Puppet 8 for DevOps Engineers

Automate your infrastructure at an enterprise scale

David Sandilands

BIRMINGHAM—MUMBAI

Puppet 8 for DevOps Engineers

Associate Group Product Manager: Preet Ahuja

Senior Editor: Athikho Sapuni Rishana

Technical Editor: Rajat Sharma

Copy Editor: Safis Editing

Project Coordinator: Sean Lobo

Proofreader: Safis Editing

Indexer: Hemangini Bari

Production Designer: Alishon Mendonca

Marketing Coordinator: Rohan Dobhal

First published: June 2023

Production reference: 2310723

Published by Packt Publishing Ltd.

Livery Place

35 Livery Street

Birmingham

B3 2PB, UK.

ISBN 978-1-80323-170-9

www.packtpub.com

This book is dedicated to my family: my wife, Linzi, for always supporting and believing in me; my sons, John and Jude, for inspiring me to push myself; and my parents, Neil and Janet, who have always been so nurturing and loving throughout my life.

– David Sandilands

Foreword

I'm not exactly sure when I first became aware of David Sandilands as a member of the Puppet Community, but it must have been about a decade ago, when I was the CTO of Puppet, and DevOps, infrastructure as code, and automated configuration management were all still things I was having to evangelize and explain to enterprises. It was immediately apparent that David was a rare individual, someone who was keenly plugged into the cutting edge of IT automation, but with extensive experience inside large, and relatively conservative, organizations.

This combination meant that David understood that automation at scale is a social activity as much as a technical one, and that driving real value and improvement means treating your Puppet installation as a platform for collaboration, not just a collection of technical capabilities.

In the decade since, we've seen DevOps explode in popularity and the emergence of the platform engineering movement, where enterprises all over the planet have realized they need to be taking the approach that David knew was necessary 10 years ago.

This foresight, concern for the human side of automation, and collaborative mindset are why, many years later, I was thrilled to have David join my team at Puppet as a principal solutions architect, helping companies big and small build out true Puppet platforms at scale.

There have been a number of books written about Puppet over the years, but this is a truly special one, distilling David's immense knowledge and experience into a pragmatic journey through the core concepts and tooling around Puppet, all with an eye toward building a collaborative platform.

I don't believe we've ever seen a Puppet book like this, and whether you're a first-time Puppet user or an enterprise IT architect, or you're looking to use your existing Puppet skills to deliver more for your organization, I know you're going to find this an invaluable reference.

Nigel Kersten

VP of product and engineering at Synadia

(Nigel served in a variety of senior technical executive roles at Puppet over 12 years, from CTO to VP of engineering to head of product.)

Contributors

About the author

David Sandilands is a principal solutions architect at Puppet, with a focus on the product management of Puppet's development ecosystem and integrations. This includes management of the Forge, supported modules, the Puppet Developer Kit, and integrations such as ServiceNow and Splunk. Before this, David worked within Puppet's solutions architect team, helping Puppet's largest customers deliver infrastructure automation at scale, and supported these customers in their DevOps working practices. He spent eight years at NatWest as a cloud infrastructure engineer delivering their IaaS platform. Based in Falkirk, Scotland, David has a Bachelor of Engineering in computer science from the University of Edinburgh.

I want to thank my wife, sons, parents, and parents-in-law for their love, support, and patience.

I would also like to thank Nigel Kersten for championing my work on this book, along with so many Puppet colleagues and Puppet Community friends who I have learned so much from on this research journey.

Finally, thanks to my editors and reviewers for their endless patience with me throughout the writing of this book.

About the reviewers

Aditya Soni is a DevOps/SRE tech professional who has taken an inspiring journey with technology and achieved a lot in a short period of time. He has worked with product- and service-based companies, including Red Hat and Searce, and is currently positioned at Forrester as a DevOps engineer. He holds AWS, GCP, Azure, Red Hat, and Kubernetes certifications. He mentors and contributes to the open source community. He is a CNCF ambassador and has been an AWS Community Builder for four years. He leads AWS, CNCF, and HashiCorp user groups for the state of Rajasthan in India. He has spoken at many conferences, both in-person and virtual. He is a hardcore traveler who loves food, exploring, and self-care and shares stories on social media as `@adityasonittyl`.

I would like to thank my parents and family who support me in numerous ways with my busy schedule. Taking the time to review the book wouldn't have been easy without them. To my friends, who always cheer me up and give me confidence whenever I feel lost. Last but not least, to the open source communities, mentors, managers, and co-workers from the early days up to now who have helped me along my path.

Jerald Sheets has been an industry system administration professional for over 30 years specializing in automation with Puppet. He owns and operates a Puppet consulting firm within the Metro Atlanta, GA, area and can be reached via `https://sscgatl.com`. He has been a Puppet Partner since 2014, has consulted for both employers and clients in the areas of Puppet Enterprise and Puppet Community, and has authored for the Puppet Blog. He can be found on the Puppet Community Slack under the handle `CVQuesty` and loves to help other Puppet Community users in any way he can. A father and step-father of 7, he lives, works, and worships with his family in rural Georgia.

Edwin Maldonado has worked in the tech industry for more than 15 years. He is a product manager at Puppet by Perforce, where he leads the Puppet Forge. Before joining the product team, Edwin worked as a solutions architect at Puppet, and as a software engineer and consulting architect in the Americas and Europe. Edwin loves learning about history and being a home barista.

Table of Contents

Preface xv

Part 1 – Introduction to Puppet and the Basics of the Puppet Language

1

Puppet Concepts and Practices 3

Puppet's history and relationship to DevOps 3
Puppet as a declarative and idempotent language 5
Key terms in the Puppet language 6
Puppet as a platform 12
Common misconceptions 15
Summary 17

2

Major Changes, Useful Tools, and References 19

Technical requirements 20
Major changes since Puppet 5 21
- Puppet 5 21
- Puppet 6 22
- Puppet 7 22
- Puppet 8 23

Legacy Puppet patterns 23
IDEs and tools to assist in Puppet development 24
How to deploy your Puppet lab and development tools 25
- Mac desktop 25
- Windows desktop 26
- Linux desktop – RPM-based 26
- Linux desktop – APT-based 27
- Configuring tools 28

References and further research 31
Summary 32

3

Puppet Classes, Resource Types, and Providers 33

Technical requirements 34

Classes and defined types 34

Including a class 35
Defined types 36
Namespaces 37

Resources, types, and providers 37

Lab 40
The package type 42
The file type 44
Service types 47
Lab 51
Core resource types 52
The exec type 53
The Augeas type 55
The notify type 56

Metaparameters and advanced resources 57

audit 57
tag 57
The resources metatype 58
Arrays of titles 59
Overriding parameters 59
Attribute splats 60
Lab 60

Anti-patterns 61

Abstract resource types 61
Defaults 61
schedule 63

Summary 64

4

Variables and Data Types 67

Technical requirements 68

Variables 68

Naming 69
Reserved variable names 69
Interpolation 70

Data types 70

Strings 71
Numbers 76
undef 80
Booleans 80
Regexp 81
Lab 81

Arrays and hashes 82

Assigning arrays 82
Accessing an array index 82
Accessing a subset of an array 83
Nested array 83
Array operators 84
Array data type 86
Assigning hashes 86
Accessing hash values 87
Nested hashes 87
Hash operators 88
Hash data type 88
Mixing hashes and arrays 89
Lab 89

Abstract data types, including Sensitive 89

Prefixes 90
Patterns 91
Arrays and hashes 92
Parent data types 94
Lab 94

Scope 94

Summary 96

5

Facts and Functions 97

Technical requirements 98

Facts and Facter 98

Custom facts and external facts 102

External facts 102
Custom facts 105
Lab 112

Functions 113

Statement functions 113
Prefix and chained functions 114
A selection of built-in functions 115

stdlib module functions 122

Lab 123

Deferred functions 124

Summary 126

Part 2 – Structuring, Ordering, and Managing Data in the Puppet Language

6

Relationships, Ordering, and Scope 129

Technical requirements 130

Relationships and ordering 130

Containment 135

Scope 140

Best practices and pitfalls 142

Lab – overview of relationships, ordering, and scope 142

Summary 143

7

Templating, Iterating, and Conditionals 145

Technical requirement 145

Templating formats in Puppet – EPP and ERB 146

EPP templates 146

ERB templates 151
EPP and ERB comparison 152

Iteration and loops 153
Iterative loops 156
Data transformation 157
Nested data 158

Conditional statements 159
If and unless statements 159
Case statement 160
Selectors 161
Capture variables 162

Lab – creating and testing templates containing loops and conditions 162

Summary 163

8

Developing and Managing Modules 165

Technical requirements 166
What is a module and what is in it? 166
Lab – reviewing the apache module 171
Roles and profiles method 171
Writing and testing a module using the PDK 175
Testing with RSpec using the PDK 182
The describe and context keywords 184
Examples, expectations, and matchers 185
Parameters and preconditions 187
Relationships 188
Data from Hiera and facts 189
Managing dependencies with fixtures 192
Coverage reports 193
Further research and tools for RSpec 193
Serverspec 194

Understanding Puppet Forge 194
Lab – creating a module and testing it 198
Summary 199

9

Handling Data with Puppet 201

Technical requirements 202
What is Hiera? 202
Using the built-in backends 203
Accessing data 206
Using custom backends 214

Hiera layers 216
Lab – add data to a module 219

Deciding when to use static code or dynamic data 220
Keeping data secure 221
Lab – use eyaml to store a secret 223
Pitfalls, gotchas, and issues 224
Lab – troubleshoot Hiera 228
Summary 229

Part 3 – The Puppet Platform and Bolt Orchestration

10

Puppet Platform Parts and Functions 233

Technical requirements 235
Puppet platform installation and versioning 235
Puppet Server 236
The embedded web server 236
The Puppet API service 237
The Admin API 239
CA 239
JRuby interpreters 243
Configuration and logs for Puppet Server 243
The Puppet agent-to-server lifecycle 246
Lab – monitoring certificate signing logging 250
PuppetDB and PostgreSQL 251
Lab – querying PuppetDB 255
Scaling with compilers 256
Lab – viewing compiler and load balancer configuration 257
Summary 258

11

Classification and Release Management 261

Technical requirements 262
Puppet environments 262
Environment directories and paths 263
Environment configuration files 264
Environment validation and deployment 265
Understanding node classification 266
Node definitions 266
Classifying nodes with Hiera 268
ENC scripts 271
PE classifier 272
Recommended approach 275
Puppet runs 276
Managing and deploying Puppet code 277
Creating a workflow 284
Lab – classifying and deploying code 285
Summary 286

12

Bolt for Orchestration 289

Technical requirements 289

Exploring and configuring Bolt 290

Connecting to clients with transports and targets 290

Running ad hoc commands with Bolt 292

Output and debugging 294

Understanding the structure of projects 296

Configuring a project 297

Configuring transports 298

System level and legacy 300

Introducing tasks and plans 301

Creating tasks 301

Creating Puppet plans 305

Constructing targets 305

Using plan functions 306

Logging and results 307

Handling errors 308

Managing data sources 309

Documenting plan metadata 310

Plan testing 311

Introducing YAML plans 311

Plugins 313

Reference plugins 314

Secret plugins 315

Puppet library 315

Lab – creating and using a Bolt project 316

Summary 316

13

Taking Puppet Server Further 319

Technical requirements 320

Logging and status 320

Exploring log locations 320

Forwarding server logs 322

Report processors 324

Accessing status APIs 325

Metrics, tuning, and scaling 327

Exploring metrics dashboards 328

Identifying and avoiding common issues 330

Lab – configuring metric dashboards 334

External data provider pattern 335

Understanding external data provider components 336

External data provider implementations 337

Lab – hands-on with Splunk and Puppet Data Service 339

Summary 340

Part 4 – Puppet Enterprise and Approaches to the Adoption of Puppet

14

A Brief Overview of Puppet Enterprise 345

Technical requirements 346
What is Puppet Enterprise? 346
Exploring the Puppet Enterprise console and services 347
Puppet Server 348
Introducing Puppet web console components 348
Using Bolt with Puppet Enterprise 351
Orchestrator services 352
Running jobs 354
Configuring performance settings 355
Automating deployment and reference architectures 356
Understanding supported architectures 356
Deployment and configuration 359
Puppet Enterprise-related projects and tooling 361
Monitoring and troubleshooting Puppet Enterprise infrastructure 361
Managing deployments and ensuring compliance 362
Lab – Puppet Enterprise extensions and configuration 364
Summary 364

15

Approaches to Adoption 367

Scope and focus 368
Managing heritage estates with no-op mode 370
A platform engineering approach 372
Adoption in regulated environments 374
Moving to the cloud 375
Summary 376

Index 379

Other Books You May Enjoy 392

Preface

As DevOps and platform engineering have pushed demand for powerful internal development platforms, the need for infrastructure automation solutions has never been greater. Puppet is one of the most powerful infrastructure automation solutions used by the largest enterprises in the world and has a strong open source community. This book comprehensively explains both the Puppet language and the platform. Starting with the basic concepts and approach of how Puppet works as a stateful language, it builds up to explaining how to structure Puppet code to scale and allow flexibility and collaboration among teams. It then looks at how the Puppet platform allows management and reporting of infrastructure configuration, showing how the platform can be integrated with other tools such as ServiceNow and Splunk. Finally, approaches will be discussed to understand how to implement Puppet to fit into heavily regulated and audited environments as well as modern hybrid cloud environments.

By the end of this book, you will have a full understanding of the capabilities of both the Puppet language and platform and be able to structure and scale Puppet to create a platform to provide enterprise-grade infrastructure automation.

Who this book is for

This book is ideal for DevOps engineers looking to automate infrastructure configuration with Puppet. It specifically focuses on Puppet's configuration management capabilities but goes on to touch on other infrastructure management practices in general. It will allow both beginners and current Puppet users to learn about the full power of the Puppet language and platform.

What this book covers

Chapter 1, Puppet Concepts and Practices, focuses on why Puppet was developed, how it has changed over time, and the core concepts and practices of Puppet. It also focuses on how Puppet assists in a DevOps transformation and our approach to it.

Chapter 2, Major Changes, Useful Tools, and References, discusses major changes such as harmful terminology, sensitive values, deferred functions, and other high-level items that have emerged since Puppet 5. It will also highlight items that have dropped out of Puppet. It will cover useful tools to assist development, such as VS Code and the **Puppet Development Kit** (**PDK**), showing how the lab and development environment will work for the book. It will also show various Puppet and community references for further learning.

Chapter 3, Puppet Classes, Resource Types, and Providers, introduces the most basic building blocks of Puppet and how to use them so you can understand the initial stages of writing Puppet code, showing how resource types and providers work together to create stateful code independent of the underlying OS implementation and how classes allow us to group these resources.

Chapter 4, Variables and Data Types, details how to assign variables with data types in Puppet, how they can be managed in arrays and hashes, the use of the sensitive data type to secure variables, and how the variable scope is managed. Then, we'll provide some best practice advice on how to use these variables and data structures well within Puppet.

Chapter 5, Facts and Functions, looks at the facts and factors that it provides, how to use them in Puppet code, and how to customize them. It will also look at functions: what they are, how lambdas can be used with them, and how the relatively recent deferred functions can be used with them

Chapter 6, Relationships, Ordering, and Scope, covers how Puppet handles relationships and order as well as scope and containment. These issues come together to help the user understand how cross-module or cross-class resources and variables will intersect.

Chapter 7, Templating, Iterating, and Conditionals, shows how to use templates, iteration, loops, and various conditional statements, such as `if` cases and selectors to affect the flow and management of code.

Chapter 8, Developing and Managing Modules, discusses the structure of modules, the use of the PDK to create them, and how we can test modules. It will also discuss how to use Puppet Forge well to both consume and share code and understand the quality of shared modules.

Chapter 9, Handling Data with Puppet, runs through how Puppet handles data, discussing what Hiera is, at what levels to store data, and some pitfalls and mistakes to avoid in structure and approach.

Chapter 10, Puppet Platform Parts and Functions, helps you understand what Puppet is as a platform, how the various components work together and communicate, and common architecture approaches to deliver scale.

Chapter 11, Classification and Release Management, discusses how Puppet manages servers and code in environments, how servers can be classified, and how the Puppet run of this classification actually runs. The tooling to deploy code into these environments will also be discussed.

Chapter 12, Bolt for Orchestration, looks at how to use Bolt as an orchestrator for procedural tasks, showing the various transport options – SSH, WinRM, and PCP – to use via Puppet agents. You will see how tasks and plans can complement Puppet code and how Puppet code can be orchestrated and deployed via Bolt itself.

Chapter 13, Taking Puppet Server Further, looks at more advanced topics to ensure you can monitor and scale your infrastructure, deal with common issues, and integrate external data sources.

Chapter 14, A Brief Overview of Puppet Enterprise, highlights the differences between Puppet Enterprise compared to open source, and the integrations and services available to help scale and tune infrastructure.

Chapter 15, Approaches to Adoption, discusses how Puppet can be adopted and worked with in real brownfield environments, highlighting lessons learned in the field and from various adoptions, and looking at correctly scoping use cases to benefit from delivering regularly. It will look at how Puppet can work within platform engineering as well as with heritage estates, and even in highly regulated and change-managed estates.

To get the most out of this book

Some background understanding of system administration for Unix and Windows systems and application deployment is required. Also, some core development concept knowledge is required, such as revision control tools (Git, virtualization, and testing) and coding tooling (such as vi or Visual Studio Code).

Software/hardware covered in the book	Operating system requirements
Puppet 7 or 8	Windows, macOS, or Linux
Bolt	Windows, macOS, or Linux
Visual Studio Code	Windows, macOS, or Linux
Azure	
Puppet Development Kit (PDK)	Windows, macOS, or Linux
The PEADM module	Windows, macOS, or Linux

The full configuration of the required software for the lab environment will be covered in *Chapter 2*.

If you are using the digital version of this book, we advise you to type the code yourself or access the code from the book's GitHub repository (a link is available in the next section). Doing so will help you avoid any potential errors related to the copying and pasting of code.

Download the example code files

You can download the example code files for this book from GitHub at `https://github.com/PacktPublishing/Puppet-8-for-DevOps-Engineers`. If there's an update to the code, it will be updated in the GitHub repository.

We also have other code bundles from our rich catalog of books and videos available at `https://github.com/PacktPublishing/`. Check them out!

Download the color images

We also provide a PDF file that has color images of the screenshots and diagrams used in this book. You can download it here: `https://packt.link/vPsXh`

Conventions used

There are a number of text conventions used throughout this book.

`Code in text`: Indicates code words in text, database table names, folder names, filenames, file extensions, pathnames, dummy URLs, user input, and Twitter handles. Here is an example: "The lookup function key, `data_hash`, accepts `yaml_data`, `json_data` and `hocon_data` as values but most Puppet implementations just use YAML data, so this book will default to the `yaml_data` backend."

A block of code is set as follows:

```
hierarchy:
- name: "YAML layers"
  paths:
    - "nodes/%{trusted.certname}.yaml"
    - "location/%{fact.data_center}.yaml"
    - "common.yaml"
```

When we wish to draw your attention to a particular part of a code block, the relevant lines or items are set in bold:

```
type { 'title':
  attribute1 => value1,
  attribute2 => value2,
}
```

Any command-line input or output is written as follows:

```
bolt --verbose plan run pecdm::provision --params @params.json
```

Bold: Indicates a new term, an important word, or words that you see onscreen. For instance, words in menus or dialog boxes appear in **bold**. Here is an example: "Select **System info** from the **Administration** panel."

> **Tips or important notes**
> Appear like this.

Get in touch

Feedback from our readers is always welcome.

General feedback: If you have questions about any aspect of this book, email us at customercare@packtpub.com and mention the book title in the subject of your message.

Errata: Although we have taken every care to ensure the accuracy of our content, mistakes do happen. If you have found a mistake in this book, we would be grateful if you would report this to us. Please visit www.packtpub.com/support/errata and fill in the form.

Piracy: If you come across any illegal copies of our works in any form on the internet, we would be grateful if you would provide us with the location address or website name. Please contact us at copyright@packt.com with a link to the material.

If you are interested in becoming an author: If there is a topic that you have expertise in and you are interested in either writing or contributing to a book, please visit authors.packtpub.com.

Share Your Thoughts

Once you've read *Puppet 8 for DevOps Engineers*, we'd love to hear your thoughts! Scan the QR code below to go straight to the Amazon review page for this book and share your feedback.

https://packt.link/r/180323170X

Your review is important to us and the tech community and will help us make sure we're delivering excellent quality content.

Download a free PDF copy of this book

Thanks for purchasing this book!

Do you like to read on the go but are unable to carry your print books everywhere?

Is your eBook purchase not compatible with the device of your choice?

Don't worry, now with every Packt book you get a DRM-free PDF version of that book at no cost.

Read anywhere, any place, on any device. Search, copy, and paste code from your favorite technical books directly into your application.

The perks don't stop there, you can get exclusive access to discounts, newsletters, and great free content in your inbox daily

Follow these simple steps to get the benefits:

1. Scan the QR code or visit the link below

`https://packt.link/free-ebook/9781803231709`

2. Submit your proof of purchase
3. That's it! We'll send your free PDF and other benefits to your email directly

Part 1 – Introduction to Puppet and the Basics of the Puppet Language

This part will establish the core concepts of what Puppet is, what you can achieve with Puppet, how it fits into a DevOps approach, and how we will approach it in this book. We will then take a high-level overview of the core components of Puppet. The development lab environment used throughout the book will be reviewed, along with useful references and further learning resources. Then, we will begin with the basics of the language by looking at classes, resources, variables, and functions.

This part has the following chapters:

- *Chapter 1, Puppet Concepts and Practices*
- *Chapter 2, Major Changes, Useful Tools, and References*
- *Chapter 3, Puppet Classes, Resources Types, and Providers*
- *Chapter 4, Variables and Data Types*
- *Chapter 5, Facts and Functions*

1
Puppet Concepts and Practices

This chapter will focus on the origins of Puppet, why it was created, and how it is used in DevOps engineering. It will look at Puppet's approach to configuration management and how its declarative approach differs from more regular procedural languages. Puppet has many features that are common in other languages such as variables, conditional statements, and functions. But in this chapter, we will cover the key terms, structure, and ideas of the language that make it different and how the underlying platform runs. We will give a clear, high-level overview of its approach and how it relates to customer needs and infrastructure environments. Finally, as there are a lot of preconceptions regarding Puppet, this chapter will finish by addressing some of the most common ones, including where they come from, and unwrap them.

This should ensure a fundamental understanding of Puppet and its approach before we build up a deeper, technical understanding of the language in upcoming chapters. It will also ensure this book is not just about technology but how genuine value can be delivered to customers using the service that Puppet provides.

In this chapter, we are going to cover the following main topics:

- Puppet's history and relationship to DevOps
- Puppet as a declarative and idempotent language
- Key terms in the Puppet language
- Puppet as a platform
- Common misconceptions

Puppet's history and relationship to DevOps

Puppet was started by creator and founder Luke Kanies, who was working as a sysadmin and consultant. He was unable to find the tooling he wanted to use and that his customers could rely on, so he created Puppet as a Ruby-based open source configuration management language in 2005. The success of this open source project resulted in the release of a commercial offering, Puppet Enterprise, in February 2011. But as the demands increased and Puppet needed to reform and expand as both a company and an open source project, Luke stood down, stating that the challenges of growing Puppet to enterprise-scale were *far from what I love to do most, and far from my core skills. We need to scale, and we need to execute.*

The new leadership that followed took a direction that saw the company develop its professional services, and focus more effort on developer tooling and education while expanding its product range both organically and via acquisitions, striking a difficult balance between the open source community and its enterprise customer demands. Puppet was acquired by Perforce Software on May 17, 2022, following the Chef (2020) and Ansible (2015) acquisitions, as the last of the standalone configuration management start-ups. Luke summed up the change that has taken place in the industry: *DevOps teams are different now. Companies are looking for a complete solution, rather than wanting to integrate individual best-of-breed vendors.*

This history has seen Puppet move from a tool that left it to the developer to decided how best to use it to solve problems to, today, a tool with patterns and solutions that users can just consume to standardize their automation and deployment. This has allowed users to focus on their solutions and not the underlying technology.

DevOps itself has become a frustrating term in the IT industry; the definition given by formal sources differs hugely from how companies actually use it, and references to it can be used as a cynical buzzword or sales gimmick. The focus of this book is on DevOps engineering, as used particularly by large companies and has been well researched and discussed in studies such as the Puppet-run *State of DevOps Report*. DevOps engineering is normally delivered as part of projects such as digital transformations, cloud-first migrations, and various other modernization projects. What is typically seen in these projects is a desire to automate self-service deployment, compliance, and remove toil. This approach follows the DevOps goal of breaking down the silos between developers and ops teams by allowing better communication and establishing shared goals. What is noticeable is that the system administrator role in which Luke worked originally has effectively been replaced by roles such as DevOps engineers.

Puppet will be used as part of a DevOps toolchain, and *Figure 1.1* shows an example set of tools and their relative functions. It is typical for Puppet to start its role at the end of a provisioning pipeline, as infrastructure is stood up in a platform and needs to be configured and enforced:

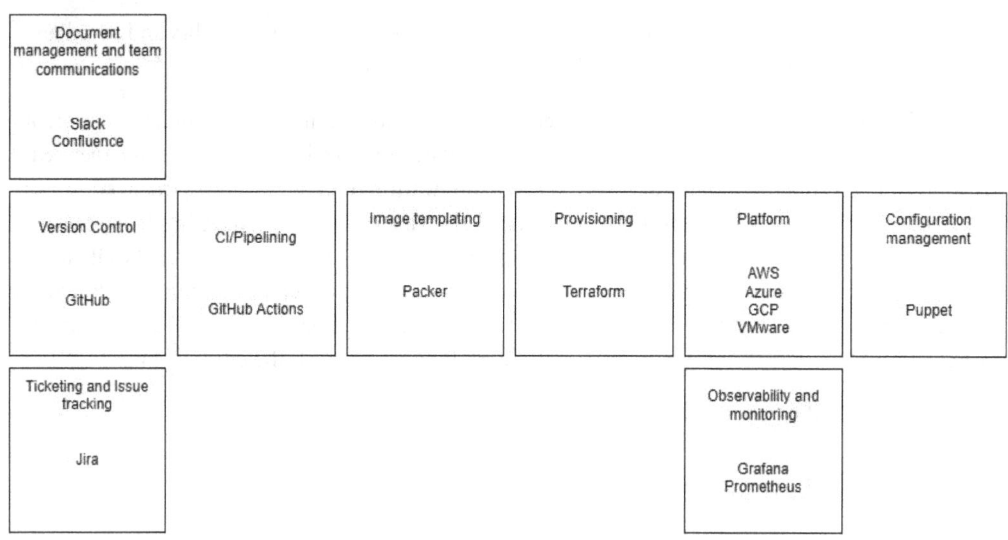

Figure 1.1 – A DevOps toolset

This book will focus not just on a technological understanding but also on how to use the maturity of the Puppet language, tooling, and platform with opinionated patterns. These approaches have been developed through years of customer engagements for Puppet and the communities' own implementations to allow users to reduce their effort in finding the right approach, focus on their solutions, and deliver immediate benefit and return to their customers.

Puppet as a declarative and idempotent language

The first important thing to understand is how Puppet differs from normal scripting or coding languages. Puppet is declarative, meaning you describe the state you want the system to be in. For example, you could describe that your system should have a user called `username` with UID `1234`, a configuration file should not exist, and a kernel setting should be at a particular value. In comparison to most languages where you have to describe the process to get to the state, Puppet's approach brings us closer to how customers request services. They don't want to know how it's done, just that it will meet their requirements. These resource definitions can be saved in your version control system. Often, this approach is described as being part of **Infrastructure as Code**.

Puppet is idempotent, meaning that it will only make the changes required to get into the declared state. Meanwhile, most procedural languages will run steps every time and, typically, require various checks such as `if` statements to be added to make checks to avoid duplication. This is particularly powerful as what is called *enforcement* can be run with the Puppet language, ensuring the state you declared has been reached, and is capable of detecting whether a change happened because of you updating the state you wished the machine to be in or whether it was a change that happened on the machine itself moving away from the desired state. This can greatly assist with audits and avoid any configuration drifts in an estate and ensure change is managed and deliberate.

Puppet is OS-independent; the language is focused on the state, not the underlying implementation of how particular OSes install a package or add a user. This gives us a universal language that is independent of any underlying implementations, allowing for less duplication of code, avoiding the need to use layers of `case/if` statements to detect differences, and allowing multiple language implementations such as PowerShell for Windows and Bash for Unix-based systems. Additionally, it makes it easier to recover after failures in applying code. If in a procedural language, a step fails, it might not be safe to run the script in full again depending on how well the check steps have been coded. In contrast, Puppet code is able to resume only performing the steps it needs to reach the correct state.

A simple example of Puppet code to create a user would look like this:

```
user { 'david'
  uid => '123'
}
```

In contrast, a shell script might have a section like this:

```
if ! getent passwd david; then
  useradd -u 123 david
elif ! $(uid david) == 123; then
  usermod -u 123 david
fi
```

In the preceding shell example, we have to check whether a user exists, and if not, create one. If it does exist, then does it have the right UID? If not, we change it. This script only covers OSes that can use `useradd` and `usermod`. To achieve compatibility with multiple OSes, we would need a test to detect the OS type and produce a section of code like this for every OS or group of OSes and their required commands. Often, it would be more practical to write in multiple languages and scripts to cover a broader base of OS flavors, that is, if we wanted to cover both Unix and Windows, for example.

This compares to the Puppet declaration, which will work on multiple OSes without change as Puppet will detect the required commands and perform all the necessary state checks as part of that.

This example is all just for a single resource with a single attribute. You can quickly see how the shell script example will not scale as it becomes increasingly complex with almost endless checks and options.

Key terms in the Puppet language

Looking at the Puppet language in more detail, the most fundamental item in Puppet is a **resource**. Each resource describes some part of the system and the desired state you wish it to be in. Each resource has a **type**, which is a definition for the Puppet language of how this particular resource can be configured, which **attributes** can be set, and what **providers** can be used. The attributes are what describe the state. So, for a user, this might be a home directory or, for a file, the permissions. **Providers** are what make the Puppet OS independent since they do the underlying commands be they for creating a user or installing a package.

So, let's take an example of a company that typically submits build request forms to an environments team to request the configuration for a server:

Build Request Form			
Project Details			
Cost code	PD8NGH		
Requester	Iain Miller		
Host details			
Servers names	Server1	Server2	Server3
Users			
User name	UID	GID	
exampleapp	1234	123	
Directories			
Directory name	Owner	Group	Permissions
/opt/exampleapp/	exampleapp	exampleapp	755
/etc/exampleapp/	example app	example app	750

Table 1.1 – An example build request form

In *Table 1.1*, the request form, we see groupings of users, groups, and directories, which are all, essentially, **types**. Each item under them is a resource, and the configuration settings are the attributes.

This request could translate to something like the following:

```
user { 'exampleapp':
  uid => '1234'.
  gid => '123'
}

group { 'exampleapp':
  Gid => '123'
}

file { '/opt/exampleapp/':
  owner => 'exampleapp',
  group => 'exampleapp',
```

```
  mode   => 755
}

file { '/etc/exampleapp/':
  owner => 'exampleapp',
  group => 'exampleapp',
  mode   => 750
}
```

The preceding example shows how Puppet translates more directly to user requests and can remain readable without even understanding any of the Puppet language.

What isn't visible, in this example, is the **providers**. Puppet has defaults, such as in the preceding example, where the user resource assumes a RedHat host will use the `usermod` provider. Instead, if I wished to use LDAP commands for user creation, I would set my `provider` attribute to LDAP.

The next important thing to note is that due to the nature of writing Puppet in a stateful way, we are not writing an ordered process that executes line by line but only declaring the state of resources that could be implemented in any order. Therefore, if we have any dependencies, we need to use the `relationship` parameter; this describes a before/after relationship, which is exactly as it sounds, or a subscribe/refresh, whereby, for example, updating a configuration file could cause a service to restart. In the previous example, Puppet automatically creates certain dependencies such as ensuring the group is created before the user, so we don't have to add a **relationship** parameter. Often, these relationships are seen as one of the most difficult parts of Puppet to adapt to, as many coders are used to writing a process to follow and mistakes can be made. This can cause a cycle of dependencies, whereby a chain of these dependencies cycles round, and there is no way to create a starting resource that isn't dependent on another.

Evidently, the resources we declare need a structure, and the first step is for this code to be in a file. Puppet calls these **manifest** files, which have an extension of `.pp`. **Classes** are blocks of Puppet code that give us a way to specifically call sections of code to be run on hosts. Normally, as a good practice, we only have one **class** in a **manifest** file. Puppet then uses **modules** as a way to group these **manifests** and **classes**. This grouping is based on the principle that a **module** should do a single thing well and represent a technical implementation, such as a **module** configuring the IIS application or configuring postfix as a mail relay. **Modules** are simply a directory structure storing the **manifests**, **classes**, and other Puppet items (which we will cover, in detail, in *Chapter 8*) and are not a keyword in the language itself. So, ideally, modules should be shareable and reusable for different users and organizations with many taken straight from the **Puppet Forge**, which is Puppet's catalog of modules with both commercial and open source offerings.

An example of one common style and practice for modules is to have a manifest file with a single class for the following:

- `install.pp` (grouping resources related to installing software)
- `config.pp` (grouping resources related to configuring software)
- `service.pp` (grouping resources related to running services)
- `init.pp` (a way of initializing the module and accepting parameters)

At a higher level, we then have **roles** and **profiles, which** are used to create the structure of your organization. While **modules** should be sharable and repeatable installations of technical implementations, such as Oracle or IIS, **roles** and **profiles** will only have context within your organization. **Roles** and **profiles** are **classes** used to group **modules** and selected parameters into logical technical stacks and customer solutions. It is common to make a **roles module** and a **profiles module** while keeping together the **classes** used.

What can be confusing, at this point, is that you can end up with an Oracle **role**, an Oracle **profile**, and an Oracle **module**. So, while the Oracle module configures and installs Oracle with various parameters available to it to customize the installation, the Oracle profile is about how your organization uses this module and what other modules it might add to this technology stack. You might specify that you always use Oracle with a cluster service and, therefore, your Oracle profile contains both an Oracle module and a cluster module. Alternatively, it might pass parameters to the Oracle module within your profile, which set default kernel settings for your organization's configuration.

You can think of a role as being what the customer actually wants when they submit a build request; they need a particular type of server, be it an Oracle or an IIS server. They don't care about the underlying implementations – only that it meets their requirements. While the Oracle role will certainly need the Oracle profile, it will expect it to meet the OS security standard and to have any agents or other supporting tools your organization defines. Therefore, a common profile for many organizations is a base OS security standard that ensures every server is compliant and that is part of almost every role.

Figure 1.2 shows an example of what has just been described as an Oracle role class in the roles module, which includes an Oracle profile class and an OS security profile class, both from the profile module. Then, the Oracle profile includes an Oracle module, while the `os_security` profile includes the DNS module:

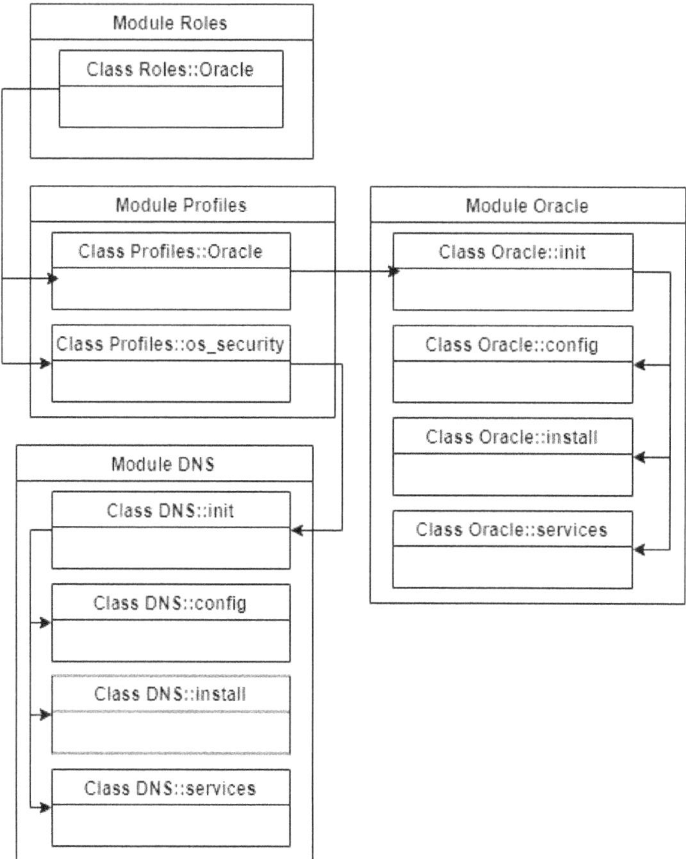

Figure 1.2 – The structure of roles, profiles, and modules

In *Chapter 8*, we will go into more technical detail, but the key takeaway from this overview is to understand that modules provide sharable and reusable single-use technical installations. In contrast, the roles and profiles pattern provides the context for your organization. Roles are for customers ordering server offerings; they don't need to understand the technical implementation, only that it meets their business requirement. The profiles in your organization's technology stack are managed by technical designers and architects, who combine and specify modules according to your organization's standards and configurations. These roles are responsible for defining how different components are integrated to create the desired technology stack. So while an Oracle module by itself can configure and install Oracle, it is the profile that defines the exact configurations that should be passed to that Oracle module and the other modules it may be dependent on such as having a NetBackup client installed.

With what we have covered in modules, roles, and profiles, going back to *Table 1.1*, instead, we can have a customer submitting the build request form but not having to specify everything they need; they could simply order an `exampleapp` role server.

What we have seen so far is fine when servers meet all the specifications and are standard, but exceptions are commonplace. **Hiera** is Puppet's data system, and it can be used to pass parameters to the roles and profiles model to handle exceptions. Hiera, as its name suggests, is hierarchical. It defines an ordered lists of data sources to access to find the most relevant setting. These data sources will typically be ordered from the default value for all nodes to a more specific group such as a particular role and specific values for an individual node.

For example, if email servers were disabled by the default OS security profile but were required for `exampleapp`, we could have the following YAML file:

exampleapp.yaml

```
profile::os_security:email_enabled: true
```

Similarly, if `server1` needed a different UID, we could have the following YAML file:

server1.yaml

```
profile::exampleapp:uid: '1235'
```

One of the most important points of creating these patterns is to avoid hardcoded values in your modules. By using Hiera, you give yourself a dynamic way to change the values in the future without modifying the code. This could evolve to access the data via a self-service portal – automating away from builds ordered via spreadsheets, emails, and discussions, which would have to be configured by the build teams instead of portals such as VMware vRealize Automation or ServiceNow:

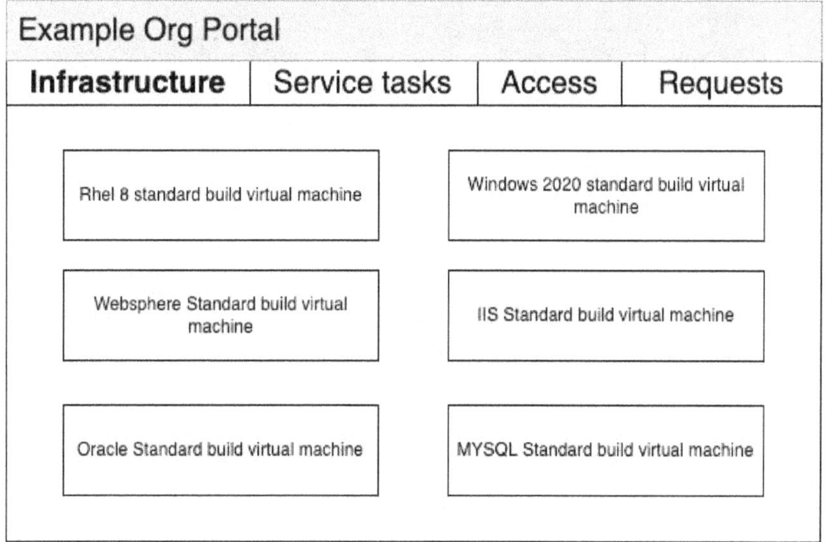

Figure 1.3 – An example portal

In *Figure 1.3*, an example portal shows how customers can be presented with simplified products. The focus of the Puppet language should be to deliver consistent products to customers and allow customers, architects, and technical staff to focus on what they care about and not have to delve into the technical requirements or coding sections themselves.

Puppet as a platform

So far, this chapter has focused on the Puppet language, but now we will look at the Puppet platform and how it applies the desired state to client servers. Puppet can be run with just an installed agent and all the files locally, which is common for testing, but this overview will focus on the client-server setup. In *Chapters 10*, *13*, and *14*, we will go into much more detail about resilience, scalability, and more advanced running options. However, for now, we will focus on how a Puppet client talks to a server to request and apply its desired state.

Every client under Puppet control will install a Puppet agent. *Figure 1.4* shows the steps of a Puppet agent run, which this section will outline:

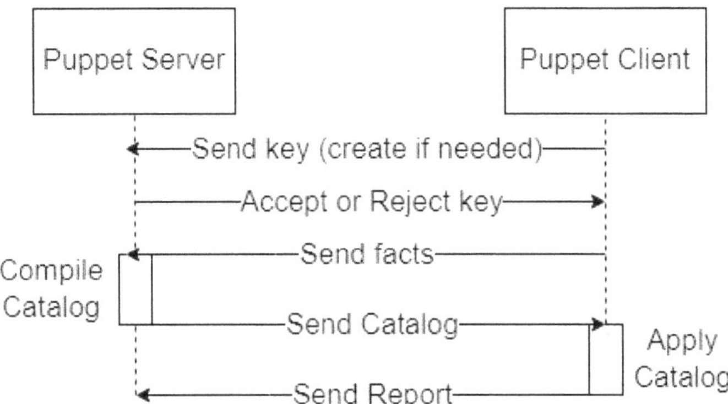

Figure 1.4 – The Puppet agent run life cycle

The first step is for the agent to identify itself to the primary server with SSL keys or to create new SSL keys for the primary server to sign. This will secure communication between the server and client.

The next action is for the client to use a Ruby library called `Facter`. This is a system profiler to gather what is known as **facts** about the system. This can be things such as the OS version or RAM size. These facts can be used in code or by Hiera to make choices about what state a host should be in, such as Windows Server 2022 having a particular registry setting.

Then, the server identifies what classes should be applied to a server. Typically, this is done by what is called an **external node classifier** (**ENC**) script, which is based on the facts and user definitions. Normally, this will apply a role class to a server, which, as we discussed in the previous section, builds up a definition of profiles and module classes.

Then, the primary server compiles a catalog and a YAML file of the resources to be applied to the node (ensuring the CPU-intensive work happens on the server and not the client).

This catalog is then sent to the client who uses the catalog as a blueprint of what the state should look like and makes any necessary changes to enforce the state on the client.

Finally, a report is sent back to the primary server confirming what resources were applied and whether these resources had to be changed due to a change in Puppet code or whether they were changed outside of Puppet control (which might be an audit or security breach).

In *Figure 1.5*, we see an example extract from a Puppet report showing the name of the resource, the type of change made, and the value it needed to change. Additionally, the report includes a record of unchanged resources highlighting what is part of Puppet's enforcement:

Message

Applied catalog in 45.89 seconds

Source: Puppet

command changed to 'GRANT CONNECT ON DATABASE "pe-puppetdb" TO "telegraf"'

Source: /Stage[main]/Puppet_operational_dashboards::Profile::Postgres_access/Pe_postgresql::Server::Database_grant[operational_dashboards_telegraf]/Pe_postgresql::Server::Grant[database:operational_dashboards_telegraf]/Pe_postgresql_psql[GRANT CONNECT ON DATABASE "pe-puppetdb" TO "telegraf"]/command
File: /opt/puppetlabs/puppet/modules/pe_postgresql/manifests/server/grant.pp **Line:** 70

Triggered 'refresh' from 1 event

Source: /Stage[main]/Puppet_enterprise::Master::Puppetserver/Puppet_enterprise::Trapperkeeper::Pe_service[puppetserver]/Service[pe-puppetserver]
File: /opt/puppetlabs/puppet/modules/puppet_enterprise/manifests/trapperkeeper/pe_service.pp
Line: 10

command changed to 'GRANT pg_monitor TO telegraf'

Source: /Stage[main]/Puppet_operational_dashboards::Profile::Postgres_access/Pe_postgresql_psql[telegraf_pg_monitor_grant]/command
File: /etc/puppetlabs/code/environments/production/modules/puppet_operational_dashboards/manifests/profile/postgres_access.pp
Line: 28

Figure 1.5 – The Puppet console server report

By default, this cycle takes place every 30 minutes. In the previous sections, the focus was on how the language can automate the building of servers. Here, we can see that, via the platform, we can ensure all our deployed servers are enforced with the state we set out to achieve; whether that be a security standard profile or whether we decided to update the settings in a particular implementation such as adding extra features to IIS. This avoids server drift, where servers on the estate are difficult to keep up to date or are vulnerable to changes made manually in error or that maliciously breach standards. *Figure 1.6* shows the dashboard view of Puppet Enterprise, giving a clear view of an estate of servers and the status of the last run. This highlights whether the servers are in compliance with our state or had to make changes in their previous run:

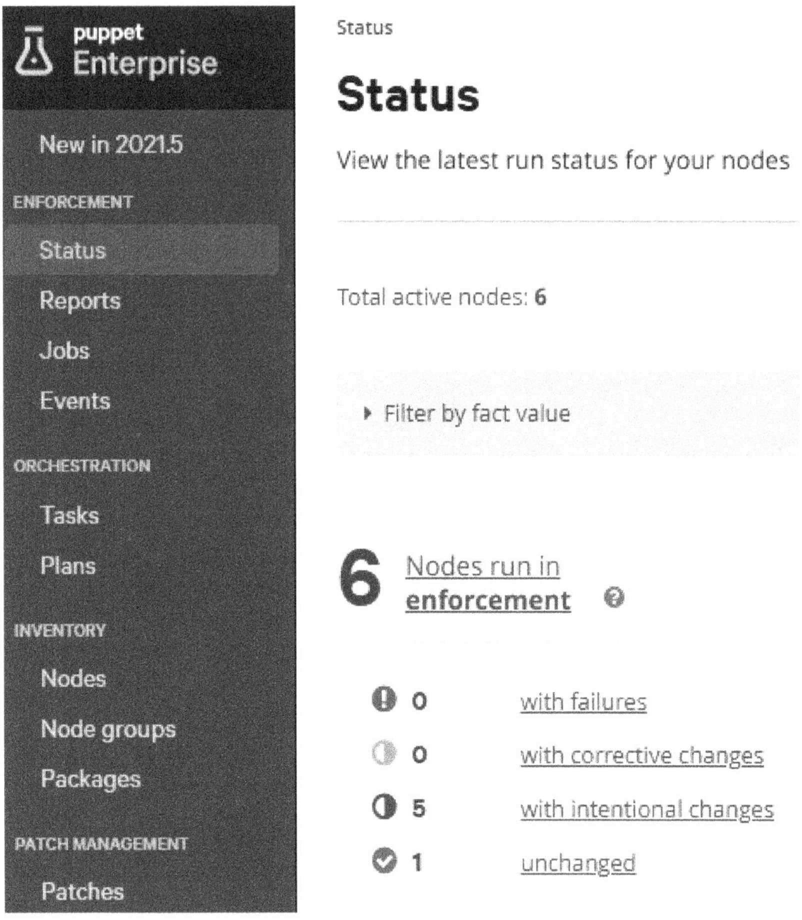

Figure 1.6 – The Puppet console status dashboard

What we have reviewed so far would presume a common code base, and when any code changes are made, all clients would have a new state enforced within the next 30 minutes as agents contact the primary server. This would clearly be problematic, as bugs will affect all servers within a brief period.

This is why Puppet has **environments**. An environment is a collection of versioned modules. This is achieved by storing the modules in revision control, such as `git`, where the version can be declared as a commit, a tag, or a branch, which we can list in a file called a **Puppetfile**.

An example module declaration would look like this:

```
mod 'apache',
   :git => 'https://github.com/exampleorg/exampleapp'
   :tag => '1.2'
```

By maintaining this **Puppetfile** in `git`, in what is known as a **control** repo, it is possible to represent multiple environments by having different branches with different versions of the Puppet file.

A common practice is to match environments against how your organization classifies server usage. Normally, this means a minimum of a development environment and a production environment. So, changes can be tested against servers in development, and then the successfully tested ones can be deployed to production. This can be taken further using canary environments to test small subsets of the server. This approach can all be customized to the change and risk setup of different organizations.

All the facts and reports we mentioned, as part of the agent cycle, are stored in `PuppetDB`, a frontend application using PostgreSQL as a backend database, which is designed to manage Puppet data such as reports and facts. This is used with the **Puppet Query Language** (**PQL**), which allows us to search the information we have gathered. This can allow for searching of facts giving `CMDB` style data and for combinations where we can check whether a certain resource for a role had changed, which could indicate a change breach had taken place.

So, in this section, we have seen that the Puppet platform gives a way to progressively deploy new code based on environments. It stores facts about the clients along with the reports generated on each run, giving a powerful view of CMDB along with audit and compliance information in the reports as we confirm what state the servers are in. This can all be searched using PQL. This can lead to huge savings in operational toil in terms of audit and compliance report generation and helps avoid building technical debt as standards and configurations change.

Common misconceptions

Isn't Puppet dead?

The focus of bleeding-edge technology has moved on to serverless and other **Software-as-a-Service** (**SaaS**)/containerized offerings, while at an **Infrastructure-as-a-Service** (**IaaS**) level, development in Puppet has reached a much greater level of maturity. 10 years ago, you might have bought this book assuming it was relevant regardless of whether you were going to work with Puppet. Today, you have a Puppet solution to implement or understand.

I need to know Ruby to use Puppet.

Some basic knowledge of Ruby would be an advantage for certain areas of Puppet code. A focus on the good use of the Puppet language to get early returns is what this book will focus on, and the reality is that the majority of Puppet professionals don't spend much time with Ruby trying to create customizations. Even specialists working for Puppet itself find that it can be some time before they need to write something custom in Ruby.

Puppet won't work with our change management.

A big fear is the idea of Puppet making changes outside the scope of governance and change management. This often reflects assumptions and a lack of communication with change management teams. Puppet will enforce the state you have described; therefore, changes will only happen if the state described in the code has changed or if it has been changed outside of Puppet's control. As previously mentioned, as long as it is agreed that Puppet is the way to define particular resources, any change to the state should be seen as outside of governance and, therefore, put back into place. Later chapters will discuss how to release code and environments to ensure that Puppet remains properly access-controlled and, therefore, within governance.

I can't make manual changes or exceptions.

This could certainly happen if users try to work around Puppet. To avoid this, it's important to define what Puppet is responsible for, what other tooling or manual processes are responsible for, and how exceptions should be requested and approved in your system. As will be discussed in *Chapters 8* and *9*, by using parameters in modules and Hiera for exceptions, a controlled method can be used for exceptions, which also keeps a record in code.

I need Puppet Enterprise to use add-ons and integrations.

There is a huge amount of confusion, particularly from industry analysts, who make comparisons about what users get with Puppet Enterprise and how open source might be limited. This book will go into more depth in *Chapter 14*, but the fundamental difference for Puppet Enterprise is you are paying for support, services, and pre-canned modules, infrastructure, and solutions. If you have the skills, developers, and time, all of these features can be replicated in open source. Ultimately, Enterprise runs on the open source components.

Everyone will need to learn Puppet.

A major focus of this book will be the importance of structuring code to allow for self-service processes. This avoids users who might wish to have small exceptions or integrations having to learn everything as a Puppet developer and only having to understand your offerings.

It will clash with other systems.

The key part is to understand what Puppet will be responsible for and what other systems will be responsible for, and to document this well. Many environments will run multiple configuration management, orchestration, and software management tools. The important thing is to use them to their strengths with clear boundaries.

Summary

In this chapter, you learned how Puppet was created by Luke Kanies as a stateful language to ease the automation of the configuration management of servers. We learned how using this stateful approach provides a language more natural for describing user requirements for configuration management and reduces the complexity involved in more traditional procedural approaches.

We looked at an overview of the core language terms and components and how they are structured via roles, profiles, and modules. This structure offers a natural way to create customer offerings, technical stacks, and reusable technical modules.

We looked at how the states described in the language are then applied to hosts via Puppet runs, and from these runs, we examined how valuable audit and compliance information can be gathered and stored in `PuppetDB`. We discussed how code can be managed in environments to allow the gradual release of state changes in a managed way in logical groups of servers that suits your organization's risk appetite and development structure.

The chapter discussed some misconceptions around Puppet along with the main themes of relevance, complexity, and flexibility. Puppet's maturity and focus on IaaS make it less fashionable, but using patterns and modules developed by Puppet and the community allows you to use Puppet to its strengths and deliver automation and self-service configuration and compliance to customers. Ensuring clear boundaries and responsibilities so that Puppet can integrate with, and work alongside, other tooling and teams avoids clashes and allows others to interact with Puppet and gain the benefits.

In the next chapter, we will review the major changes that have taken place in Puppet since version 5 and in the latest version, 7. Recommendations of the tooling to use to create an effective development environment will be made, and the creation of lab environments will be outlined and demonstrated. Additional reference sites will be outlined to allow readers to continue their research and stay up to date with developments in Puppet. This will ensure that as we start on the technical details in the following chapters, you will have the capability to test and experiment in your own environment and follow up in more detail on your points of interest.

2

Major Changes, Useful Tools, and References

This chapter will set out the major changes that have taken place since Puppet 5 up to the current versions, Puppet 7.24 and 8.0. This is viewed as the modern era of Puppet, where in the previous chapter, the change of focus was highlighted in the history of Puppet. This summary of changes will also cover some redundant patterns and approaches that might still be seen from earlier versions of Puppet, as these can still be visible in code and various sources. The chapter will then go on to discuss tooling to create a productive developer environment for Puppet, which will be used for the lab environment throughout this book. The aim will be to give an opinionated view of how to develop Puppet code and tooling that can assist with this. These tools can be installed in an environment of the readers' choice. The lab environment itself will then be demonstrated by standing up a simple setup and logging in. The chapter will finish by looking at what resources are available for you to keep up to date with Puppet and research further topics of interest.

In this chapter, we're going to cover the following main topics:

- Major changes since Puppet 5
- Legacy patterns before Puppet 5
- IDEs and tooling to assist in Puppet development
- How to deploy your Puppet lab and development tools
- References and further research

Technical requirements

The development environment will require an OS with access to the internet, which can be any of the following:

- macOS using Homebrew to install the software
- Windows 10/11 or Windows Server using Chocolatey to install the software
- A Linux environment using package managers such as apt for Ubuntu or RHEL-based using Yum

The following software is required for the development environment:

- Puppet agent (`https://www.puppet.com/docs/puppet/8/install_agents.html`)
- Bolt (`https://puppet.com/docs/bolt/latest/bolt.html`)
- Visual Studio Code (`https://code.visualstudio.com/`) with the following extensions:
  - JSON for Visual Studio Code
  - Puppet
  - rest client
  - Ruby
  - ShellCheck
  - Thunder client
  - VSCode Ruby
  - YAML
  - PowerShell
  - The puppet enterprise cloud deployment module (pecdm) (`https://github.com/puppetlabs/puppetlabs-pecdm`)
- The GitHub CLI (`https://github.com/cli/cli`)
- The Puppet development toolkit (`https://puppet.com/try-puppet/puppet-development-kit/`)
- The Azure CLI (`https://docs.microsoft.com/en-us/cli/azure/install-azure-cli`)
- An Azure account
- A GitHub account (free account)
- The SSH keys created to communicate with GitHub

The pecdm module (`https://github.com/puppetlabs/puppetlabs-pecdm`) will create resources as specified via the `bolt` command. The cost of running the labs in Azure should be carefully watched via the Azure cost analysis tools to avoid unexpected bills. Labs not in use should be destroyed or at least deallocated to reduce charges.

All of these components have equivalents that you might use in your organization. However, the purpose of this development and lab setup is to make it as simple and automated a setup as possible. It might well be an exercise you wish to do as the book progresses to test out your own components. The pecdm module itself supports AWS, Azure, and GCP with instructions on the module to configure the necessary CLI.

The code for this section can be found at `https://github.com/PacktPublishing/Puppet-8-for-DevOps-Engineers/tree/main/ch02`.

Major changes since Puppet 5

Puppet 5 reflects the change in direction of Puppet as an organization, which was highlighted in the previous chapter. Its focus is on performance and scaling for infrastructure and stability in the language. This section will cover the changes that have taken place between Puppet 5 and 8; these versions reflect the versions of Puppet in use, which you are likely to see in code bases you are working with and in modules you would take from the Puppet forge. It will also cover some old patterns and issues you might see in code that reflect how Puppet was before version 5.

Puppet 5

Puppet 4 had a large number of deprecated features, which were almost all removed in Puppet 5. It is not worth listing all of these features, but just to set the context of the release, it was more about finishing what had been started in Puppet 4 by introducing more new features. It standardized package numbering, with all the Puppet packages starting at 5.0.0 instead of the mismatches of various packages, such as Puppet 4 requiring Puppet Server 2.x and Puppet agent 1.x.

Puppet 5 as a server platform delivered big boosts in performance: agent runtimes were 30 percent lower, CPU utilization was at least 20 percent lower, the catalog compile times reported by Puppet Server were 7 to 10 percent lower, and Puppet 5 was able to scale to 40 percent more agents. Puppet Server metrics were introduced to give greater observability of the Puppet platform. In addition to this greater performance and scalability, Puppet Enterprise 2017.4 onward had capabilities to allow for disaster recovery along with package inspection, which stored information about software installed across the estate regardless of whether Puppet managed it or not. Full technical details of the Puppet Enterprise features will be discussed in *Chapter 14*.

Although not affiliated directly with Puppet 5, the **Puppet Development Kit** (**PDK**) was released around the same time, which automated a lot of tool installation, testing, linting, and the creation of module directories (this will be covered in detail in *Chapter 8*). Previously, this had to be done by hand or by individual developer automation. Additionally, Hiera 5 integrated with EYAML (a mechanism of encrypted data covered in *Chapter 9*), which was introduced and massively simplified how data could be secured and still used.

Puppet 6

Puppet 6 came with a substantial change when a lot of types that had been included with the core Puppet install were removed and put into modules, which users could choose to download from the Puppet forge. This was to narrow down the installation, as the number of core types had grown over time, and it was far more efficient for users to choose what they wanted. A review of what functions were being consistently used took place, and a number of string and math functions were moved from the `stdlib` module into core Puppet to reflect their core use. The trusted external command was also introduced; this allowed for external data sources to be queried and brought in like facts, so an API on a Satellite server or a database server could be called and brought in for use in Puppet code. This will be covered in detail in *Chapter 13*. Additionally, the deferred data type was introduced, which allowed variables to run deferred functions locally on machines at deployment time. This is particularly useful for use cases such as secret management, for example, a vault, where a traditional function would make its call from the Puppet primary server and send the secret to the agent over the Puppet infrastructure. Parametrized execs were introduced in 6.24, which allow for the separation of commands from parameters when using the `exec` resource type – a powerful security measure to prevent commands from being passed instead of parameters.

On the platform side, the Puppet certificate commands were changed from `puppet cert` commands to `puppet server ca`, which were more complete and more powerful commands. Also, PuppetDB was included on Puppet compiler servers to better manage the load of requests on PuppetDB. The full details of the platform are covered in *Chapter 10*.

Puppet 7

One of the most noticeable changes in Puppet 7 was the removal of harmful terminology following a review and work beginning in 2014. This focused on phrases such as master/slave and blacklists/whitelists. For Puppet, this means that master servers became primary servers, master services became server services, and in modules, the master branch became the main branch. It also means blacklist/whitelist terminology was replaced with allow list/block list.

The parameterized execs mentioned in the Puppet 6 updates were available with the 7.9 Puppet language. Factor was upgraded to version 4, a re-write in Ruby, which provides features such as benchmarking, timeouts, and user caching, which will be discussed in *Chapter 5*. The option to not include legacy facts via the `include_legacy_facts` option was included as of 7.21.

The platform upgraded to Postgres 11 and Ruby 2.7, which further increased performance.

The reporting mechanism also had the option to not include unchanged resources in its reports via the `exclude_unchanged_resources` option.

Again, although not affiliated with Puppet releases directly, PDK 2.0 was released close to Puppet 7, dropping support for Puppet 4.

Puppet 8

Puppet 8 primarily focused on upgrading internal components, including moving to Ruby version 3.2. While this release had significant implications for developers extending Puppet, such as immutable string literals, we won't delve into those details in this book.

For Puppet users, the most noticeable change was in event reporting, which now only reports changed resources and excludes unchanged ones. We will explore this further in *Chapter 10*.

Another important change is that Facter, by default, no longer includes legacy facts. This will be discussed in *Chapter 5*.

Additionally, Puppet no longer requires Hiera 3, although it's worth mentioning that Hiera 3 should not have been used since Puppet 4, as discussed in *Chapter 9*.

One potentially breaking change is the strict mode being enabled for variables. Accessing an undefined variable or attempting actions like adding a number and string will now result in errors. This topic will be covered in *Chapter 4*.

Deferred functions, which will be discussed in *Chapter 5*, are now lazily evaluated by default. This means they are evaluated at catalog application time, allowing dependencies to exist in the catalog for deferred functions.

As of writing, the latest version of PDK, 2.7.1, does not support Puppet 8, but PDK 2.6.0 has deprecated Puppet 5 support.

In this book, we'll explore these changes and their impact on Puppet users and practitioners. Let's dive into the world of Puppet 8 and its exciting features!

Legacy Puppet patterns

This section will highlight some old patterns and their reason for use in old versions of Puppet. This will help you to understand code that can be commonly found in older, unmaintained modules or code that simply has never been refactored over time. Puppet 4 introduced data types, but before this, all variables were strings, and a lot of comparisons and other functions could have quite strange and inconsistent results. To understand the full extent of this, you can view `www.youtube.com/watch?v=aU7vjKYqMUo`. Therefore, you might see in historic code the odd handling of variables and checks for undefined variables. Originally, **facter** facts were also just called **top-level variables**, which could be very confusing with normal variables and created the opportunity for accidental overrides. This changed to the **facts hash**, which we will show in more detail in *Chapter 5*.

The platform infrastructure was more complicated and varied with the options of using Rack or WEBrick configurations. In very early versions of Puppet code, which predate the `file_line` function of the Puppet `stdlib` module, there was no way to manage single lines of a file. This resulted in the overuse of Augeas (a tool that can parse files to allow for manipulation) and templates (which allow for the creation of file using conditional logic and variables). Augeuas is very advanced but often over-complicated and a drain on performance, while the overuse of templates resulted in whole files being enforced instead of just the individual line or setting required. Therefore, when working with Puppet code that was developed in earlier versions, it is worth reviewing to make sure any code you inherit really does need to control a whole file and does not overuse Augeas when simpler solutions now exist. The `params.pp` pattern was heavily used in modules before Hiera offered the ability for class parameters to be overwritten. The sensitive data type was not introduced until 4.6, which made it hard to handle any secret data securely in code. Finally, the original Puppet versions had no concept of loops, as provided by lambda functions, which were introduced in Puppet 4. So, you might find, in old code examples, obscure patterns being used to achieve a similar effect.

IDEs and tools to assist in Puppet development

One of the greatest issues with early Puppet development was the lack of a consensus around how to develop and a lack of integration. As discussed in *Chapter 1* this changed greatly around the time Puppet 5 was released. This section highlights some tools as opinionated recommendations based on usage and experience in Puppet, and most of them will be used in the labs and demonstrations. This is certainly not the only way to develop Puppet code, and your organization might require the usage of different tools depending on the environment.

The **PDK** is central to how Puppet development has changed for the better and will be discussed, in full, in *Chapter 8*. It bundles various elements of tooling for creating modules, linting, and testing and allows them to be run from the `pdk` command. Previously, Puppet developers had to gather the tools, install dependencies, and then run each of the various commands that `pdk` is made up of.

Visual Studio Code has become an incredibly powerful and popular source code editor. It is free and multiplatform, with a vast selection of extensions including the Puppet extension (`https://marketplace.visualstudio.com/items?itemName=puppet.puppet-vscode`). It creates powerful shortcuts that allow all your work to take place in the IDE, which will be demonstrated throughout this book.

I will not be using it as part of the lab directly, but since many prefer a command-line editor as opposed to Visual Studio Code, it should be noted there are Vim modules (`https://github.com/rodjek/vim-puppet`) that can provide linting and syntax checking within VIM.

A particularly useful web page for development is the `https://validate.puppet.com/` site, which can be used to quickly paste in Puppet code to validate and parse it and create relationship graphs.

At an even more advanced level is the Puppet debugger (`https://github.com/nwops/puppet-debugger`), which allows for running Puppet code and taking breaks in the code, which allows you to see the state of variables. This will become useful as more advanced code is authored.

How to deploy your Puppet lab and development tools

This section will run through how to install and configure your desktop environment and then use that environment to stand up the Puppet infrastructure in Azure, configure it with a control repo, deploy some modules to an environment, and test logins to the web console. This will confirm the lab environment functions as expected and should give you the confidence to start up and shut down the labs as required to avoid paying for unecessary virtual machine running time costs on Azure.

In *Figure 2.1*, the final result of this exercise is shown. The device you use as a development environment will have Visual Studio Code installed to edit the code that has been cloned from GitHub. A PowerShell or shell session, depending on the OS, will use Bolt with Terraform to stand up the infrastructure in Azure and then apply the configuration to that infrastructure, configuring a Puppet Enterprise server and an attached instance to that server. The web console of the Puppet Enterprise server will be accessible via HTTPS to a web browser:

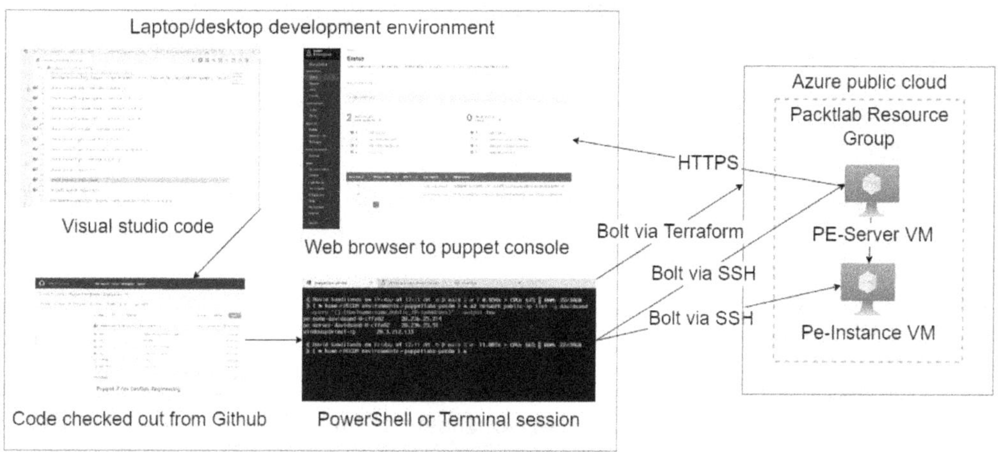

Figure 2.1 – The lab setup

Mac desktop

The Mac installation will rely on Homebrew to automate the installation process, for which Puppet has created its own repos (`https://github.com/puppetlabs/homebrew-puppet`). Run the following commands to install the desktop tooling that was highlighted in the *Technical requirements* section:

```
/bin/bash -c "$(curl -fsSL https://raw.githubusercontent.com/Homebrew/
install/HEAD/install.sh)"
brew update
brew install azure-cli
brew install --cask puppetlabs/puppet/puppet-agent
```

```
brew install --cask puppetlabs/puppet/pdk
brew install --cask puppetlabs/puppet/puppet-bolt
brew install --cask visual-studio-code
brew install gh
brew install shellcheck
brew install puppetlabs/puppet/pe-client-tools
brew install git
```

Windows desktop

The Windows installation relies on Chocolatey for installation. Run the following code in a PowerShell session; note that administrative rights are only required for the first command:

```
Set-ExecutionPolicy Bypass -Scope Process -Force; [System.
Net.ServicePointManager]::SecurityProtocol = [System.Net.
ServicePointManager]::SecurityProtocol -bor 3072; iex ((New-Object
System.Net.WebClient).DownloadString('https://community.chocolatey.
org/install.ps1'))
choco install pdk -y
choco install puppet-agent -y
choco install vscode-puppet-y
choco install puppet-bolt -y
choco install vscode -y
choco install git -y
choco install pe-client-tools -y
choco install gh -y
choco install azure-cli -y
choco install shellcheck -y
Install-Module PuppetBolt
Add-WindowsCapability -Online -Name OpenSSH.Client~~~~0.0.1.0
```

Linux desktop – RPM-based

This RPM-based Linux desktop installation was tested with Rocky Linux 8. So, some localized adjustments need to be made depending on your specific OS version and the difference in flavor. However, running the following code will add necessary the Yum repositories from the vendors and install the packages:

```
release=$(rpm -E '%{?rhel}')
sudo rpm --import https://packages.microsoft.com/keys/microsoft.asc
sudo sh -c 'echo -e "[code]\nname=Visual Studio Code\nbaseurl=https://
packages.microsoft.com/yumrepos/vscode\nenabled=1\ngpgcheck=1\
ngpgkey=https://packages.microsoft.com/keys/microsoft.asc" > /etc/yum.
repos.d/vscode.repo'
sudo rpm -Uvh https://yum.puppet.com/puppet7-release-el-${release}.
noarch.rpm
sudo rpm --import https://packages.microsoft.com/keys/microsoft.asc
```

```
echo -e "[azure-cli]
name=Azure CLI
baseurl=https://packages.microsoft.com/yumrepos/azure-cli
enabled=1
gpgcheck=1
gpgkey=https://packages.microsoft.com/keys/microsoft.asc" | sudo tee /
etc/yum.repos.d/azure-cli.repo
sudo dnf config-manager --add-repo https://cli.github.com/packages/
rpm/gh-cli.repo
sudo rpm -Uvh https://yum.puppet.com/puppet-tools-release-el-8.noarch.
rpm
sudo yum -y install epel-release
sudo yum check-update
sudo dnf install gh
sudo yum install code
sudo dnf install azure-cli
sudo yum install ShellCheck
sudo yum install puppet-bolt
sudo yum install https://pm.puppetlabs.com/pe-client-
tools/2021.7.0/21.7.0/repos/el/8/PC1/x86_64/pe-client-tools-21.7.0-1.
el8.x86_64.rpm
```

The client tools are at a specific version and should be adjusted to match your installation. Check out `https://puppet.com/try-puppet/puppet-enterprise-client-tools/` to find the `curl` command.

Linux desktop – APT-based

The APT-based Linux desktop was tested with Ubuntu 20.04, so some localized adjustments need to be made depending on your specific OS version and the difference in flavor. However, running the following code should add the necessary APT repositories and install the desktop development software required:

```
release=$(lsb_release -c | awk '{print $2}')
wget -qO- https://packages.microsoft.com/keys/microsoft.asc | gpg
--dearmor > packages.microsoft.gpg
sudo install -o root -g root -m 644 packages.microsoft.gpg /etc/apt/
trusted.gpg.d/
sudo sh -c 'echo "deb [arch=amd64,arm64,armhf signed-by=/etc/apt/
trusted.gpg.d/packages.microsoft.gpg] https://packages.microsoft.com/
repos/code stable main" > /etc/apt/sources.list.d/vscode.list'
wget https://apt.puppet.com/puppet7-release-${release}.deb
wget https://apt.puppet.com/puppet-tools-release-${release}.deb
curl -fsSL https://cli.github.com/packages/githubcli-archive-keyring.
gpg | sudo dd of=/usr/share/keyrings/githubcli-archive-keyring.gpg
echo "deb [arch=$(dpkg --print-architecture) signed-by=/usr/share/
keyrings/githubcli-archive-keyring.gpg] https://cli.github.com/
packages stable main" | sudo tee /etc/apt/sources.list.d/github-cli.
```

```
list > /dev/null
sudo apt-get update
sudo dpkg -i puppet7-release-${release}.deb
sudo dpkg -i puppet-tools-release-${release}.deb
rm packages.microsoft.gpg
rm puppet7-release-${release}.deb
rm puppet-tools-release-${release}.deb
sudo apt install apt-transport-https
sudo apt update
sudo apt install code
sudo apt -y install puppet-agent
sudo apt-get install git
sudo dpkg -i puppet-tools-release-${release}.deb
sudo apt-get install puppet-bolt
sudo apt install gh
curl -sL https://aka.ms/InstallAzureCLIDeb | sudo bash
sudo apt install shellcheck
curl -JLO ' https://pm.puppetlabs.com/pe-client-tools/2021.7.0/21.7.0/
repos/deb/focal/PC1/pe-client-tools_21.7.0-1focal_amd64.deb'
sudo apt install ./pe-client-tools_21.7.0-1focal_amd64.deb
```

The client tools are at a specific version and should be adjusted to match your installation. Check out `https://puppet.com/try-puppet/puppet-enterprise-client-tools/` to find the `curl` command.

Configuring tools

Now that you have the core tools installed on whichever desktop environment you are using, the core steps will be the same for running and managing the applications.

First of all, we need to register with GitHub (`https://github.com/join`) and register with Azure (`https://azure.microsoft.com/en-gb/free/`). Once these registrations are complete, log in to the CLIs for both. Run the following and log in to the web page that will appear:

```
gh auth login
az login
```

The next step is to generate keys that will allow for communication with GitHub. You can do this by running the following:

```
ssh-keygen -t rsa -b 4096 -P ''
```

Then, we upload the key we have created using the GitHub CLI. For Mac or Linux, run the following:

```
gh ssh-key add ~/.ssh/id_rsa.pub
```

For the equivalent location of the SSH key in Windows, run the following:

```
gh ssh-key add %USERPROFILE%\.ssh\id_rsa.pub
```

Then, extensions for Visual Studio Code can be added by downloading the `extensions.list` file from the Packt GitHub repo at `https://github.com/PacktPublishing/Puppet-8-for-DevOps-Engineers/blob/main/ch02/extensions.list` and looping through the lines to install.

For Mac or Linux, you can achieve this by running the following:

```
cat extensions.list | xargs -L1 code --install-extension
```

For Windows, you can run the following:

```
foreach($line in get-content extensions.list) {code --install-
extension $($line)}
```

The next step will be to create an area for you to have a code workspace and then download the `pecdm` module into it. For Linux and Mac, in your home directory, we will create a workspace and then clone `pecdm` into this directory by running the following:

```
mkdir ~workspace/pecdm
git clone git@github.com:puppetlabs/puppetlabs-pecdm.git ~workspace/
pecdm
cd ~workspace/pecdm
```

For Windows, we will assume the equivalent directory in the user profile, creating a `workspace` directory there, and then cloning it by running the following:

```
mkdir %USERPROFILE%\workspace
git clone git@github.com:puppetlabs/puppetlabs-pecdm.git
%USERPROFILE%\workspace\pecdm
cd %USERPROFILE%\workspace\pecdm
```

Now that we have everything installed and have a work area with the cloned module, we can configure the module and run the following Bolt plan to create the Puppet infrastructure in Azure. This will stand up a Puppet 2021.7.3 primary server and create a single client registered to it. The SSH user allows you to use the SSH keys created earlier to connect to the hosts. For this example, the `params.json` file should be downloaded to the pecdm directory from `https://github.com/PacktPublishing/Puppet-8-for-DevOps-Engineers/blob/main/ch02/params.json`, I have used the UK south region and allowed an open-to-anything firewall, but you will want to choose the cloud region closest to you and set a firewall with rules to allow only your desktop environment and Azure region to access it. The following links can help you to work out this choice:

- `https://azure.microsoft.com/en-gb/global-infrastructure/geographies/#geographies`
- `https://www.azurespeed.com/Azure/Latency`

The code is as follows:

```
bolt module install --no-resolve
bolt --verbose plan run pecdm::provision @params.json
```

This should take around 20 to 30 minutes to complete.

You can then run the following Azure CLI command to return the list of hostnames and public IPs:

```
az network public-ip list -g packtlab --query "[].
{Hostname:name,Public_IP:ipAddress}" --output tsv
```

The output will look similar to this:

```
pe-node-packtlab-0-cffe02      20.108.156.266
pe-server-packlab-0-cffe02    20.108.156.67
```

Copy the IP address listed for the entry that starts with `pe-server` into a web browser to reach the Puppet Enterprise console screen. Then, you can use the login details with the username of `admin` and the password as `puppetlabs`.

To destroy this infrastructure and ensure no unnecessary costs are incurred, run the following command:

```
bolt plan run pecdm::destroy provider=azure
```

Alternatively, if labs are to be kept for periods of time, it is possible to stop and deallocate each virtual machine to minimize the charge and then restart them later using the following commands:

```
az vm deallocate --resource-group packtlab --name <VM name>
az vm start --resource-group packtlab --name <VM name>
```

This section has fully run through the creation of your developer desktop and then standing up and destroying Puppet infrastructure. It ensures you are ready for the labs in future chapters. In this lab, the `pecdm` and `peadm` modules are used to configure a standard architecture, which is one of Puppet's supported architectures: `https://puppet.com/docs/pe/latest/supported_architectures.html`. In *Chapter 14,* we will discuss, in more detail, the different architecture options. But for now, it is important to understand that the standard is the base level providing a single Puppet Server. In this scenario, `pecdm` configures the necessary infrastructure using Terraform, while `peadm` installs the Puppet Enterprise components. Both modules will be used as examples of using Bolt projects, tasks, and plans and will be reviewed in *Chapter 12.*

You may ask why Puppet 7 has been installed for your lab desktop and PE 2021.7.3 (using Puppet 7) for labs has been chosen but at point of publication Puppet 8 was not available and using the LTS version would provide greatest stability. Puppet 8 is expected to be available in PE 2023.3 and at this point you could update the `params.json` file to use a more up to date PE version and adjust the desktop version installs to 8 but the behavior difference seen in the labs should be minimal.

References and further research

This section will cover further resources and references that can be used alongside this book. They go into further depth and enable you to learn about Puppet from both Puppet and the community.

The general page (`https://puppet.com/docs/`) is the core doc page, where you can find all the products of Puppet and sections such as patterns and tactics. We will highlight different sections of the docs to refer to as we progress through the book.

Puppet runs through various media forms where a variety of articles are published covering new product releases, security updates, and guides for implementations. Their handles are listed as follows:

- Blog: `https://puppet.com/blog`
- Podcast: `https://pulling-the-strings.simplecast.com/`
- Dev.to articles: `https://dev.to/puppetecosystem` and `https://dev.to/puppet`
- Twitter: `https://twitter.com/puppetize` and `https://twitter.com/PuppetEcosystem`
- YouTube: `https://www.youtube.com/channel/UCPfMWIY-qNbLhIrbZm2BFMQ`

Puppet has its own learning site (`https://training.puppet.com/learn`),this site includes various elements such as the Puppet practice labs, which are online labs you can run entirely from a web browser and tackle boxes, which are guides on achieving small focused tasks. Puppet's support knowledge base was made public in April 2022, allowing anyone to search and view the troubleshooting guides, best practices, and FAQs, which are available at `https://support.puppet.com`, without the need for a login. Archived articles for the older version of Puppet can be found at `https://github.com/puppetlabs/docs-archive/tree/main/supportkb#readme`.

Puppet previously run two, instructor-led training courses, which had to be paid for and lasted 3 days (*Getting started with Puppet* and *Puppet Practioner*). During 2022 the *Fundamental Core Training* modules replaced *Getting started with Puppet*, and the *Advanced Core Training* modules replaced *Puppet Practioner*.

The key difference is that the `Fundamental Core Training` modules have a free self paced version, and both training sets are broken up into three module sets that are each a day long. More details can be found on the Puppet Compass site.

Fundamental Core Training:

- PE101: Deploy and Discover
- PE201: Design and Manage
- PE301: Develop and Maintain

Advanced Core Training:

- PE401: Extend Capability
- PE501: Continuously Deliver
- PE601: Automate at Scale

Enterprise modules that produce commercially licensed Puppet modules on the Puppet forge have a blog discussing various Puppet topics at `https://www.enterprisemodules.com/blog/` and a Twitter account at `https://twitter.com/enterprisemodul`.

Two other noted Puppet consultancy and development groups were formed after the split of Example42 GmbH into Example42, which is now a brand of Lab42 with a blog at `https://blog.example42.com/blog/` and a Twitter account at `https://twitter.com/example42`, and Betadots, which has a blog at `https://dev.to/betadots` and a Twitter account at `https://twitter.com/betadots`. Both provide insights into their Puppet development work and approaches.

To ask questions about Puppet or talk with people in the community, you can join `https://slack.puppet.com/` and `https://www.reddit.com/r/Puppet/` to ask questions about Puppet and the community.

This section is not supposed to be an exhaustive list of references. It is to give a view of some of the better-known and long-lasting sources of information and communities to view and follow to get to know Puppet better.

Summary

In this chapter, we discussed the changes in the modern versions of Puppet 5 to 8 and some antipatterns to look out for that could be left over from legacy Puppet code. It might be more practical to come back to this section if you aren't familiar with Puppet and read through the changes again once completing the book.

We covered the available tooling to use in a developer environment and the IDE to automate and quicken your Puppet development environment, and we have installed these tools to introduce the lab. We have learned how to stand up both the reader's development environment and the Puppet infrastructure in Azure.

At the end of this chapter, we covered the various sources and communities that can be used to further learn about Puppet, keep up to date with ongoing developments, and signpost where to ask questions and discuss Puppet with the community.

In the next chapter, we will begin to look at the Puppet language, covering the fundamental building blocks of resources, types, and providers. We will look at the basic syntax and style of coding in Puppet and how to use various references and commands to make it easier to generate code and find documentation. We will look at the core types to start coding in Puppet and how to use them well. Then, we will look at how to use defined types for repeatable patterns of resources, use classes to contain and include resources in catalogs, and finally, finish by looking at the more advanced feature of exporting and collecting resources to share resource declarations across multiple clients.

3
Puppet Classes, Resource Types, and Providers

This chapter will cover how classes and defined types provide structure and a way to group resources, allowing code to be modular and reusable. You will learn about the components that make up resources; types, providers, and the attributes applied to them. You will be shown how to use Puppet commands to understand the current state of the system and by looking at three of the most common resource types – packages, files, and services. You will see how to find out the attributes that are available to a resource and how to declare a state.

Using these three resource types, you will see how a simple installation of a package, configuration file, and service can be quickly used to start up an application with Puppet code, such as Apache or Grafana. The other core resource types will then be discussed, highlighting the best practices and approaches. A number of metaparameters (attributes that can be applied to any resource) will be discussed, along with some advanced patterns for resource declaration.

You will then come across some anti-patterns, which, although still documented Puppet language features, are not recommended for use. This will help you understand any legacy code you may encounter and consider where code needs to be refactored.

In this chapter, we're going to cover the following main topics:

- Classes and defined types
- Resources, types, and providers
- Core resource types
- Metaparameters and advanced features
- Anti-patterns

Technical requirements

Provision a standard sized Puppet server with a Windows client and a Linux client by downloading the `params.json` file from `https://github.com/PacktPublishing/Puppet-8-for-DevOps-Engineers/tree/main/ch03` using the following command:

```
bolt --verbose plan run pecdm::provision --params @params.json
```

Classes and defined types

As discussed in *Chapter 1*, Puppet code is stored in manifest files ending with `.pp`. It is possible to just write resources into a single manifest file and then, using the `apply` command, `puppet apply example.pp`, enforce the code locally. It can also be done without the manifest file using the `execute` flag with the Puppet code in the field of the command, such as `puppet apply -e 'package { 'vscode': }'`.

> **Note**
>
> `puppet apply` can also be run against a directory of manifests and it will parse every file in order, descending a directory structure. In *Chapter 11*, node definitions will allow us to utilize this.

While both of these approaches are useful for testing and learning purposes, they have a clear limitation in terms of lacking any structure, which will result in both having to run a lot of large static commands or files and having no way to pass data. **Classes** are named sections of code that provide this structure, offering a way of grouping resources together and assigning data, which we can apply to servers. A class definition goes into a manifest file and within the class definition, we put our resource definitions. The syntax is as follows:

- The `class` keyword.
- The name of the class.
- Optional parameters within `( )`.
- Puppet code with { }

```
class  example_class (
  String example_parameter
)
{
   <code block>
}
```

Parameters will be discussed in greater detail in *Chapter 4*, but for now, it should be understood that `class` parameters allow classes to be supplied with external data. For example, a class might have a resource that installs a package, and a parameter can be used to specify the version of that package to be installed.

> **Note**
>
> An optional `inherit` keyword can be added to a class to allow class inheritance, whereby you can create a general base class and then extend it in an inheriting class or classes. This pattern is no longer used and is no longer discussed in the Puppet documentation as of Puppet 6, beyond saying it exists as a keyword. There are better ways to achieve this behavior using data, which we will cover in *Chapter 9*.

A common early source of confusion with classes is that this structure only defines a class; it does not declare it to be included in the catalog compiled from the Puppet code. This contrasts with the resource statement in manifests, which by being written and then applied are added to the catalog.

This means running `puppet apply` on a manifest containing a class will do nothing. To add the classes to a catalog, we must declare the class using the `include` function, make a class resource declaration, or we must use an **External Node Classifier** (**ENC**). ENCs will be covered in *Chapter 11*, but for now, they can be understood as Puppet server scripts that identify the classes to be included in a node.

Including a class

The `include` function is the simplest way to add classes via the declaration in the code block of a class in a manifest file of `include class_name`. It can be used multiple times across multiple classes and will result in only one entry. To declare a class with `puppet apply` directly, we can instead run `puppet apply -e "include class_name"`, which will test a manifest file with a class. Following the module structure, this would apply the manifest from the `class_name/manifest/init.pp` path.

The class resource declaration

In the next section, resource declaration will be covered in more detail, but declaring a class such as a resource allows us to pass in the attributes we have defined or looked up. It looks like this, but can only be used once in a catalog:

```
class {'class_name':
  paramter1 => 'value1',
}
```

Defined types

A defined type is a block of Puppet code, which, in contrast to a class, can be declared multiple times in a catalog by passing in parameters and a unique name. Like a class, it is by best practice defined in a manifest file by itself.

The syntax is as follows:

- Starts with a `define` keyword
- The type name
- Open brackets (( )
- List of parameters
- Close brackets ())
- Open braces ({)
- The resource body
- Close braces (})

In addition to the parameter list defined, the `$title` and `$name` variables are available to be used within the definition. This ensures the resources we declare are unique. A very simple example could take a name and a group and ensure a user and a group are created and a file is placed in the `user` home directory owned by the user and group we have created:

```
define exampledefine (
  String user = "${title}",
  String group
) {
user { ${user}: }
group { ${group}: }
file { '/export/home/${user}/.examplesetting':
  user => ${user},
  group => ${group},
  content => "User is ${user} and group is ${group}",
}
}
```

Defined types are the same as classes; applying the manifest file will not produce anything. A defined type resource declaration must be made in a class, which can then be included:

```
exampledefine {'user1':
  group => 'group1',
}
exampledefine {'user2':
```

```
    group => 'group2',
  }
```

This example has its dangers since if the second declaration for `user2` also used a group of `group1`, this would result in a duplicated resource declaration.

Namespaces

Namespaces are segments that identify the directory and file structure for classes in manifest files. These namespaces are separated with two colons (`::`), so, for example, the following directories would translate as follows:

File path name	Namespace
/manifests/base.pp	base
/manifests/windows/grafana.pp	windows::grafana
/manifests/linux/apache.pp	linux::apache
/manifests/linux/ubuntu/landscape.pp	linux::ubuntu::landscape

Table 3.1 – Namespace directory translation

If we wanted only to apply the `windows::grafana` class, we could therefore run `puppet apply -e "include windows::grafana"` from within the `manifest` directory.

There is no limit to the depth a namespace can have, but the best practice would be to stick to a couple of levels.

In *Chapter 8*, we will see modules that have namespaces where the module name is the root level for all classes except one.

Resources, types, and providers

Resources are the fundamental basic unit of the Puppet language; every stateful item we wish to describe is a resource. Resources must be unique in terms of what they manage since Puppet has no way of managing or prioritizing conflict between resources. It will simply call out that a clash exists and fail to compile a catalog.

Each resource will have a type, which is a description of what we are configuring, such as a file or a registry setting; parameters, which are variables containing the settings we can customize for the resource; and a provider, which is the underlying implementation allowing Puppet to be OS independent. This provider is often a default based on the OS but can be added as an attribute if required. So, a resource declaration has the following syntax:

- Opens with the type name, such as `file`, with no quotes and in lowercase
- A curly brace (`{`)
- The title of the resource in quotes
- A colon (`:`)
- A list of attribute names and the value of that named attribute with `=>` between, ending with a comma (`,`)
- A closing curly brace (`}`)

> **Note**
> Everything between the two curly braces is known as the **resource body**. It is possible to have multiple bodies in a single resource declaration, essentially declaring multiple resources of the same type, but for clarity, I would generally advise against this.

As pseudocode, this syntax looks like the following:

```
type { 'title':
   attribute1 => value1,
   attribute2 => value2,
}
```

Here's a real example of ensuring a package named `vscode` is at the latest version on the system:

```
package { 'vscode':
  ensure => 'latest',
}
```

What was given in the syntax list for both the resource and class declarations/definitions was the minimum required, while the code examples were spaced and broken over several lines for stylistic reasons and following best practices. It is possible to write declarations and definitions as a single line but Puppet has developed a style guide – `https://www.puppet.com/docs/puppet/latest/style_guide.html` – that we will use throughout this book, along with other opinionated best practices to create readable, maintainable, and simple code.

Here are some examples of the style guide being applied in the code examples:

- Use a two-space indent
- No trailing whitespace

- Attribute names should align
- Attribute => symbols should align
- Attribute values should align
- Include trailing commas after all attributes

Although there are no limits or syntactical meanings for whitespace, the Puppet language style guide's recommendations aim to make the code more readable and consistent. The style guide states all attributes should have trailing commas; this ensures adding a new attribute will only show a single change in a Git diff, but you may find some code follows a pattern of having no comma on the last attribute, which would make it clear it was the last element. This will pass linting checks but may cause issues for not meeting Puppet style guides if you wish to get code approved by Puppet for module use.

As there are a number of syntactic and stylistic rules, the best way to learn is to use style guide linting, made available via the Ruby gem, `puppet-lint`, with syntax validation made available via the `puppet parser validate` command. The Puppet extension on Visual Studio Code has these commands integrated into its checks, so it highlights syntax and lint issues as you edit. In the screenshot in *Figure 3.1*, the warning output of the lab is visible with some stylistic and syntactical errors:

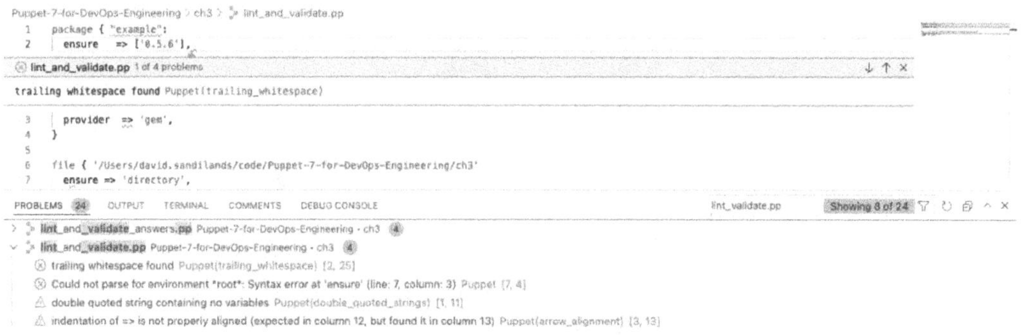

Figure 3.1 – Visual Studio Code showing syntax and lint issues

Similar effects can be achieved in `vim` using `https://github.com/rodjek/vim-puppet`.

> **Important note**
>
> Throughout this book, advice will be given on best practices and approaches to coding, with a lot of this advice taken from sources such as the Puppet style guide. One of the best things an organization can do to develop clear and consistent Puppet code is to write its own best practices and style guidelines, building on top of the foundation provided by the Puppet style guide and ensuring it is followed when reviewing code. This can equally disagree with points raised in the style guide or this book, as long as it is best for your organization and developers and it is agreed to.

Resources of each type must be uniquely **titled**, so you could have file, service, and package resources all titled `ntp`, but not two service type resources both titled `ntp`. There are no other limitations on how they are named in terms of characters or spacing, but for performance purposes, titles should be kept short and never be longer than 140 characters. This **title** is what identifies the resource to Puppet itself when it generates a catalog.

The `namevar` attribute (also known as the **name attribute**) is what the target system uses to identify the resource and confirm uniqueness. `namevar` by default is the same as the title unless attributes are assigned. In some cases, types will use multiple attributes to define `namevar`, such as a package using the command and name together. This is used in cases where multiple copies of the same configuration can be installed via different mechanisms, such as installing a package of the same name as a Ruby gem and as a **Red Hat Package Manager** (**RPM**).

Installing the Apache package can demonstrate the difference between `namevar` and **title**. For example, the `apache_package` name variable is set based on the operating system. For Fedora, the package name will be `httpd`, while for all other operating systems, it will be `apache2`. This means our title for this package resource is `apache`, and when referring to this resource in Puppet code, we can always refer to it as the `apache` resource package, while the target system will refer to it by the appropriate package name, ensuring it is a uniquely managed installation:

```
$apache_package_name = $facts['os']['name']? {
  'Fedora' => 'httpd',
  default  => 'apache2',

}

package { 'apache':
  ensure => 'latest',
  name   => "$apach_package_name",
}
```

Let us now move on to some practical examples.

Lab

To practice what has been learned so far, look at the file at `https://github.com/PacktPublishing/Puppet-8-for-DevOps-Engineers/blob/main/ch03/lint_and_validate.pp` and try to correct the errors highlighted in VS Code. Alternatively, use the `puppet-lint -f` (`-f` automatically fixes issues where possible) and `puppet parser validate` commands from the VS Code integrated terminal or a separate terminal session.

`https://validate.puppet.com/` can also be used to do validation checks online.

Examining the current system state

This chapter so far has discussed how resources should be structured and styled and, with all these rules, it can be intimidating when starting to write your own resources. The `puppet resource` command allows us to produce Puppet code from the state of a current machine; this command is supplied parameters of a type and a `namevar` variable. To give an example, looking at the directory on which a Windows desktop puppet has been installed would produce something like the following:

```
C:\ProgramData\PuppetLabs>puppet resource file "c:\Program Files\
Puppet Labs"
file { 'c:\Program Files\Puppet Labs':
  ensure   => 'directory',
  ctime    => '2022-01-31 22:01:02 +0000',
  group    => 'S-1-5-18',
  mode     => '2000770',
  mtime    => '2022-01-31 22:01:02 +0000',
  owner    => 'S-1-5-18',
  provider => 'windows',
  type     => 'directory',
}
```

From this example, it should be noted that certain attributes are returned only for the information we refer to as properties and cannot be managed by Puppet, such as `mtime` and `ctime`. Other attributes such as `provider` do not need to be declared, as `windows` would be the presumed provider on a Windows machine. Apart from this, with minor adjustments, this output can just be directly put into Puppet manifests and run. (Later in this chapter, we will show you how to review type attributes.)

> **Note**
>
> Visual Studio Code allows you to run Puppet commands via the command palette (*Ctrl* + *Shift* + *P*, or for Mac, *Command* + *Shift* + *P*). Type `puppet resource`, then the resource type, and optionally name `var`. It will then paste the output into your open file.

In the previous example, we ran `puppet resource` against a single `namevar` attribute. For certain types, you can discover what the state of every resource of that type would be on a machine, such as running `puppet resource package` for packages. This clearly will not work for the likes of files, as recursively going through every file on a host would produce too much information, but you can quickly produce information on your host's setup.

In VSCode, try opening a new file, running the command palette with `puppet resource`, and just entering `package`. This will list all the packages recognized by Puppet and available Puppet providers. An example of this output is available at `https://github.com/PacktPublishing/Puppet-8-for-DevOps-Engineers/blob/main/ch03/puppet_resource_package.pp`.

Introducing types with the package, file, and service pattern

Having discussed the structure and style of declaring resources, the next step is to introduce the core types available to Puppet and how you can discover the attributes and features of a type.

The core types are documented online at `https://www.puppet.com/docs/puppet/latest/type.html#puppet-core-types` and can be viewed on the command line with the `puppet describe` Puppet command. Using `puppet describe --list` will list all the types available in your environment; you can then review a type by passing the type name, for example, `puppet describe package`. This documentation is also visible in VS Code when you hover the mouse pointer over the types and attribute names in a resource declaration.

Starting with the combination of package, file, and service types, you will be able to install, configure, and start an application.

The package type

Running `puppet describe package` or viewing the web contents at `https://www.puppet.com/docs/puppet/latest/types/package.html`, we can view the description of what the type is for and a list of attributes and available providers.

A package used at its simplest level can just be declared as a package resource with a title:

```
package { 'vscode': }
```

This sets several attributes to defaults, resulting in using the default provider for the underlying operating system, such as `yum` for Red Hat or, for Windows, the Windows provider, which handles `.exe` and `.msi` files. It will also install at the latest package version available but, when enforced, will only ensure the package is installed and not maintain it at the latest version.

This versioning behavior is controlled by the `ensure` parameter and the example defaulted to a value of `present`, which can also be declared as `installed`. The `latest` value, just as it sounds, ensures the package is at the latest version available to the provider. For more flexible versioning, it is possible to set a value as a string version, such as `1.2.3`, and, depending on the support of the provider, to use ranges, such as `> 1.0.0 < 2.0.0`. Using the value of `absent` is an important part of Puppet, where resources don't just ensure what is present in the server state but also what should not be there.

Related to using the `absent` value for `ensure` is the `purged` value, which is a provider-dependent option. If set to `true`, it removes configuration files on the removal of packages.

The `providers` attribute is often left as the default, but if it is required to be installed via another package management system such as `pip` or `rubygems`, can be assigned an appropriate provider's name as its value.

To see what providers can be used, the `-p` flag can be used on the `describe` command: `puppet describe package -p`.

Taking the example of Windows, it is important to note that it tells us the Windows provider is the default provider and it lists the supported features, which are attributes that will work with this provider. This difference in attributes reflects the different underlying commands used by the provider.

The `source` attribute is a URL to the package file; this allows for remote calls to web sources such as JFrog Artifactory or locally downloaded files and is a required parameter for certain providers, such as Windows, which requires a location of the `.bin` or `.exe` file.

The `command` attribute, new since Puppet 6, allows you to select which command the provider should run. This is necessary for situations where you have multiple versions of an installer command available on a machine.

The `name` attribute, which should be the name of the package, will be set as the title by default and combined with the command attribute; since Puppet 6, this is what makes the `namevar` attribute for a package. In Puppet 5, the provider attribute is used instead of the command attribute.

> **Note**
>
> Sometimes, it may be necessary due to dependency issues to run install commands such as yum with multiple packages in a single command. There isn't a way to do this under the package type; the best approach would be to use an `exec` type, which we will talk about later in this chapter.

So, as an exercise, write a manifest for the following; create a new file for each platform example, `package_rhel8.pp`, in `vscode` or a terminal.

On RHEL 8, do the following:

- Install `rubygem activerecord` so that it is greater than version 7
- Install the latest `cowsay` from yum
- Ensure the `pinball` package is absent from the system in a resource titled `no games`

View the suggested solution at `https://github.com/PacktPublishing/Puppet-8-for-DevOps-Engineers/blob/main/ch03/package_rhel8_answer.pp`.

On Windows Server, do the following:

- Install `ruby and devkit from the .exe file already downloaded to c:\tmp\rubyinstaller-devket-3.1.1-1-x64.exe` with the `/VERYSILENT` install option
- Install `rubygem activerecord` so that it is greater than version 7 but less than 9
- Ensure the `pinball` package is installed at version `2005-xp` in a resource titled `fun games`

View the suggested solution at `https://github.com/PacktPublishing/Puppet-8-for-DevOps-Engineers/blob/main/ch03/package_windows_answer.pp`.

> **Note**
>
> For more advanced Windows package management, it is worth looking at Chocolatey, which will be covered in *Chapter 8* (`https://forge.puppet.com/puppetlabs/chocolatey`).

The file type

Having installed packages, it is then common to add application configuration files and directories to contain them. The file type is ideal for creating files and making directory structures. It can handle the content, ownership, and permissions of files, links, and directories.

The simplest declaration of a file type is the title as a fully declared path:

```
file { '/var/tmp/testfile' : }
```

Looking at the file type via `puppet describe file`, in this case, there are only two **providers** – a Windows file or a POSIX file, which will match whichever operating system family you are configuring.

For the `ensure` attribute, there are `present` and `absent` options. Selecting `present` will default to the file value, ensuring the created resource is a normal file but only enforcing that the file path exists regardless of whether it is a symbolic link, a file, or a directory.

To create and enforce a resource, we must select the value of a file and use `direct` to create a directory or directory nest or `link` to create a symbolic link.

The path is the `namevar` attribute for this type and should be a fully qualified path, or it can default from the title.

For example, a resource titled `Puppet directory`, which creates `ensure` for the existing `directory` at `C:\ProgramData\PuppetLabs`, is as follows:

```
file {'Puppet directory' :
  ensure => 'directory',
  path   => 'C:\ProgramData\PuppetLabs',
}
```

For resources we ensure as files, the `content` attribute gives us multiple ways of putting content into the file. The simplest version is simply to put a string of the text into the file but using the functions, file, and template, we can copy the contents of whole files stored in Puppet modules or use templated files, allowing us to substitute values into pre-parsed files. These functions will be covered in detail in *Chapter 7* and *Chapter 8*.

Three attributes are then used to manage ownership and permissions: `user`, `group`, and `mode`. For `user` and `group`, this is as simple as entering the UID and GID or the username and group name. If this is not set, this will default to the user and group Puppet is running under. `mode` deals with permissions using the Unix 4-digit-style permissions mode, but for Windows systems, entering this gives a very rough and limited translation and it is better to leave `mode` undeclared and supplement files with the ACL module: `https://forge.puppetlabs.com/puppetlabs/acl`.

To give an example of the attributes we have covered, the following declaration creates a file called `config.test` with both `owner` and `group` set and the content of two lines of text:

```
file {'Example config':
  ensure => 'file',
  path   => '/app/exampleapp/config.txt',
  owner => 'exampleapp',
  group => 'examplegroup',
  content => "verbose = true\nselinux=permissive",
}
```

The `recurse` parameter allows recursive management of the contents of a directory. If set to `true` when ensuring a directory and using `source`, it will copy the directory contents recursively. It is important to note Puppet is not a file synchronization tool, so do not put too many files under Puppet management, or files that are too big. There is no specific number documented, but a common recommendation is 10 or fewer files in a recursive file resource and no greater than 25 MB. This is due to the comparative nature of Puppet, which uses `md5` checksums for content, which are expensive to run over large-sized files or large numbers of files.

Information

In the case of large numbers of files and directory structure, the module archive – `https://forge.puppet.com/modules/puppet/archive` – can be used to download and extract it into place. Alternatively, when auditing and versioning files, it is better to build a package and manage it with the package resource we spoke of previously.

Several parameters can provide protection with `recurse` using `max_files`, which can warn or error if a command is going to go over a certain limit. `recurselimit` can be used to limit how many levels of recursion will be performed.

There are only two scenarios in which it is advised to use this parameter – when you have a small number of files and the content of the files should be enforced, or when also using the `purge` parameter, which, when set to `true`, will ensure no files outside of Puppet's control will remain in the directory.

Note

We will review data types and variables in detail in the next chapter, but for now, note a parameter that takes `true` or `false` can take a value without quotes, and that is the style this book will use.

The purge parameter can only be used with ensure set to directory and recursive set to true and provides a powerful way to ensure the directory only contains files under Puppet management, removing any other files it finds. In the following, we give an example of recursion, ensuring the /etc/httpd/conf directory only contains files under Puppet's control:

```
file {'Remove apache config files outside of puppet control' :
  ensure  => 'directory',
  purge   => true,
  recurse => true,
  path    => '/etc/httpd/conf',
}
```

> **Note**
>
> There is a recursive_file_permissions module (https://forge.puppet.com/modules/npwalker/recursive_file_permissions), which can assist in managing recursive permissions over a large number of files in a performant way. This can be combined with the archive module we previously mentioned.

The validate_cmd parameters can be particularly useful with configuration files, where there is a known way to check the file we are putting in place. If the validation command fails, the old file will be left in place, preventing issues.

The target parameter is required if ensuring a link. By combining it with the path value, we get a symlink, as demonstrated in the following code:

```
file {'Picking a python on Rhel 8' :
  ensure  => link,
  path    => /usr/bin/python3,
  target  => /usr/bin/python,
}
```

The source parameter can be of several types: URIs, local files, NFS shares, or web or Puppet modules. This can also be presented as an array to provide multiple choices depending on the hostname or operating system, where it would use the first file it could find. In the following code block, we show an example, where host would be substituted with the applicable hostname and operatingsystem with the locally installed operating system:

```
file {'/etc/exampleapp.conf':
  source => [
    "nfsserver:///exampleapp/conf.${host}",
    "nfsserver:///exampleapp/conf.${operatingsystem}",
    'nfsserver:///exampleapp/conf'
  ]
}
```

In this example, on a Windows server called `server1`, applying this resource declaration would look on `nfsserver` under the `exampleapp` share to find the first match, looking for `conf.server1`, then `conf.windows` if it could not find it, and finally `conf`.

The `backup` parameter is not recommended, as managing and scaling file buckets to store these backups proves difficult, and as we will see in *Chapter 11*, there are better approaches we can consider, managing our code in Git to allow for back-out scenarios.

The `replace` parameter should be used sparingly, but if set to `true`, allows for a file to have content enforced only if it does not exist. If the file exists, the state is met. This can be useful for applications that require an initial configuration file but then overwrite it.

Having discussed a lot of attributes, try practicing constructing examples by writing a manifest file to meet the requirements listed:

1. On a Unix-based system, ensure only Puppet-controlled files are in the `/etc/sudoers.d` directory.
2. Add a `/etc/sudoers.d/mongodb` file with `robin All=(ALL) NOPASSWD: su - mongo` content and a validation command, `visudo -c`, owned by `root`, a group of `root`, and permission `0660`.
3. Create a symlink from `/opt/mongodb/mongos /home/robin/mongos`.
4. View the suggested solution at `https://github.com/PacktPublishing/Puppet-8-for-DevOps-Engineers/blob/main/ch03/file_unix_answer.pp`.

For Windows, see the following:

1. On a Windows-based system, ensure only Puppet-controlled files are in the `c:\inetpub\wwwroot` directory but subdirectories are untouched.
2. Add a `c:\inetpub\wwwroot\page with source nfsshare1:\\publish\page.html` and a validation command `c:\program files\httpvalidator\httpvlidate.exe` file.
3. Create a symlink from `c:\program files\httpvalidator\httpvlidate.exe C:\Users\david\Desktop` and use the `replace` option to replace the file if it exists.
4. View the suggested solution at `https://github.com/PacktPublishing/Puppet-8-for-DevOps-Engineers/blob/main/ch03/file_windows_answer.pp`.

Service types

Having installed software and created configuration files, the next common step is to start services with the service type. Since system services can vary widely in terms of what they support and provide, we must be careful to provide all the necessary parameters. Some services lack proper status commands but can be provided via the parameters of service.

Running and reviewing the output of `puppet describe service -p`, you will see various providers, although in most cases, the default service provider is what will be required. On certain occasions, such as legacy software on a modern Red Hat system only providing `init` scripts, we may expect to select a different provider.

The first two parameters to consider are `enable` and `ensure`. `ensure` accepts the values `stopped` or `running`, which can also be represented as `false` or `true`, respectively. This is a simple binary of whether the service should be running or not. `enable` defines in the service whether it should start on boot and is only provided by certain providers. This can be `true` or `false` to be enabled or disabled, and then there are several provider-dependent options; for example, on Windows, `false` means the service is disabled and cannot be started, and `manual` means the service is set to a manual startup type, which doesn't start with Windows but does allow the service to be started manually. `true` is an automatic startup type and `delayed` means the service is set to the automatic (*delayed*) startup type, which starts the service a couple of minutes after Windows has started up.

One final parameter to highlight for Windows would be `logonaccount`, which specifies an account for the service to run as.

To give examples of the attributes we have covered, see the following code for a Windows service, `wuauserv`, a running service with a delayed startup service and running as the `localsystem` user. The `bam` service is stopped and disabled:

```
service { 'wuauserv':
  ensure       => running,
  enable       => 'delayed',
  logonaccount => 'LocalSystem',
}
service { 'bam':
  ensure => stopped,
  enable => 'false',
}
```

Comparing this to `systemd`, the default provider for RHEL 8 and other Linux systems, we can see in the description under supported features that `systemctl` does not have delayed login or `manual` but does have `mask`, which, in system terms, means it disables the service so not even services that are dependent on it can activate it.

> **Note**
>
> Beware that the defaults for `ensure` and `enabled` are entirely dependent on the underlying provider implementation.

In cases where there are no startup scripts provided for an application, combining the `start` and `stop` parameters, you can use Puppet to bridge this gap, defining which commands start and stop the service in these parameters. The `pattern` parameter would by default take the name of the service and look for the name in the process table to confirm a running status, or you can supply a regular expression, strings, or any permissible Ruby pattern to search the process table. Alternatively, the `status` parameter can be used to point at a status script, which should return a zero-exit code if the service is running.

The following shows an example of a legacy service with scripts for starting, stopping, and checking the status of the server pulled together in this service resource:

```
service {'legacy service':
ensure     => running,
  enable   => true,
  start    => '/opt/legacyapp/startlegacy -e production',
  stop     => '/opt/legacyapp/stoplegacy -e production' ,
  status   => '/opt/legacyapp/legacystatus -e production',
}
```

It can be seen based on the nature of implementation that a careful parameter choice must be made and that this varies by scenario. Later in this chapter, we will show methods for how to cover these differences while declaring resources using a splat (`*`).

Running Puppet locally with multiple resources

In *Chapter 8* and *Chapter 10*, we will cover using Puppet agents and classification to apply Puppet code, but to test the code developed just now, as mentioned at the start of the chapter, `puppet apply` can be used to run code locally. In our labs, we will use Bolt to automatically copy our manifest files to our remote labs and run `puppet apply`.

> **Note**
>
> An additional way of applying resources is via the `resource` command we reviewed earlier. Adding parameters and settings to the command will cause it to be applied to the resource. The Puppet service could be enforced as enabled and running with the `puppet resource service puppet ensure=running enable=true` command. You will often see this command in Puppet knowledge base articles when performing fixes to Puppet services since it can usefully start/restart services without having to think about which operating system it is running on.

Relationships will be covered in detail in *Chapter 6*, but to allow for resources that are dependent on one another, as the package, file, and service pattern requires, the basics of the `require`, `before`, `subscribe`, and `notify` metaparameters need to be known. `require` and `before` mirror one another, creating a relationship between two resources so that when Puppet runs one resource, it will run before the other. It is not semantically important which way you define the relationship, although it may prove more logical where there is a many-to-one relationship to apply the dependency metaparameter to many resources.

Similarly, the `subscribe` and `notify` metaparameters allow a resource to not only have this dependency but also to send refresh events to types that support them if the resource state changes (this can be confirmed in the type documentation using `puppet describe`). This is particularly useful in service resources where updating a configuration file should result in the service restarting.

The syntax for these metaparameters is a resource reference, which comprises a resource type with a capital letter and a resource name in square brackets. To give some examples of this, the following shows examples of `before`, `notify`, and `require` being used to make the package, file, and service pattern:

```
package { 'example app package':
  ensure => latest,
  name   => 'exampleapp'
  before => File['example app configuration'],
}
file { 'example app configuration':
  content => 'attribute=value',
  notify  => Service['example app service'],
}
Service {'example app service':
  name    => 'exampleapp',
  enable  => true,
  ensure  => running,
  require => Package['example app package'],
}
```

In this example, `package` is installed first, the configuration file is then added, and the service should start. If the configuration file changes state, this will cause the service to restart. In the next section, we will talk in more detail about resource references.

Shorthand can be used to create an array of the same resource types in a dependency:

```
file { ' C:\Program Files\Common Files\Example':
  require => Package['package1', 'package2',
}
```

Running Puppet will generate reports that describe how resources, if not in the desired state, were changed into the correct state and, if the server was in the correct state, will produce little output beyond the time it took to run checks. The code can also be run in **noop mode**, which will list out what changes are needed to get to the correct state but without applying them. In *Chapter 10*, we will discuss in more detail around reporting and using `noop`.

Lab

So, use the lab environment to apply some Puppet code to our client servers.

For CentOS, we will install `httpd` and serve a web page displaying *Hello World*. Create an `apache_linux.pp` file; this will require the `httpd` package to be installed and a file to be created at `/var/www/html/index.html` with the following content:

```
<html>
<head>
</head>
<body>
    <h1>Hello World<h1>
</body>
</html>
```

We have a `/etc/httpd/conf/httpd.conf` configuration file with content sourced from `https://raw.githubusercontent.com/PacktPublishing/Puppet-8-for-DevOps-Engineers/main/ch03/httpd.conf` and validated by running `httpd -t -f` and an `httpd` service, which is enabled on boot and running.

For Windows, create a `grafana_windows.pp` file; we will install the Grafana server from `https://dl.grafana.com/oss/release/grafana-8.4.3.windows-amd64.msi`, ensuring the service is running and enabled and, in the `C:\Program Files\GrafanaLabs\grafana\conf\grafana.ini` configuration file, ensuring the content contains the following:

```
[server]
Protocol = HTTP
Http_port = 8080
```

Updating the configuration file should restart the service.

You can apply the code you have written using Bolt, which will be covered in *Chapter 12*. Using the `bolt apply apache_linux.pp --targets linuxclient.example.com` or `bolt apply grafana_windows.pp --targets windowsclient.example.com` command will copy the manifest to the server and run `puppet apply` on the client. For both Linux and Windows examples, test your solution by navigating to `http://hostname:8080` and confirming **Hello World** for Linux or the Grafana login page for Windows is visible.

Example solutions are available at `https://github.com/PacktPublishing/Puppet-8-for-DevOps-Engineers/blob/main/ch03/apache_linux.pp` and `https://github.com/PacktPublishing/Puppet-8-for-DevOps-Engineers/blob/main/ch03/grafana_windows.pp`.

To test the runs in `noop` mode, you can apply the `_noop => true` option to the Bolt command.

While it would be impractical to discuss every core type in detail, the next section will cover at a high level the other core types, which are useful for creating more advanced configurations.

Core resource types

In this section, we will discuss the core resource types.

User and group types

The user type and group type are core to most configurations, allowing the `ensure` attribute to be set to `present` or `absent`. With a Unix platform as the provider, the user would normally have a minimum set of attributes of `uid` and `gid`, with the group having a minimum of `gid`. The user can be further enforced via the `password` attribute, which can ensure the limits for any password set, passing an encrypted password and enforcing the home directory and shell. For Windows Server, it is important to note only local users and groups can be managed, although a group resource can manage adding domain accounts to the membership of that group via the `members` parameter. The names are case sensitive in Puppet but case insensitive in Windows. The case should match so we do not lose any of the auto requirements that are formed. Windows also uses multiple types of names, so it can be `<name of computer\<user name>`, `BUILTIN\<username>`, or just `<username>`.

So, for example, `'DESKTOP-1MT10AJ\david, 'BUILTIN\david'` and `david` are all treated the same by Puppet.

The following code shows examples in Windows and Unix of an account and a group:

```
user { 'david':
  ensure   => 'present',
  groups   => ['BUILTIN\Administrators', 'BUILTIN\Users'],
}

group { 'Users':
  ensure   => 'present',
  members  => ['NT AUTHORITY\INTERACTIVE', 'NT AUTHORITY\Authenticated
Users', 'DESKTOP-1MT10AJ\david'],
}

user { 'ubuntu':
  ensure               => 'present',
  comment              => 'Ubuntu',
```

```
  gid                  => 1000,
  groups               => ['adm', 'dialout', 'cdrom', 'floppy', 'sudo',
'audio', 'dip', 'video', 'plugdev', 'lxd', 'netdev'],
  home                 => '/home/ubuntu',
  password             => '!',
  password_max_age     => 99999,
  password_min_age     => 0,
  password_warn_days => 7,
  shell                => '/bin/bash',
  uid                  => 1000,
}

group { 'ubuntu':
  ensure    => 'present',
  gid       => 1000,
}
```

We see here that Windows user David is a member of the administrator's group and user's group. We see the user's group and its list of members. We can then see the detailed setup of a Ubuntu user on Unix with password settings, a home directory, and group settings. Similarly, certain users and groups can be added as resource declarations, ensured absent, and removed from the system.

The exec type

The `exec` type is quite different from most Puppet types and can be dangerous if not used correctly. While most Puppet types try to describe the state a server should be in, `exec` provides a way of running scripts or commands on servers. This means declaring an `exec` type takes effort to make sure the resource will be **idempotent**. We can achieve this if the command itself is already idempotent, such as `apt-get update` (the command for updating package sources in Ubuntu), if we use the `onlyif` attribute, `unless`, or `creates`, or if the `exec` has a refresh-only attribute.

In the first case, if the command is **idempotent**, it will do no harm, but it will log in each Puppet run that it has run, and therefore using the other two methods is better to avoid the `exec` reporting runs.

With the `onlyif` attribute, we can declare a command that if it returns `true`, then our `exec` will run. `unless` is the opposite of `onlyif`, using a command that if it returns `true`, then our `exec` will not run. Finally, `creates` looks for a file to be created to show the script has run.

This first example looks at disabling public Chocolatey access unless the command finds in the sources that it is already disabled:

```
exec { 'disable_public_chocolatey':
    command => "C:/ProgramData/chocolatey/choco.exe source disable
-n=chocolatey",
    unless   => "\$sourceOutput = choco.exe source list; if
(\$sourceOutput.Contains('chocolatey [Disabled]')) {exit 0} else {exit
```

```
1}",
    provider => powershell,
}
```

This second example shows an example command, which generates a file using the `cowsay` command unless that file has already been created:

```
exec { 'Cowsay file':
   command => '/bin/cowsay Hello world > /etc/cowsaysays',
   creates => '/etc/cowsaysays',
}
```

> **Note**
>
> There is an optional PowerShell provider to allow `exec` to run PowerShell scripts: `https://forge.puppet.com/puppetlabs/powershell`.

The third scenario uses the `refreshonly` attribute, so using the `notify` and `subscribe` attributes, we can set the `exec` to only run if another resource is refreshed. The following `exec` can be useful when scripts are simply not going to be replaced by Puppet code:

```
exec  { 'refresh exampleapp configuration' :
  command      => '/bin/exampleapp/rereadconfig',
  refreshonly  => true,
  subscribe    => File['config file'],
}
file {'config file':
  path      => '/etc/exampleapp/configfile',
  content   => 'setting 1 = value',
}
```

This might be the case if the script/command is vendor-provided or simply a heritage script that works, and the effort of refactoring it into Puppet code would not be worth it.

On Unix platforms, a recent feature called parametrized execs was introduced with Puppet 6.24+ and 7.9+, allowing you to pass a `command` attribute as an array, the first part of the array being the command and the second part being the arguments. This uses the secure method of parametrized system calls to ensure code cannot be injected. In the following example, a traditional `exec` with just the command would run all the commands separated by the semi-colon, in our simple example echoing `real parameters` and running `rm`, while with the improvement of parametrized execs, it will take the second argument as a string to be passed and echo it, ensuring the original purpose of the command and preventing command injection:

```
exec  { 'parametrized command'
  command => ['/bin/echo', 'real parameters; rm -rf /'],
}
```

This example using `echo` is obviously simplified and it will become clearer where this plays a part when we look at *Chapter 8* and *Chapter 9*. There, we will see how user data can be fed into Puppet code and that we must code defensively.

The Augeas type

Augeas is a type only available on Linux; it was used more historically in earlier versions of Puppet when the options for manipulating files were much more limited, but in more advanced situations, it can have its uses. It can be computationally more expensive, so you should be careful in how you use it. Augeas can parse files in their native formats into a tree, which you can then manipulate. It uses lenses to perform these translations.

To give an example, if we want to manipulate the `access.conf` file, we can view the file using `augtool` (the CLI interface for Augeas) and print it using the following command:

```
augtool print /files/etc/security/access.conf
```

Let's say our file contains the following lines:

```
+ : john : 2001:4ca0:0:101::/64
+ : root : 192.168.200.1 192.168.200.4 192.168.200.9
- : ALL : ALL
```

This would result in the following being printed using the default lens:

```
/files/etc/security/access.conf/access[1] = "+"
/files/etc/security/access.conf/access[1]/user = "john"
/files/etc/security/access.conf/access[1]/origin =
"2001:4ca0:0:101::/64"
/files/etc/security/access.conf/#comment[83] = "All other users should
be denied to get access from all sources."
/files/etc/security/access.conf/access[2] = "+"
/files/etc/security/access.conf/access[2]/user = "root"
/files/etc/security/access.conf/access[2]/origin[1] = "192.168.200.1"
/files/etc/security/access.conf/access[2]/origin[2] = "192.168.200.4"
/files/etc/security/access.conf/access[2]/origin[3] = "192.168.200.9"
/files/etc/security/access.conf/access[3] = "-"
/files/etc/security/access.conf/access[3]/user = "ALL"
/files/etc/security/access.conf/access[3]/origin = "ALL"
```

This allows you to make programmatical references to individual sections and values in the syntax, so if in the client state, you wanted to remove any entry with the user `john` from all entries, `augtool` could run the following:

```
augtool rm /files/etc/security/access.conf/*[user="john"]
```

To use this in Puppet, Augeas only has one **provider**, and the core attributes are `changes`, which is the Augeas command you wish to run, `lens` if you wish to use a different translation from the default, and `onlyif`, which can perform a check of the content of the tree to see whether the change needs to be run. Creating the previous example as a Puppet resource would look like the following:

```
Augeas { 'remove John from access.conf' :
  changes => 'rm /files/etc/security/access.conf/*[user="john"]',
}
```

> **Note**
>
> Augeas is a powerful tool but should be used sparingly. More details on the syntax can be found at `http://augeas.net/docs/` and `https://forge.puppet.com/modules/puppetlabs/augeas_core/reference`.

The notify type

The `notify` type is used to send messages to the logs. This is more likely to be used for debugging purposes than production use, as it is not idempotent, and it will cause the Puppet report to see changes on every run. Using the `message` parameter as a string of what to print will take a default from the title. A simple example would be as follows:

```
notify { 'print a message to logs':}
```

> **Note**
>
> The `notice` function can be more practical for printing messages, as they will not show up in the Puppet report change logs. See *Chapter 5*.

There are more core types, but the commands demonstrated in this chapter to list types available, view the attribute, and provide documentation should give you the ability to understand how to go on and investigate other types that you may find useful, including types installed from `puppet forge`, which will be covered in *Chapter 8*.

> **Information**
>
> Throughout this chapter, we have highlighted resources coming under Puppet's control by being added to the catalog, whether they enforce presence or absence. Puppet has no concept of back-out, so removing a resource from Puppet's control will just leave it unmanaged as it was set in the last Puppet run. This should therefore always be considered in your back-out process for a code change.

Metaparameters and advanced resources

This section will start by looking at metaparameters, which are attributes that work on any resource type. For the lab work, we covered `before`, `required`, `notify`, and `subscribe`, which were used to create dependencies between resources. To follow this, there are several other useful attributes with a range of effects on resources. To see the full documentation of metaparameters on types and providers, the `meta` flag can be added to the `describe` command: `puppet describe <file type> --meta`.

audit

The `audit` metaparameter allows us to monitor unmanaged Puppet parameters; this could either be an array list of attributes or all for monitoring all undeclared attributes. In the following example, we declare this:

```
file {'/var/tmp/example'
  mode     => '0770',
  audit    => [owner,group],
}
```

This creates a `/opt/puppetlabs/puppet/cache/state/state.yaml` file on Puppet Enterprise or `/var/lib/puppet/state/state.yaml` in the open source version of Puppet, which records the audit state. Applying the preceding resource would produce the following output:

```
Notice: /Stage[main]/Main/File[/var/tmp/example]/owner: audit change:
previously recorded value 'absent' has been changed to 'root'
Notice: /Stage[main]/Main/File[/var/tmp/example]/group: audit change:
previously recorded value 'absent' has been changed to 'root'
```

As the resource was created, its state will be recorded as changing from `absent` to `present`, and it will then be reported whether the previously recorded value was found to have changed on Puppet runs. The `state.yaml` file would update to this new value, so it's important to take action on this change if it is required.

tag

The `tag` parameter allows us to apply tags to our resource, which can be a single string or multiple tags with an array of strings. By default, several tags are applied to a resource: the title, resource type, and the class the resource is contained in. Tags are particularly useful in scenarios where we only want to run parts of our manifests since both Puppet local and agent-based runs can take a `--tag` flag to run only resources with a particular tag.

For example, let's look at the Puppet resources in a manifest called `example.pp`:

```
class example::access {
  group {'ubuntu':
    ensure   => 'present',
    gid      => 1000,
    tag      => ['pci','sox'],
  }
  user {'ubuntu':
    ensure  => 'present',
    tag     => 'pci',
  }
}
```

The group will have the `group`, `ubuntu`, `pci`, and `sox` tags while the user will have the `user`, `ubuntu`, and `pci` tags. Additionally, both would have a tag of the class name, `example::access`. With the `puppet apply --tags pci example.pp` command, both resources would be applied similarly; `ubuntu` would apply both while running with a tag of `sox` would just run the group.

There are further metaparameters, such as `alias` and `loglevel`, that are simply not in common use although they have no risks worth discussing in detail; they can be read about at `https://www.puppet.com/docs/puppet/latest/metaparameter.html` or by running `puppet describe <any type> -m`.

The resource declarations shown before now have followed the same simple declaration pattern, but there are several other methods to allow more flexibility and advanced features.

The resources metatype

Puppet has a `resources` metatype, which can be used to ensure unmanaged resources of a type are removed. If it is thought of like the output of the `<type>` Puppet resource, finding anything with no matching `namevar` attributes from your code to mark as `absent`. It uses four attributes; a **name**, which is the type you want to apply to, the `purge` attribute, which can be `true` or `false`, and two attributes relevant when you are using resources on the user type – `unless_system_user`, which accepts `true`, `false`, or a specified minimum UID and ensures the system definition, or you can define integers or an array of integers in the `minimum_uid` parameter, which will be protected from the purge. To generate a list of numbers, the `range()` function from the `stdlib` module can make this easier. We will discuss functions in *Chapter 5*, to make it clear how functions work. As with all resources, the metaparameters can be used and **noop** is advisable here, as purging all users may be too aggressive, so seeing which users will be removed may initially be the best reporting to see:

```
resources {'user':
  purge => true,
  noop  => true,
}
```

> **Note**
>
> The `ssh_authorized_key` type should be managed on the *user* type via the `purge_ssh_keys` attribute.

Arrays of titles

When declaring several resources with the same attributes, the title can be declared as an array of resources, acting like multiple resource declarations. We will cover arrays in *Chapter 4*, but for now, understand an array of titles can be used with opening square brackets and a separating comma, so the title for a resource would be like the following example:

```
file{ ['/opt/example1','/opt/example1/etc','/opt/example1/bin'] :
  owner => 'user',
  group => 'user',
  mode  => '0750',
}
```

Overriding parameters

Here's the syntax for a resource reference:

- Type starting with a capital
- Title in square brackets
- Opening curly brace ({)
- Attributes to override
- Closing curly brace (})

It is possible to override attributes of a declared resource. In this example, we set `audit` to `true` and `group` to `other_group` on the `resource /opt/example/bin` file:

```
File['/opt/example/bin/'] {
  group  => 'other_group',
  audit  => true,
}
```

This is best used combined with the array of titles so that common defaults can be defined and then particular attributes set for a named resource. In this book, we recommend using this sparingly to avoid confusion when everything is declared.

Attribute splats

The attribute splat (*) is a mechanism of using a hash to fill out attributes of a type; this can be useful in situations where we want to cover the differences in attributes used by different providers. In a resource using the normal syntax, you can have set of the attributes as * and then create a hash of the attributes you would use. We will cover hashes, variables, and case statements in *Chapter 4* and *Chapter 7*, but for this example, it should be clear that we are setting the package options hash to contain a name attribute equal to apache2 for Debian and httpd as a default:

```
case $facts['os']['name'] {
  /^(Debian|Ubuntu)$/: {
  $package_options = {
    "name"  => "apache2"
    }
  }
  default:  {
  $package_options = {
    "name"  => "httpd"
    }
  }
}
package { 'http' :
  ensure => latest,
  *      => ${package_options},
}
```

This results in the package http resource using the name http2 for Ubuntu and Debian systems and httpd by default for any other systems. This feature should be used carefully so as not to detract from readability.

Lab

To practice a little of what we have discussed, let us follow up on our previous example and have a single manifest, all_grafana.pp, which can install, configure, and run Grafana on both Linux and Windows. As we have not covered facts yet, understand that as in our previous example, a case statement could use $facts ['os']['family'] to look for Red Hat or Windows to distinguish between our two clients. Note the rpm install file is available at https://dl.grafana.com/enterprise/release/grafana-enterprise-8.4.3-1.x86_64.rpm and the configuration file for Linux is at /etc/grafana/grafana.ini.

As a second exercise, create a separate manifest to create some users on the Linux client, linux_users.pp create 3 users exampleappdev, exampleapptest, exampleappprod, and a group, exampleapp, with all the users using this group as their primary group. exampleappprod should purge ssh keys from authorized. Finally, it should check whether there are any other non-system-level users on the client (but not enforce anything).

As per the previous lab, you can test your manifests by running a bolt command with your manifest name and client name listed: bolt apply manifestname.pp --targets servername.example.com.

You can find solutions at https://github.com/PacktPublishing/Puppet-8-for-DevOps-Engineers/blob/main/ch03/all_grafana.pp and https://github.com/PacktPublishing/Puppet-8-for-DevOps-Engineers/blob/main/ch03/linux_users.pp.

Anti-patterns

In this section, we will talk about some resource features you will find documented and useable with Puppet but that this book strongly recommends you do not use and make it part of your best practices to avoid. The resource features we highlight here are powerful but make resource declarations harder to read and require more translation and calculations required to see the state we are attempting to get the server into.

Abstract resource types

An abstract resource type is used for declaring a resource when we do not want to predefine a type and may decide which resource we will use based on the client. In this simple example, a variable is set to the type and the resource is then declared using the Resource[<TYPE>] { <RESOURCE BODY>} syntax:

```
$selectedtype = exec
resource[$mytype] { "/bin/echo 'don't use this' > /tmp/badidea":
creates => '/tmp/badidea', }
```

A simple translation of this statement would be as follows:

```
exec {"/bin/echo 'don't use this' > /tmp/badidea":
  creates  =>  '/tmp/badidea',
}
```

This book recommends against using abstracts, it is not commonly used, and it makes the code a lot less readable, particularly for less experienced Puppet users. The best approach is to use case statements or if statements, which we will cover in *Chapter 7*. If there is too much divergence in the code, it becomes best to separate the resources into separate classes and not force platforms that share little in resource types together.

Defaults

There are two methods of declaring defaults, but this book advises against using either. A default body with multiple bodies in a resource declaration breaks good practices around single purposes for a declaration, and a default resource statement can be dangerous in terms of understanding its scoping.

A default resource body

Here, a resource can have a default body, following the same syntax as a normal resource declaration but starting one of the bodies with `default:`; the ordering of the bodies does not matter.

This example shows two sets of arrays of titles, taking the defaults and changing the default mode for the second array of the titles set:

```
file {
  default:
    ensure => directory,
    owner  => 'exampleapp',
    group  => 'exampleapp',
    mode   => '0660',

  ;
  ['/opt/example','/opt/example/app','/etc/exampleapp']:
  ;
  ['/var/example','/var/example/app',]:
    mode => '0644',
  ;
}
```

As discussed when declaring resources, this book strongly recommends keeping a clear single-purpose resource declaration as grouping multiple bodies together makes the code harder to read. The recommended method for comparable results is to use arrays of titles and override parameters where appropriate.

Resource default syntax

The second method is to use a default resource statement syntax:

- The type starting with a capital and {
- A List of attributes and default values
- Ending with }

If the type has multiple namespaces, such as `concat::fragment`, then each namespace section should be capitalized.

In this example, we use a file to set the default for all files resources that do not declare a value for an attribute's owner group or mode:

```
File {
  owner  => 'exampleapp',
  group  => 'exampleapp',
  mode   => '0660',
}
```

Commonly used in early versions of Puppet, this is now considered only suitable for use on the `site.pp` file (a global settings file we will cover in *Chapter 11*). This is a result of Puppet no longer using a dynamic scope for variable lookup and default resources still being dynamically scoped, which can result in scope creep and unintentionally affect other resources in your catalog. (Scope will be discussed in detail in *Chapter 6*). Since having defaults in `site.pp` makes them unexpected and less visible, this book recommends against using resource defaults.

schedule

`schedule` is a metaparameter used in conjunction with the schedule resource type. This allows us to describe a specific schedule with a resource type, which defines when a particular resource can be run so that if Puppet is applied outside of this time, it will ignore the resource and how many times it can run in this period. The schedule resource type uses various attributes to describe ranges, repetition, or days: a simple example would be to cover the hours 6 p.m. to 9 a.m. over Friday night and Saturday morning:

```
schedule { 'Friday Night':
  day     => 'Friday',
  range   => '18 - 9',
}
```

That could then be applied to a resource:

```
exec {'/bin/echo weekend start > /tmp/example':
  schedule = > 'Friday Night',
}
```

This book advocates against this use case. It may seem tempting, particularly for highly regulated environments that have restricted windows for changes, but Puppet should enforce the expected state, so diverging from this state should be an issue. Creating schedules makes it obscure as to what will be applied and opens the state to periods of vulnerability where servers are only partially enforced by Puppet.

Exporters and collectors

Exporting and collecting Puppet resources happens when Puppet tries to allow information to be exchanged between nodes for interdependency. It allows a resource to be declared and run on one node and then other nodes to also apply these resources. This is done by exporting the information to the `PuppetDB` database, which Puppet runs will consult with when collecting. This means it can only be run via a Puppet agent setup and not via local Puppet runs.

Exporting a resource just involves adding `@@` in front of a normal resource declaration. The exported resource must be unique in `PuppetDB`, so commonly the hostname fact (a variable containing the hostname) is used in the declaration. In this example, a host entry is being exported to be put in the host file of each collecting server:

```
@@host { "Oracle database host entry ${::hostname}" :
  name   => 'dbserver1',
  ip     => '192.168.0.6',
  tag    => 'oracle',
}
```

Collecting the resource then involves declaring a collector, which is the type starting with a capital and a *spaceship* (<< | | >>) declaration; inside this, tags can be declared to filter the collection. Completing this example, this collection would ensure all exported host resources tagged `oracle` would be applied to the server:

```
Host <<| tag = oracle |>>
```

Exporting and collecting have two key issues; the first is it becomes harder to read the code and understand the resources that may be applied to a node. The second is it complicates the scalability and high-availability considerations for your Puppet infrastructure setup. As a result, by best practice, this book recommends avoiding any use of exporters and collectors.

Summary

In this chapter, you learned about declaring resources and the syntax and styling checks that can be performed to develop consistent code. Classes were shown to be a way to group resources and allow us to call classes and apply these groups of resources to servers. Defined types were then shown as a way to create repeatable patterns of Puppet code, which can vary by parameters.

We showed how to explore and use types and providers and saw some of the most commonly used core types and how to use them well. The file, package, and service types were shown to provide a great foundation for installing, configuring, and starting an application. It was seen how Puppet resources can relate to each other to ensure an order and how to then apply these resources written locally to servers for testing.

The chapter covered the core resource metaparameters to understand how to use various features of resources – tagging to allow filtered runs of resources; auditing to monitor changes, which happen to unmanaged attributes on a resource, and using `noop` to allow a resource to be declared as non-executable but reported on.

Finally, various anti-patterns were covered – default resources, which have scoping issues; default bodies, which result in overloaded resource statements; schedules, which make understanding Puppet runs complex; and export and collectors, which have issues both in terms of scalability and availability and in terms of abstracting data away from the code.

In the next chapter, we will cover variables and data types, which will allow us to assign values to variables and control what those values are and how they can be interacted with. This will allow us to reduce duplication and make our resources easier to update and manage, as well as providing a way to pass in data to our classes.

4

Variables and Data Types

This chapter will cover how Puppet handles variables and, in particular, how Puppet differs from most declarative languages in terms of how they are used and declared. We will look at the core data types that are used to define what the value of a variable can contain and how it can be interacted with. Then, we will look at how data types and variables allow the classes we discussed in *Chapter 3* to receive external data and handle default values.

Arrays and hashes will be discussed in detail, including how to declare them, access values, and manipulate them with operations. The `Sensitive` data type will be shown, which you can use to secure values in logs and reports while making the limitations of this data type and what it does not secure clear. We will also cover abstract data types and show you how to allow more complex and flexible definitions of variables and values. The chapter will finish by covering how variable scopes and namespaces work with variables. We will also discuss the scope of variables and how variables from different scopes can be accessed and which scopes can access which levels of data.

In this chapter, we are going to cover the following main topics:

- Variables
- Data types
- Arrays and hashes
- Abstract data types, including `Sensitive`
- Scope

Technical requirements

For this chapter, you will need to provision a Puppet server standard architecture with a Windows client and a Linux client by downloading the `params.json` file from `https://github.com/PacktPublishing/Puppet-8-for-DevOps-Engineers/blob/main/ch04/params.json` and then using the following command from your `pecdm` directory:

```
bolt --verbose plan run pecdm::provision --params @params.json
```

In various sections of this chapter, examples will be given of using the `notify` function, which outputs to the agent command line. These examples can just be run in the local development environment by putting all the code into a manifest file – for example, `example.pp` – and then running `puppet apply example.pp`.

Alternatively, any variables that are required can be set using the environment variable format of `FACTER_variable_name` and running `puppet apply -e '<example_code>'`. To run one of the substring examples, you can run the following code:

```
export FACTER_example_string='substring'
puppet apply -e 'notify{ "${example_string[3]}": }'
```

Variables

In this section, we will cover how to use variables in Puppet and how this differs from other procedural languages. The key thing to understand about Puppet variables is that they are only assigned once during compilation in a given scope. In a traditional procedural language, it's common to use variables throughout your code, where you might gather the current state as your code runs, use variables to keep track of it, and update it to act and make procedural decisions at various stages of your code. The following is an example of a simple PowerShell script that runs a command several times and adds the output to a single variable. It does so by using `select-string` to search for files containing ? in the `.sh` and `.pp` files in the user's code directory:

```
$Matches = Select-String -Path "$PSHOME\code\*.sh" -Pattern '\?
...
...
$Matches = $Matches + Select-String -Path "$PSHOME\code\*.pp" -Pattern
'\?
```

This state check is not done in Puppet since all evaluation takes place at the start of the catalog, based on the state the server sent to be compiled. This, in turn, provides the steps required to get it into a desired state. In Puppet, we assign variables for repetitive uses such as file paths or conditional logic such as `if` or `case`. In these cases, a value must be chosen to be assigned, depending on the initial state.

> **Note**
>
> We are simplifying the process slightly here as there are now deferred functions that can run after complications. However, this still does not allow us to reassign variables. We will cover this in more detail in *Chapter 5*.

Puppet variables are declared with a dollar sign (`$`) followed by the variable's name, an equals symbol (`=`), and the value to assign. For example, a variable called `example_variable` that's assigned a value of `'this is a value'` would look like this:

```
$example_variable = 'this is a value'
```

Note that, unlike resources, variables depend on the order of evaluation and must be declared in code before they are called.

> **Note**
>
> There are several variables known as built-in variables that return server information. However, since these are more about infrastructure and the environment, they will be covered in *Chapter 10* and *Chapter 11*, where they will be relevant.

Naming

Variable names are case-sensitive, and they can include upper and lowercase letters, numbers, and underscores but must start with an underscore (_) or lowercase letter. The exception is regex capture variables, which are variables only named with numbers such as `$0`, `$1`, and so on. We will cover these in *Chapter 7* and *Chapter 11*, where we will use them as part of conditionals and node definitions.

> **Note**
>
> Starting a variable name with an underscore will limit it to local scope use, as discussed in detail at the end of this chapter.

Reserved variable names

There are several built-in variables that you cannot use in your code. These are as follows:

- `$facts`
- `$trusted`
- `$server_facts`

These are all built-in variables generated from Facter, which cannot be used or reassigned. We will discuss these variables in detail in *Chapter 5*, as well as what values they provide.

As we covered in *Chapter 3*, the `$title` and `$name` variables are used by classes and defined types and should not be used.

The full list of reserved words can be viewed in the Puppet documentation at `https://www.puppet.com/docs/puppet/latest/lang_reserved.html`.

Interpolation

Puppet variables can be evaluated and resolved into their assigned values when called without quotes or as part of a variable mixed with our data in double quotes. The style guide enforced by the lint checks will ensure single quotes are used when an assignment does not contain a variable and no quotes are used when an assignment only contains a variable. A mix of values and variables can be written with double quotes. When you should use a mix is stated in the style guide. This is checked by the linter to ensure it uses curly braces { }. This can be seen in the following example of assigning `mixed_variable` to the variable's declaration when using double quotes for interpolation:

```
$database_id = $dbname
$base_directory = '/opt'
$database_directory = "${base_directory}/database/${database_id}"
```

As we described in the previous chapter, the `notify` function can be used to check the values of variables:

```
notify{'debug variable':
  message => "The database directory is ${database_directory}",
}
```

This section discussed how Puppet variables differ compared to other programming languages due to statefulness and how it can declare, access, and name variables.

Data types

Every value in Puppet has a data type; in the previous section, for example, the variables had `String` values assigned. A data type, when used as a capitalized unquoted string, such as `Integer`, can be used to specify what parameters in a class, defined type, or lambda should contain, allowing the data to be validated:

```
class example (
  String example_string = 'hello world',
  Integer example_integer = 1
) {
}
```

Data types can also be used to compare a variable's value, conditionally check values, and take different actions, depending on the result. For example, to confirm that a variable contains an integer, the following match expression can be used. Here, we are confirming that the `example_integer` variable contains an integer:

```
$example_integer =~ Integer
```

Conditional statements and comparisons will be covered in full in *Chapter 7*.

The next section will run through the most commonly used core Puppet data types. Unfortunately, Puppet has no equivalent to the `puppet describe` command for data types .so all references must be taken from the web and GitHub documentation at `https://www.puppet.com/docs/puppet/latest/lang_data_type.html`. If you're using types provided by modules from the forge, which will be covered in detail in *Chapter 8*, the documentation should be on the reference page of the module. Various functions work with data types, but we will not cover this here. We will look at functions in detail in *Chapter 5*.

Strings

Strings are the most common data type used in Puppet and, as discussed in *Chapter 1*, were originally the only type of data used in early Puppet. Strings are pieces of unstructured text of any length, encoded in UTF-8. There are four ways to declare strings in Puppet:

- Unquoted
- Single quoted
- Double quoted
- Heredocs

Unquoted strings

Unquoted strings are single words starting with a lowercase letter and containing only letters, digits, hyphens (`-`), and underscores (`_`) and must not be reserved words. Reserved words are typically keywords such as class or other language functions; the full list can be viewed at `https://www.puppet.com/docs/puppet/latest/lang_reserved.html#lang_reserved_words`.

Unquoted strings are used for resource attributes that accept a limited set of words, such as the Puppet service resource type, which accepts `running` or `stopped` for its `ensure` attribute:

```
service { 'defragsvc':
  ensure => stopped,
}
```

Single-quoted strings

Single-quoted strings can contain multiple words but, as previously discussed, cannot interpolate variables. However, they can contain line breaks and escape sequences using a backslash (`\`), an escape backslash, or a single quote (`'`). This allows the use of single quotes within the string itself and for a backslash to be used at the end of a string. Additionally, a line break can be achieved via the *Enter* key.

The following example shows the `sed_command` variable with the single quotes needed as part of the `sed` command to escape in the single-quoted string and then the `install_dir` variable for a Windows file path with an ending backslash:

```
$sed_command = '/usr/bin/sed -i \'s/old/new/g\''
$intall_dir = 'c:\Program Files(x86)\exampleapp\\'
```

Double-quoted

Double-quoted strings can fully interpolate variables and more available escape characters. In addition to the single quote, backslash, and escape characters, double quotes can interpret the following:

- `\n`: Newline
- `\r`: Carriage return
- `\s`: Space
- `\t`: Tab
- `\$`: Dollar sign, to prevent variable interpolation
- `\uXXXX`: Here, `xxxx` is a four-digit hex number for a Unicode character
- `\u{X}`: Here, `X` is a hex number between two and six digits between curly braces, { }
- `\"`: Double quote
- `\r\n`: Window line break

As per single quotes, a line break can also be used within text (that is, just by hitting *Enter*).

> **Note**
>
> If a backslash is used in double quotes without you escaping it and without a valid escape character after, it will continue and treat it as normal characters but result in the following message in the logs: *Warning: Unrecognized escape sequence*. Here is an example:
>
> ```
> Warning: Unrecognized escape sequence '\T' at C:/Users/david/
> code/test.pp:1:50
> ```
>
> This commonly affects Windows user paths that are used in double quotes.

A simple example of a double-quoted string making use of new lines and tabs (which are important to the syntax of Makefile content) is as follows. This creates a string to then be used in the content of a file:

```
$make_file_content = "hello:\n\techo \"hello world\""

file '/home/david/makefile' : {
  content => $make_file_content,
}
```

Heredocs

Puppet's implementation of heredocs involves using a tag to indicate the start and end of the heredoc file's content. The starting tag is typically composed of the following elements:

- `'@('`
- A string, known as the end text, that may be surrounded by double quotes to enable interpolation
- An optional escape switch (or switches) that begin with a forward slash to enable escape switches in the text
- An optional colon (`:`) followed by a syntax name check
- `')'`

To use a heredoc in Puppet, the content should be entered on the lines immediately following the starting tag, with the exact formatting that is desired. The end of the heredoc is indicated by an end tag, which should include the following elements:

- An optional vertical bar (`|`), which indicates how much indentation should be stripped from the lines of the text
- An optional hyphen (`-`), which removes the final line break from the heredoc
- The same end text tag that was used in the starting tag

The end text in a Puppet heredoc is a string that can consist of mixed-case letters, numbers, and spaces, but cannot include line breaks, slashes, colons, or parentheses. By default, the content of the heredoc will not interpret escape characters, so optional escape switches must be declared if they are needed. The following escape switches are available and mirror the same escape sequences for double-quoted strings, but do not require double quotes (since they have no special meaning in heredocs):

- `n`: New line
- `r`: Carriage return
- `t`: Tab
- `s`: Space

- `$`: Dollar sign, to prevent interpolation if the end text is double-quoted
- `u`: Unicode characters
- `L`: A new line or carriage return
- `\` : : All of the previously mentioned escape sequences are available

When any escape sequence is selected, you can use `\\` to escape a backslash.

Variable interpolation is disabled by default so, as discussed, the end text should be surrounded in double quotes if needed.

Syntax checking is available for various content, such as Puppet manifests via `pp` or Ruby files via `ruby`:

```
@(END:pp)
@(END:ruby)
```

Syntax checking will only run if variable interpolation is not turned on; if a type unavailable to Puppet is entered, it will be ignored. The full details of available syntax checkers can be found in the Puppet specification, which also contains details on creating custom syntax checkers.

Heredoc declarations can be placed anywhere a string declaration can be, so, for example, a long command in an `exec` command could be declared as follows:

```
exec { 'create databases':
  command => @("Database Commands"/L)
    sudo -u postgres psql \
    -c "CREATE DATABASE ${database1} ENCODING 'utf8' LC_COLLATE 'en_
US.UTF-8' LC_CTYPE 'en_US.UTF-8'" \
    -c "CREATE DATABASE ${database2} ENCODING 'utf8' LC_COLLATE 'en_
US.UTF-8' LC_CTYPE 'en_US.UTF-8'" \
    -c "CREATE DATABASE ${database3} ENCODING 'utf8' LC_COLLATE 'en_
US.UTF-8' LC_CTYPE 'en_US.UTF-8'"
    |-"Database Commands"
}
```

This book recommends using heredocs sparingly. For long commands in exec, as shown in the preceding example, this may be suitable, but particularly for file content, it typically clutters and confuses the code with text and is better placed in templates, as covered in *Chapter 7*, or as files in modules, as covered in *Chapter 8*. The topic of how best to store data will be discussed in *Chapter 9*.

Accessing substrings in variables

To call a string in Puppet, the simplest method is to use the `$` symbol, followed by the variable name. However, if the variable name contains invalid characters, such as a space, Puppet will assume that the variable name has ended. Therefore, to ensure proper interpolation of variables within strings, it is safest to enclose the variable name in curly braces, `{}`.

To access a particular character or substring within a string, Puppet allows you to specify a range of indices using `[<start index>, <stop position>]`, which can include support for negative numbers to count back from the end of a string or change the order of characters to be returned. For example, the following code sets a variable named `'example_string'` to the `'substring'` string:

```
$example_string = 'substring'
```

Various combinations can be used; for example, a single character can be called by taking an index from the start, such as 3, to return `s` (we start at 0).

To extract a single character from a string variable in Puppet, you can specify the index of the desired character starting from `0`. For example, to extract the third character of a string variable, you would use an index of `3` (since indexing starts at `0`). In Puppet, this can be expressed as follows:

```
notify { "${example_string[3]}" :}
```

This would return the character at index `3`, which in this case would be `'s'`.

A negative index can go from the end and return the same `s` character with `-6`:

```
notify { "${example_string[-6]}" :}
```

To extract a specific portion of a string variable in Puppet, you can use the square bracket notation to indicate the start index and the stop position of the substring. For example, if you have a string variable named `'example_string'` with a value of `'substring'`, and you want to extract a substring that starts at the third character and includes the next five characters, you can use the following syntax in Puppet:

```
notify { "${example_string[3,6]}" :}
```

This would return the substring that starts at index `3` (which corresponds to the letter `'s'` in `'substring'`) and includes the next five characters, which in this case would be `'string'`.

To extract a substring starting from a negative index position, you can specify a negative value for the stop position. For example, to extract the substring that starts from the fourth index from the end and includes the next three characters, you can use the following syntax:

```
notify { "${example_string[-4,-1]}" :}
```

This would return the substring that starts at the fourth character from the end (which corresponds to the letter `'t'` in `'substring'`) and includes the next three characters, which in this case would be `'tri'`.

Finally, to extract a substring that starts from a negative index position and includes a positive number of characters, you can use the following syntax:

```
notify { "${example_string[-4,4]}" :}
```

This would return the substring that starts at the fourth character from the end (which corresponds to the letter `'t'` in `'substring'`) and includes the next four characters, which in this case would be `'ring'`.

This sort of substring work can be particularly useful when package names, application versions, or other consistent name strings need to be broken up into different variables. As a more practical example, an organization has hostnames that start with a location code and contain a role, their environment, and a server ID:

```
$hostname = flkoracprd00034
$location = $hostname[0,3]
$role =$hostname[3,3]
$environment = $hostname[6,3]
$id = $hostname[-5,5]
```

String data type parameter

When setting the type of a parameter as a string, the capitalized keyword of `String` is used, along with optional minimum and maximum lengths of the string:

```
String[<Minimum length>, <Maximum Length>] $variable_name
```

The default for the minimum is 0 and the maximum is infinity. To use the default implicitly, you can use the default unquoted string keyword.

Let's look at a class called `database`, which accepts a database ID string of four characters, a username between six and eight characters that defaults to `dbuser` if it's not provided, and a description of any length:

```
class 'database': {
  String[4,4] database_id,
  String[6,8] username = 'dbuser' ,
  String description,
}:
```

Numbers

This section will cover the two types Puppet uses for numbers: integers and floating points. We will also look at what arithmetic operations can be performed on them, how numbers can convert to and from strings, and the variations on these types.

Both types of numbers are declared without quotation marks. Here, casing does not matter where letters are used. The following patterns are available:

- **Integers and octal integers**:
  - An optional negative, – (positive is presumed in absence)
  - Numeric digits (starting with a `0` for octal)
- **Hexadecimal Integers**:
  - An optional negative, – (positive is presumed in absence)
  - `0x` or `0X` (case is not important)
  - A mix of numeric digits and upper or lowercase letters
- **Floating point**:
  - An optional negative, – (positive is presumed in absence)
  - Numeric digits (a `0` is required if using a number between -1 and 1)
  - A decimal point
  - Numeric digits
  - An optional `e` or `E` preceded by digits (for scientific float)

The following are some simple and appropriately named examples of each of the preceding types:

```
$integer = 42
$negative_integer = -84
$float = 32.3333
$scientific float = 3e5
$octal = 0678
$hex = 0x
```

It is important to note that an octal or hexadecimal number cannot be expressed as a floating-point number and will result in an error as it is not a valid octal or hexadecimal number, as applicable.

Arithmetic operators

We cannot perform operations to reassign a variable but can assign new variables based on operations between assigned variables. The following expressions can be used between variables:

- `+`: Addition
- `-`: Subtraction
- `/`: Division

- `*`: Multiplication
- `%`: Modulo, the remainder of dividing left by right
- `<<`: Left shift
- `>>`: Right shift

Left shift and right shift are less familiar and need further explanation. A left shift is the first variable multiplied by two to the power of the second variable. Taking an example of `5 << 3`, this would translate into $5 * 2^3$, which would result in 40.

A right shift is the first variable divided by two to the power of the second variable. Taking an example of `32 >> 2`, this would translate into $32 / 2^2$, which would result in 8.

> **Note**
>
> For both left shift and right shift, floats will round down to an integer.

Additionally, the negative symbol (`-`) can be used as a prefix to negate a variable and brackets can be used to manage the priority of operations, where **Brackets, Orders (powers/indices or roots), Division, Multiplication, Addition, and Subtraction** (**BODMAS**) rules apply. Shifts are essentially treated as multiplication and modulo division in this priority.

Any operations between an integer and a float will result in a float and an operation on an integer, which would result in a float being rounded down to an integer.

The following are some examples of using these operators:

```
$a = 5
$b = 3
$addition = $a + $b
$subtraction = $a - $b
$division = $a / $b
$multiplication = $a * $b
$modulo = $a % $b
$shift_left = $a << $b
$shift_right = $a >> $b
$negate = -$a
```

To further show brackets enforcing BODMAS rules, the following example will be equal to negative 40:

```
$bodmas_example = ($a + $b) * -$a
```

String to numeric conversion

If a string is used in a numeric operation, it will automatically convert, but this will not happen in any other context. To convert a string into a number, an object can be declared as an integer, float, or numeric (we will cover numeric objects in the *Abstract data types, including Sensitive* section). An example of conversion is taking a string, `1`, to an integer and a string, `1.1`, to a float:

```
$string_integer='1'
$string_float='1.1'
$converted_integer=Integer($string_integer)
$converted_float=Float($string_float)
```

Numeric to string conversion

Numeric types automatically convert into strings when interpolated in a string; the automatic conversion uses base 10 notation. The `String` object declaration can also be used to convert, as follows:

```
$string_from_integer = String(342)
```

Integer data type

When setting the type of a parameter as an integer, the capitalized `Integer` keyword is used, along with optional minimum and maximum values of the integer:

```
Integer[<Minimum Value>, <Maximum Value>]
```

The defaults are technically negative infinity and positive infinity but as Puppet uses 64-bit signed integers, this is in the region of –9,223,372,036,854,775,808 to 9,223,372,036,854,775,807.

Float data type

When setting the type of a parameter as a float, the capitalized `Float` keyword is used, along with optional minimum and maximum values of the integer:

```
Float[<Minimum Value>, <Maximum Value>]
```

The defaults are technically plus and negative infinity but in practical terms, this is the range of a double precision float of -1.7E+308 to +1.7E+308 in Ruby implementation.

As an example, consider the following code block, which defines `Class application::filesystem` for assigning a percentage of a volume group within known limits of `100` to `10000`:

```
class application::filesystem (
Float[0.1, 99.9] percentage_application,
Integer[100, 10000] volume_group_size,
) {
}
```

undef

`undef` is considered equivalent to nil in Ruby and represents the absence of a value assigned to a variable. By default, the `strict_variables` setting is set to `true` for Puppet 8 and `false` for Puppet 7 and below, which means variables that have not been declared will `error` in Puppet 8 but have a default value of `undef` in Puppet 7 and below. In *Chapter 10*, we will see that this can be set in the `puppet.conf` configuration file.

As a simple example, the following line will `error` in Puppet 8 but for Puppet 7 and below will notify `Print` that `test1` has not been declared:

```
notify {"Print $test1":}
```

The only value an `undef` data type has is the unquoted `undef` and it is not used for parameter data typing by itself. This is because enforcing the absence of a value would have no purpose.

In the *Abstract data types, including Sensitive* section of this chapter, we'll see how `undef` values can be accepted for parameters as a part of a selection of feasible options.

When interpolated into a string, `undef` is converted into an empty string (`''`).

Callout

In *Chapter 5* and *Chapter 8*, we will learn about functions such as `delete_undef_values` and `filter`, which can be used to trim the arrays and hashes of `undef` values.

Booleans

Booleans in Puppet represent `true` or `false` and in *Chapter 7*, when looking at `if/case` statements, you'll see that all Puppet comparisons return a Boolean type. A Boolean variable should simply contain an unquoted `true` or `false` value. As a result, this makes the data type quite simple with no parameters – just the capitalized `Boolean` keyword.

As an example, the following code is for an `exampleapp` class that has a parameter to manage users that are set to `true` by default and a couple of variables hard-coded:

```
class exampleapp (
  Boolean manage_users = true,
) {
  $install_ssh = true
  $install_telnet = false
}
```

Conversion

Automatic conversion into Boolean values will occur in most cases unless an explicit data type has been specified. For example, in an `if` statement, a variable can be used as if it were a Boolean by simply writing `$variable_name`. However, automatic conversion can be confusing because only `undef` will result in a conversion to `false`. This means that the `'false'` string, an empty string (`''`), an integer of `0`, and a float of `0.0` will all convert into `true`.

When using a Boolean declaration, an empty string will fail to convert, as will `undef`, while a string of `'false'`, an integer of `0`, or a float of `0.0` will convert into `false`.

Since this is confusing, it is safer to use the `num2bool` and `str2bool` functions from the `puppetlabs-stdlib` module, which will be covered in *Chapter 8*.

Regexp

The `regexp` type is different from the types we've seen so far. It represents a valid regular expression in Puppet, which are expressions contained between forward slashes based on Ruby's regex implementation: `http://ruby-doc.org/core/Regexp.html`.

Regex use will be covered in more detail in *Chapter 7*, where it will be more practically applied. However, it is worth noting that, later in this chapter, several abstract types that combine multiple types, including `regexp`, will be covered.

Lab

In the previous chapter, a combined `all_grafana` manifest was created and a solution was provided at `https://github.com/PacktPublishing/Puppet-8-for-DevOps-Engineers/blob/main/ch04/all_grafana.pp`. Adjust this file so that it is within a class called `all_grafana` and instead of using Facter, parameters are used.

These parameters should include the following:

- **Source download**: A string variable defaulting to `https://dl.grafana.com/enterprise/release/grafana-enterprise-8.4.3-1.x86_64.rpm`
- **Port**: The port for the service to listen on as an integer
- **Service enabled**: By a Boolean

To achieve class assignment, write a class declaration that assigns the variables to ensure the class is included in a catalog run. When you run `bolt` against your manifest, it will ensure you have included your variables. Solutions are available at `https://github.com/PacktPublishing/Puppet-8-for-DevOps-Engineers/blob/main/ch04/all_grafana.pp`.

Arrays and hashes

This section will cover the two core collections of data in Puppet: arrays and hashes. You will learn how to create, access, and perform operations to manipulate the values into a new variable.

Assigning arrays

Puppet arrays are created by surrounding comma-separated lists of values with square brackets. An optional comma can be added after the last element, but this book recommends against that for styling. For example, an array called `example_array` containing the `first`, `second`, and `third` strings, and would be declared as follows:

```
$example_array = ['first','second','third']
```

Arrays can contain any data type, as well as a mix of data types. A Puppet variable cannot be reassigned in terms of individual values or in terms of any other manipulation such as the addition or removal of values. The following code shows how to assign the `mixed_example_array` array with the integer of `1`, a Boolean value of `false` from the `example_boolean` variable, and the `example` string:

```
$example_boolean = false
$mixed_example_array = [ 1, $example_boolean , 'example']
```

Arrays can also be empty with nothing between the brackets, `[]`. They will not be identified as `undef` but as an empty array. It would be unlikely to declare an array empty directly; this normally occurs as a result of interpolated variables and operators causing it to become empty.

Accessing an array index

To access an array variables element, a specific element can be specified by index, which will return that element. For example, to take the second string in the second index of `example_array` and assign it to a variable, you can use the following code:

```
$example_array = ['first','second','third']
$second_index = $example_arrary [1]
```

The following code shows a `notify` resource outputting the string interpolating the third element from `example_array` as negative numbers counting back from the end of the array:

```
notify{ "The first element is ${example_array[-1]}":}
```

Accessing an element that does not exist will result in `undef` being returned. You mustn't put any spacing between the square bracket and the variable name; otherwise, it will be interpreted as a variable and the square brackets will be separate.

Accessing a subset of an array

When accessing a subset of an array, a second number is used to indicate the stopping point. This is different from how substrings are handled. In the case of arrays, a positive number represents the number of elements to return. For example, using a count position of 1 will return an array with a single element. To extract a sub-array containing only the 'second' element from example_array, you could use the following code:

```
$sub_array =  example_array[1,1]
```

This will assign the ['second'] sub-array to the $sub_array variable.

Choosing a length beyond the length of the array will simply return the available elements.

A negative length will count back from the end of the array. Importantly, unlike accessing substrings, you cannot reverse the order by going past the starting index; this will simply return an empty array. In the following example, the negative_sub array will return the whole array since its starting position is 0 and its finishing position is the first element from the end of the array. The empty_sub_array variable will be assigned an empty array since the ending position would be before the starting position. The second_element_array variable would be assigned an array with the second element:

```
$negative_sub_array = example_array[0, -1]
$empty_sub_array = example_array[1, -3]
$second_element_array = example_array[1, -2]
```

Nested array

A nested array can be declared by inserting an array value within an array as many times as needed. The value can then be accessed by using multiple sets of square brackets to access the desired level. As an example, if a nested array is created with the first element as a string, a second value as an array of three strings, and a third value of a string, attempting to access the first element as a nested element results in returning a string, i, since the element at index 0 returns a string. A second set of brackets are then used on the first string.

The nest_second variable returns the nest_second string since it returns the nested array at element 1; then, with the second set of brackets accesses the second element:

```
$nested_array= ['first',['nest_first','nest_second','nest_
third'],'third']
$sub_string = $nested_array[0][1]
$nest_second = $nested_array[1][2]
```

To interpolate nesting in an array, curly brackets must surround the variable name and square brackets. For example, the following notify resource will print the first element of nested_array:

```
notify {"Print ${nested_array[1][0]}":}
```

It is possible to use the subset approach within nested brackets, but this can create confusing and hard-to-follow accesses and is not a recommended style in this book.

Array operators

Arrays cannot be manipulated once assigned but operators can manipulate the content of arrays to assign new array variables. The following operators are available:

- `<<`: Append
- `+`: Concatenate
- `-`: Remove
- `*`: Splat

Append

Append takes any type of value and adds it as a new element at the end of the array. This includes adding an array as a nested array. To combine two arrays, concatenate (+) must be used. To demonstrate this, let's look at an example of an array with integers `1` and `2` as elements that appends `3` into a new array. This will produce a new array, `[1,2,'three']`; appending an array of `[3,4]` to `example_array` will produce a new nested array, `[1,2,[3,4]]`:

```
$example_array=[1,2]
$new_array=$example_array << 'three'
$append_nest=$example_array << [3,4]
```

Concatenate

Concatenate takes an array and essentially combines its content with another array. If the first value is not an array, the compiler will assume this is a numerical operator. For numbers, strings, Booleans, and regular expressions, it will essentially work the same as an append and add the value to the end. To achieve a nested array entry, you must supply a nested array. So, to show some examples, `combined_1` will become an array of `[1,2,1]`, `combined_2` will be assigned `[1,2,1,2]`, and `combined_3` will result in a nested array being assigned `[1,2,[1]]`:

```
$combined_1 = $example_array + 1
$combined_2 = $example_array + [1,2]
$combined_3 = $example_array + [[1,2]]
```

If a hash must be concatenated, it will be converted into an array unless it is turned into an array with a single hash element. So, in the following code, the converted variable would be assigned a nested array with elements of `test` and `value`, giving it an array of `[1,2, [test, 'value']`, while the `nested_hash` variable would add a nested hash that assigns `[1,2,{test => 'value'}]`:

```
$converts = $example_array + {test => 'value'}
$nested_hash =$example_array + [{test => 'value'}]
```

Remove

Remove assigns an array after removing all matching elements from a source. The first variable must be an array; otherwise, it will be assumed to be a numeric operator. For the second variable, if it's a number, string, Boolean, or regular expression, it will search each element of the first variable array and remove it if there is a match. For example, removing the one string from `another_example_array` will match the first element and the third element and remove them, but not the first element of the nested array, assigning `['two','three','four','three',['one','three','four']]` to the `remove_string` variable:

```
$another_example_array = ['one','two','one','three','four','three',['o
ne','three','four']]
$remove_string = $another_example_array - 'one'
```

When you have an array as a second variable, it will iterate through each element in that array, removing them as if they had been presented directly, as in our previous example. In this example, it will remove one as per the previous example and then perform the searches for matching strings of `three` and `four`, removing the fourth, fifth, and sixth elements while assigning `['two',['one','three','four']]` to the `remove_array` variable:

```
$another_example_array = ['one','two','one','three','four','three',['o
ne','three','four']]
$remove_array = $another_example_array - ['one','three','four']
```

When a nested array is used as the second variable, it will match any elements with the same array and remove them. So, in this example, the `remove_nested_array` variable will be assigned `['one','two','one','three','four','three']`:

```
$remove_nested_array = $another_example_array -
[['one','three','four']]
```

As with concatenation, hashes must be placed in an array; otherwise, they will remove any matching element of a translated nested array.

Splat

Splats are different from the other operators as they are used to make an array provide comma-separated lists as an argument in a function call. This is true for both case and selector statements. Using array splats will be covered in detail in *Chapter 5* and *Chapter 7*.

Array data type

When setting the data type of a parameter to an array, the capitalized `Array` keyword must be used with a data type for elements of the array, the minimum size of the array, and the maximum size of the array:

```
Array[<Data Type>, <Minimum Size>, <Maximum Size>]
```

The defaults for data types are data, which will be covered in the *Abstract data types, including Sensitive* section of this chapter, but this means that numbers (both integers and floats), strings, Booleans, and regular expressions, as well as arrays and hashes of these types apply. If you select a more specific data type, such as `String`, it will expect every element in the array to contain a string. In the *Abstract data types, including Sensitive* section, other mixed types will be covered that provide more flexibility.

The minimum size is 0, while the maximum size is infinite.

As an example, the `database` class could accept a variable of `db_uids`, where at least one element is expected in the array but could contain up to six elements. The `user_names` variable can be an empty array or up to five elements but most only contain strings. Finally, the `extra_flags` variable is an array with default values, so it can be an empty array up to an infinite size with the contents matched against data types:

```
class 'database': {
  Array[default,1,6] db_uids,
  Array[string,0,5] user_names,
  Array extra_flags,
}
```

Assigning hashes

Hashes are written as comma-spaced key-value pairs separated by `=>` and the list is surrounded by curl braces, `{ }`. A trailing comma can be added after the last pair, but this is not a recommended style by this book. For example, the following hash pairs could be declined to assign the `make` key with the `skoda` string, the `model` key with the `rapid` string, and the `year` key with the `2014` integer:

```
$my_car = { make => 'skoda', model => 'rapid', year => 2014 }
```

For style purposes, a new line is often taken with each key to ensure the start of the keys line up and the arrows line up. Taking a final new line for the closing curly brace and lining it up with the opening curly brace is what this book recommends when writing arrays:

```
$my_car = { make  => 'skoda',
            model => 'rapid',
            year  => 2014
          }
```

Hash keys and values can be any type, but it rarely makes sense for the keys to be anything but strings. Just like arrays, hashes are variables in Puppet and can only be assigned once and not manipulated unless a new hash is assigned.

> **Note**
>
> Puppet can only serialize string hash keys into a catalog. Therefore, you cannot assign a hash with non-string keys to a resource attribute or class parameter.

Accessing hash values

Similarly to arrays, hash values can be accessed using square brackets with the key value to access. As an example, the following would print the `rapid` value:

```
notify {"Print ${my_car[model]}":}
```

Nested hashes

As with arrays, by declaring a hash within a hash, a nested hash can be created, which can be accessed with chained keys. The following example shows a variable package list containing the `packages` and `services` keys. The `packages` key contains the `httpd` key, with a string value of `latest`, and the `cowsay` key, with a float value of `4.0`. The `services` key contains the `httpd` key with a string value of `running` and the `nginx` key with a string value of `stopped`:

```
$package_list = { packages =>  { httpd   => 'latest',
                                 cowsay  => 4.0
                               }
                  services =>  { httpd   => 'running',
                                 nginx   => 'stopped'
                               }
                }
```

To print both of the nested `httpd` keys, a `notify` resource can be declared, as follows:

```
notify {"Print ${package_list[packages][httpd]} ${package_
list[services][httpd]}":}
```

Hash operators

There are two operators for hashes – merging (+), which can assign a new hash by adding key pairs to an existing hash, and removal (-), which can assign a new hash by removing key pairs from an existing hash.

Merging

Merging is performed by taking a hash variable, a + symbol, and a hash or an array with an even number of values. Note that this is done while looking to add a new key; if a key already exists, it will not be added. In the following example, merging a hash with the `database` key with an `oracle` string and a `version` key with an integer of `11` with the `app_web` hash with a `web_server` key with a string of `httpd` and a `version` key with a value of `12` will result in the `combined_app` variable containing the `database` key and value and the `web_server` key and value. However, it will ignore the `app_web` key version as a key already existed in `app_db`:

```
$app_db    = { database => 'oracle', version => 11}
$app_web = { web_server => 'httpd', version  => 12 }
$combined_app = $app_db + $app_web
```

Removal

The removal operator takes a hash variable, a – symbol, and a hash, an array of keys, or a single string key. If giving a hash, the values in the hash will not matter as the removal is simply removing any matching keys. In the following example, a hash of `software_versions` with the `oracle` key and an integer of `11`, the `httpd` key and a value of `12`, and the `cowsay` key with a value of `9` can be seen. When a single key is removed to create the `no_cowsay` variable, the key-value pair of `cowsay` and `9` is removed. When `only_cowsay` is assigned, the values of `oracle` and `httpd` in the hash to be removed do not matter and it will simply remove the key and value. For the `only_oracle` variable, removing an array will make the removal operator run through each matching key and remove matches:

```
$software_versions = { oracle => 11, httpd => 12, cowsay => 9}
$no_cowsay = $software_versions - cowsay
$only_cowsay = $software_versions - { oracle => 'anything' , httpd =>
'anything' }
$only_oracle = $software_versions - [httpd,cowsay]
```

Hash data type

The hash data type accepts an optional key type and value type; if a key type is specified, a value type must be specified. A minimum size and maximum size can be specified for how many key pairs there should be:

```
Hash[<Key type>, <Value type>, <Minimum size>, <Maximum size>]
```

For example, the following class has a `tunables` parameter, which must contain a hash with `1` to `10` key-value pairs of strings and integers:

```
class kernel_overrides (
  Hash[String,integer,1,10] tunables,
)
```

Mixing hashes and arrays

Since the value of a hash key value or an array value can be any data type, nesting can be performed. Care should be taken not to overcomplicate the structure.

The following example shows the `server_cmdb` hash containing a hash of `nfs_share_servers`, with the `prod` and `dev` keys containing arrays of strings:

```
$server_cmdb = {
  nfs_share_servers => {
    prod => ['prdnfs01','prdnfs02','prdnfs02'],
    dev  => ['devnfs01','devnfs02','devnfs03'],
  }
}
```

To access the first `prod` array's third value, `prdnfs02`, the following call could be made:

```
$server_cmdb[nfs_share_servers][prod][2]
```

Lab

To practice what we've covered, write a class that takes an array of packages and installs the packages with a hash-defining standard parameter for installing the provider and version. Remember to declare the class with variables, as per the previous lab. As an example, you could install the latest version of the RubyGems webrick, puma, and sinatra. The suggested solution can be found at `https://github.com/PacktPublishing/Puppet-8-for-DevOps-Engineers/blob/main/ch04/packages_array_hash_paramters.pp`.

Abstract data types, including Sensitive

Abstract data types give you the flexibility to mix the core data types for parameter enforcement and particular patterns, as well as provide some more advanced features in terms of parameter checking. There are a large number of abstract types, so this section will cover the most commonly used ones. Other types can be found at `https://github.com/puppetlabs/puppet-specifications/blob/master/language/types_values_variables.md` and `https://www.puppet.com/docs/puppet/8/lang_data_abstract.html#variant-data-type`.

Prefixes

Although not Puppet terminology, the types we'll review will be described as prefixes, where a type is prefixed in front of another type with no options.

Sensitive

The `Sensitive` data type is used to mark strings as sensitive, which means the value will be displayed in plain text in the code and the catalog, but not in any Puppet reports or logs. By prefixing the `Sensitive` keyword to parameters and assignments with brackets, the strings' contents are made sensitive. This affects both string types and resources that can contain strings or can be converted into strings. In the following example, we are showing a string, a string in an array, and an array that can be assigned. The output will print `[value redacted]` over the section that has been marked as sensitive:

```
$secret_string = Sensitive('password')
notify {"Print ${secret_string}":}
$single_sensitive_array = [Sensitive('password'),'password']
notify {"Print ${single_sensitive_array}":}
$secret_array = Sensitive(['password','password'])
notify {"Print ${secret_array}":}
```

When the value must be used in code, the `unwrap` function allows us to view the sensitive value. This example shows how it could be unwrapped to print with a `notify` resource:

```
notify {"Print ${secret_string.unwrap}":}
```

This is purely an example and would defeat the purpose of hiding a value from logging and reporting; more likely, it would be passed to another resource. Resources such as user-recognized sensitive values for attributes such as `password` do not need to be unwrapped, but resources such as `exec` do not interpolate, so the values must be unwrapped. To avoid leaking data resources such as `exec`, which cannot interpolate, you can wrap it as `Sensitive` to ensure no part is exposed in logging. The following example shows passing the sensitive string to `user` and passing the sensitive string as a password to a `curl` command:

```
user { 'max':
  id => 7,
  password => $secret_string,
}
exec {'secure curl':
  command => Sensitive("C:\\Windows\\System32\\curl.exe -u
david:${secret_string.unwrap} http://example.com"),
}
```

If only the unwrap is performed when running Puppet with `debug`, the command and password would be fully visible.

In *Chapter 7*, we will cover templates, including how to use sensitive values. However, as of Puppet 7.0 and 6.20, you no longer need to unwrap sensitive values before using them in templates.

> **Note**
>
> Securing data fully from end to end will be discussed in *Chapter 9*.

`Enum` and more advanced pattern data type patterns, which will be covered in the next section, will not work with `Sensitive` and should be avoided. Here, you should only use basic types such as `string`.

Optional

The `Optional` data type allows `undef` to be used as an acceptable input for a data type, in addition to the types allowed by the data type it prefixes:

```
Optional <type> <variable name>
```

For example, to allow an `Integer` parameter or `undef` to be assigned to the `oracle_uid` variable, simply add the `Optional` keyword in front of the `Integer` type:

```
class oracle (
  Optional Integer orace_uid,
)
```

The `notundef` type has the opposite effect but is of much more limited and exceptional use.

Patterns

Pattern types allow for combinations of types, such as regular expressions or specific choices of strings, to be enforced on attributes.

Enum

The `Enum` data type allows you to enumerate strings, allowing multiple options to be used on a `class` parameter. The following code declares `Enum`, followed by an array of strings as options with a minimum of one string or more:

```
Enum[<string>,*<string>]
```

The following example shows how to use this in a class called `regional` with a parameter of `uk_region` accepting one of the available UK regions:

```
class regional (
  Enum['Scotland,'England','Wales','Northern Ireland'] uk_region,
)
```

Variant

The `Variant` data type allows you to combine any other data types as an array. The following code uses the `Variant` keyword and declares the list of allowed types on a parameter:

```
Variant[<type>,*<type>]
```

For example, the following class accepts Booleans of `true` and `false` or the `true` or `false` strings for the `create_user_home` variable. It will also take a string or an array of strings for the `user_names` variable:

```
class user_accounts(
  Variant[Boolean, Enum['true', 'false']] create_user_home,
  Variant[String,Array[String]] user_names,
)
```

Pattern

The `Pattern` data type is similar to `Variant` but is a way of providing a list of regular expressions where the parameter can match any of them. The syntax is as follows:

```
Pattern[<regexcp>*<regexcp>]
```

Here, we are declaring with the `Pattern` keyword, followed by an array of `regexp` types. For example, the following defined type, `server_access`, takes a hostname that must have a string starting with `edi`, `gla`, or `abe`:

```
define server_access (
  Pattern[/^edi/,/^gla/,/^abe/] hostname,
)
```

Arrays and hashes

In this section, we will cover the various arrays and hashes types.

Tuple

In the previous section, we discussed that the array type could have one type declared for all of its content. Tuple allows any number of types to be used at specific indexes within an array and optional minimum and maximum sizes. The minimum size, if smaller than the number of types assigned, makes those types optional, while a maximum size allows for the last type to be repeated if the maximum size is greater than the number of types declared. A maximum size requires a minimum size to be declared:

```
Tuple[ <type>, *<type>,  <minimum size>, <maximum size>]
```

To provide an example of this, let's consider three variables: `user_declaration`, `calculation`, and `file_download`. The `user_declaration` variable requires a string for the username, an integer for the UID, and at least one string up to eight characters in length, which represents the groups that a user can be assigned to. The `calculation` variable requires an integer, a float, and an integer. The `file_download` variable requires a URI and a string, and, optionally, an integer. Please note that the integer is optional and is not required:

```
class exampleapp (
Tuple [ string, integer, string, 3 , 10 ] user_declaration,
Tuple [ integer, float, integer] calculation,
Tuple [ uri, string , integer, 2] file_dowload,
)
```

Struct

`Struct` provides a similar type to `Tuple` for hashes. In the `Hash` data type, a single key type and value type was declared, while a struct allows for a particular order string keys with the option to have `optional` or `undef` and value types to be declared. Unlike `Tuple`, there is no minimum or maximum size:

```
Hash[<*optional *undef String name>, <Value type>, *(<*optional *undef
String name >,<value type>)
```

To illustrate how the use of optional keys and values can affect variable assignments, let's consider three examples: `config_file`, `application_binary`, and `application_startup`. The `config_file` variable requires key pairs, including the `mode` key with a string value of either `file` or `link`, and a `path` key with a string value. The `application_binary` variable is similar to `config_file`, but it allows for an optional `owner` key with a string value. If present, the `owner` key must have a string value. The `application_startup` variable requires an `owner` key that can either be undefined or a string. Additionally, the value for each key must match the expected data type:

```
class skeleton (
Struct[{mode => Enum[file, link],
        path => String}] config_file

Struct[{mode                => Enum[file, link],
        path                => String,
        Optional[owner] => String}] application_binary

Struct[{mode                => Enum[file, link],
        path                => String,
        owner               => Optional[String] }] application_startup
)
```

Parent data types

The following data types allow you to group multiple data types into a single parameter. Using them directly can make code shorter and clearer:

- `Any`: The `Any` type matches any Puppet data type, making it useful when the exact data type is unknown or does not matter.
- `Collection`: The `Collection` type matches any array or hash data type, making it useful when an array or hash can have multiple data types.
- `Scalar`: The `Scalar` data type matches strings, Booleans, regular expressions, and numerics. It is useful when a single value with any of these data types is required.
- `Data`: The `Data` type matches scalar, undefs, and arrays containing values that match data, and hashes with keys matching scalars and values that match data. It is useful when complex data structures are required.
- `Numeric`: The `Numeric` type matches float and integer data types, which is useful when a numerical value is required.

Lab

Continuing our work on the `all_grafana` class, create an `all_grafana_data_types` class and add to it so that it accepts a `file_options` parameter. This must have a name but can optionally have a mode, a user, and a group as a hash. Ensure each of those resources has restricted data types. Add a Grafana user and a sensitive parameter password that is passed to the user.

To achieve class assignment, write a class declaration before you assign the class some variables. When you run `bolt` against your manifest, it will include your variables. The solutions are available at `https://github.com/PacktPublishing/Puppet-8-for-DevOps-Engineers/blob/main/ch04/all_grafana_data_types.pp`.

Scope

In Puppet, a scope is a level of code that has limited access to variables and default settings for resources. The three levels of scope are top scope, node scope, and local scope. Top scope variables reflect variables that are declared globally, most commonly in the `site.pp` manifest file. Node scope variables are assigned in node definitions, which are typically also made in `site.pp` or via an **External Node Classifier** (**ENC**). For example, variables can be declared in the `site.pp` manifest file within a Puppet environment to make them globally available to all nodes. Alternatively, variables can be declared in a node definition in `site.pp` or the ENC to be made available at the node level for a particular server or group of servers. `site.pp` is a special manifest file in Puppet that contains the main configuration for a Puppet environment. A resource default is a default setting for a resource, which can be overridden in a more specific scope, such as node scope or local scope. The full use of `site.pp`, ENC, and node definitions will be explained in detail in *Chapter 10*.

When accessing a variable, by default, the server will access the lowest level first and essentially override variables of the same name at higher levels. Other local scopes can be accessed by using the namespace but cannot be assigned values.

Here's an example of how these concepts work together in a single Puppet manifest file. We can define a global variable called `'global'` with a string value of `'world'`, and a node definition that, by default, assigns all nodes a variable called `'node'` with a string value of `'mynode'`. The node definition includes two classes, `'local'` and `'also_local'`. In the `'local'` class, we assign a variable called `'global'` with a string value of `'override'`, which has a local scope and overrides the global value. We will use two notify resources to demonstrate how variable scope works. The first notify resource prints `'Print override'`, showing that the `'global'` local variable has overridden the global value. The second notify resource uses the `::` syntax to reference the global variable, so it prints `'Print world'`. The third notify resource prints `'Print node'` because there are no local variables with that name. In the `'also_local'` class, we define a new variable called `'another_global'` with a string value of `'another world'`. The first `notify` resource in this class uses the directly-accessed variable to print `'Print another override'`. The second `notify` resource uses the `::` syntax to reference the global variable and prints `'Print another world'` because no local variable called `'global'` is declared. A `notify` resource is a Puppet resource type that simply logs a message to the console or system log. It's commonly used for debugging or informational purposes:

```
$global = 'world'

node default {
  $node = 'mynode'
  include local
  include also_local
}

class local
{
$global = 'override'
  notify {"Print ${global}":}
  notify {"Print ${::global}":}
  notify {"Print ${node}":}
}

class also_local {
  notify {"Print another ${local::global}":}
  notify {"Print another ${global}":}
}
```

Resource titles or references to resources are not limited by scope as they must be unique to the whole catalog. As shown in the preceding example, the `notify` resources that were used in the `also_local` class had their titles adjusted so that they contained `another`. This helps us avoid resource title clashes when the variables are interpolated. Otherwise, both the `local` and `also_local` classes would have contained `notify` resources called `Print override` and `Print world` and would fail to compile with duplicate resources.

As discussed, the `also_local` class can call the `global` variable from the `local` class but cannot assign it to that local scope.

Summary

In this chapter, we learned that Puppet variables are different from those in normal procedural languages as they can only be assigned once. We saw that certain words are reserved and cannot be used in naming variables. We also saw that Puppet variables can be interpolated, depending on how and where strings are placed.

We covered various core data types and how they can be used to both restrict parameters and assign variables. We also looked at `undef` and Booleans, which need to be carefully managed when translating values to get the expected results.

Next, we looked at arrays and hashes and how to assign them. Although they can't be changed, we learned how operators can manipulate them into new assignments. We also covered how arrays and hashes can be nested and mixed as hashes of arrays and arrays of hashes.

Then, we looked at abstract data types and how they apply restrictions to parameters more flexibly with the `Sensitive` type, which provides scoped protection for logs and reporting.

After, we reviewed how Puppet variables can be declared at different scopes and how variables can be shared/seen in different scopes.

In the next chapter, we will cover facts and functions. We will look at the system profiling tool, Facter, the information it gathers, and how it can be customized to gather user-specific data on system profiles. Functions provide Ruby code plugins, allowing code to be run at compile time, which can perform actions such as data manipulation or affect the catalog run. We will cover built-in functions and functions from the standard `lib` module, from Puppet Forge, which can be used to manipulate data types into the variables we discussed in this chapter.

5

Facts and Functions

This chapter will cover facts. We will show you how the Facter tool gathers them to show the profile of systems, how to interact with Facter, and how to use them in Puppet code. We will also cover how custom and external facts can be added to the provided core facts, to allow for more user-specific facts to be gathered.

Then, we will cover functions. We will explain what functions do and the three types of functions – statement, prefix, and chained. We will examine a selection of the core provided functions to show you their capabilities. A selection of functions will also be shown from the `stdlib` module, where we will explain the module's approach and uses.

Deferred functions, which were introduced in Puppet 6, will also be covered. Here, we will show you how deferred functions differ from normal functions, how to make a function deferred, and pitfalls to avoid while using deferred functions.

In a nutshell, the following topics will be covered in this chapter:

- Facts and Facter
- Custom facts and external facts
- Functions
- The stdlib module functions
- Deferred functions

Technical requirements

For this chapter, you will need to provision a Puppet server standard architecture with a Windows client and a Linux client by downloading the `params.json` file from `https://github.com/PacktPublishing/Puppet-8-for-DevOps-Engineers/blob/main/ch05/params.json` and then using the following command from your `pecdm` directory:

```
bolt --verbose plan run pecdm::provision --params @params.json
```

Facts and Facter

Facter is Puppet's system profiler, a set of Ruby libraries that work cross-platform and gather information, known as **facts**, about clients. This tooling provides the necessary information to evaluate the profile of the client and allows configuration decisions to be made based on the pre-existing state of the host in Puppet code.

Puppet 5 and 6 use Facter 3, while Puppet 7 and 8 uses Facter 4. Only a handful of new features are available in Facter 4, which will be highlighted, and a small number of facts have changed, but most users will find no difference. These differences can be viewed by running the `puppet facts diff` command. In *Chapter 8*, we will highlight how module testing can ensure code is compatible across different versions.

The output of Facter can be seen by running the `facter -p` or `puppet facts` command on the command line or VSCode terminal. Running either of these commands without any further options will return all the core facts. The `-p` flag ensures that Puppet-specific facts are gathered. Due to a circular dependency being created between Facter and Puppet, it had been previously planned to depreciate the `-p` flag and replace it with the `puppet facts` command. This approach was abandoned with the release of Facter 4. This book will use the `facter` command for its examples, which follows the documentation and community practices.

> **Note**
>
> By default, the `facter` command outputs in a Puppet hash format, while `puppet facts` outputs in JSON format. Both of these commands accept options for choosing the appropriate format.

We will now look at some examples of Facter output. The simplest type of fact is a simple key-value pair, such as the `Kernel` fact, which in this case tells us that the kernel is Windows-based:

```
"Kernel": "windows"
```

There are also hashes, known as structured facts, which can be broken into nested levels. The `os` fact is commonly used. The following example of a Windows 10 laptop shows the various levels that are available:

```
os => {
```

```
  architecture => "x64",
  family => "windows",
  hardware => "x86_64",
  name => "windows",
  release => {
    full => "10",
    major => "10"
  },
  windows => {
    display_version => "21H2",
    edition_id => "Core",
    installation_type => "Client",
    product_name => "Windows 10 Home",
    release_id => "21H2",
    system32 => "C:\WINDOWS\system32"
  }
}
```

The full list of core facts can be found at `https://puppet.com/docs/puppet/latest/core_facts.html`; running `facter -p` on your client system and reviewing the output is recommended. An individual fact can be accessed by running the `facter` command against the fact's name, such as `facter -p kernel`, to return the `kernel` fact. To access a specific nested level value in a structured fact, dot notation is used, which separates each key level name with dots (`.`). So, to access the `family` fact within the `os` structured fact, the `facter -p os.family` command can be run.

As Facter has gone through several iterations and structured facts were not in earlier versions, Facter 3 hid several legacy facts such as architecture, which was put into the `os` structured fact as `os.structured`. The `--show-legacy` flag allows these facts to be made visible in the Facter output; they are documented in the core fact documentation.

When Puppet is run, either via the agent or by running `puppet apply` on the command line, Facter will run, with legacy facts, disabled in Puppet 8 or enabled in Puppet 7 and below and the output will be assigned to global variables. As of Puppet 7.21 and Puppet 8 the `include_legacy_facts` setting was added to allow overriding the default behavior on each client in the `puppet.conf` file.

These variables can then be accessed in Puppet manifests in two ways – either directly by the name of the fact on a global variable or via the facts array. It is strongly recommended to access facts only via the facts array since this makes it clear that facts are being accessed and not potentially other global variables.

For example, in the following code, the `notify` resources would access the `kernel` and `os family` variables and print logging messages containing the `kernel` and `os` families of the host it was run on:

```
notify { "This clients kernel is ${facts[kernel]}": }
notify { "This client is a member of the os family ${facts[os]
[family]": }
```

Note that not all facts will appear on all clients. Facts often have a context on which to filter themselves, such as which operating system is running or if a particular underlying hardware is being used.

> **Note**
>
> As you'll see in the next section, functions use dots for chained functions, so the dot-separated access syntax of the `facter` command cannot be used to call the `facts` variable directly. However, the `getvar` function can be used.

Facter can be customized and tuned on a host-by-host basis by configuring a `facter.conf` file. This file is not created by default and should be created at `/etc/puppetlabs/facter/facter.conf` on Nix-based systems and `C:\ProgramData\PuppetLabs\facter\etc\facter.conf` on Windows. For testing purposes, the `facter` command can be run with the `-c` flag to select a configuration file to be run.

An example `facter.conf` group looks like this:

```
facts : {
    blocklist : [ "disks", "dmi.product.serial_number", "file system"
],
    ttls : [
        { "processor" : 30 days },
    ]
}
global : {
    external-dir      : [ "/home/david/external1", "/home/david/
external2" ],
    custom-dir        : [ "/home/david/customtest" ],
    no-exernal-facts : false,
    no-custom-facts  : false,
    no-ruby           : false
}
cli : {
    debug     : false,
    trace     : true,
    verbose   : false,
    log-level : "warn"
}
fact-groups : {
 custom-exampleapp : ["exampleapp1", "exampleapp2"],
}
```

The first section, `facts`, includes a blocklist, which allows us to list facts and fact groups that will not be run. This can be useful in situations where calculating the fact can be computationally expensive. For example, in the preceding example, we block the `disks` and `file system` groups since, in some legacy UNIX systems, SAN storage can be configured with thousands of paths. It also disables `dmi.product.serial_number`, which might be decided as something secure that should not be visible in Puppet. To see a full list of blockable groups, you can run the `facter --list-block-groups` command, which will list the group names and a list of the facts inside it. For example, the `disks` group looks like this:

```
disks
- blockdevices
- disks
```

The other part of the facts section is `ttls`, which allows caching to be configured. Cached facts are stored as JSON in `/opt/puppetlabs/facter/cache/cached_facts` on UNIX-based systems and `C:\ProgramData\PuppetLabs\facter\cache\cached_facts` on Windows. In the preceding example, the `processor` group will only be refreshed every 30 days. To see a full list of cacheable groups, you can run the `facter --list-cache-groups` command, which will display a similar format to the block groups.

The `global` section allows an array of directories to be passed to `external-dir` so that you can define where `facter` should look for external facts. Similarly, an array directory can be passed to `custom-dir` to define where `facter` should look for custom facts. Custom and external facts will be covered in the next section.

The `global` section has three Boolean values:

- `no-external-facts`: To disable external facts if set to `true`.
- `no-custom-facts`: To disable custom facts if set to `true`.
- `no-ruby`: To prevent Facter loading via Ruby. Any facts that use Ruby and custom facts are set to `true`.

All these settings are more likely to be used for debugging and development purposes.

The `cli` section sets a log level with a string of (`none`, `trace`, `debug`, `info`, `warn`, `error`, `fatal`) and has three options: `verbose`, `trace`, and `debug`. Each of these three options is enabled or disabled with a `true` or `false` Boolean. The `trace` option will show a backtrace if an exception occurs in a custom fact. This should not be confused with the trace log level; a better name for this option might have been stacktrace. The `verbose` option enables `verbose` information output from Facter, while the `debug` option enables debug-level output from Facter.

The `fact-group` section is new to Facter 4 for Puppet and allows you to define custom groups for caching and blocking. Core facts and custom facts can be specified but not external facts.

> **Note**
>
> As the `facter.conf` file uses a HOCON format, it can be easier to manage it via the HOCON module from Puppet Forge (`https://forge.puppet.com/modules/puppetlabs/hocon`), where it can be classified on an individual node or group of nodes basis as required.

Facter 4 in Puppet 7 and above, also reintroduced benchmarking of facts, which had previously been in Facter 2. To run benchmarks on a particular fact, you can run the `facter -t <fact name>` command. For example, running `facter -t os` will produce an output similar to the following:

```
fact 'os.name', took: (0.000007) seconds
fact 'os.family', took: (0.000006) seconds
fact 'os.hardware', took: (0.000007) seconds
```

If a structured fact is selected, it will time each part of the fact and will return it to the normal output of the facter call after.

Having covered what core facts are and how to run and configure Facter to test and manage them, the next stage is adding customizations via custom and external facts.

Custom facts and external facts

In this section, you will learn how to add to the facts provided by core via custom facts. These are written in Ruby, similar to core facts or external facts, which are either hard-rigged values or client-native executable scripts. While it can be tempting to gather everything, the extra burden facts put on the Puppet infrastructure, particularly with many agents, should be considered and balanced with the need for data.

External facts

External facts are executables that can set facts based on the logic within the scripts or facts set statically by the structured data of the file.

External facts can be stored in the following directories for Unix-based operating systems:

- `/opt/puppetlabs/facter/facts.d/`
- `/etc/puppetlabs/facter/facts.d/`
- `/etc/facter/facts.d/`

For Windows systems, they can be stored in `C:\ProgramData\PuppetLabs\facter\facts.d\`.

In *Chapter 8*, you will learn how external facts can be distributed to clients from modules via the plugin sync process, where facts are added from modules contained in a `facts.d` folder within the module.

> **Note**
>
> Puppet can be run as a non-root user on UNIX-based systems, while external facts can be stored in `~/.facter/facts.d/`. However, this book will not cover running as a non-root user.

Static external facts

Static external facts must be in JSON, YAML, or TXT format. As an example, let's set an `Application` fact to `exampleapp`, a `Use` fact to `production`, and an `Owner` fact to `exampleorg`. In a YAML file, this can be created like so:

```
---
Application :  exampleapp
Use : Production
Owner : exampleorg
```

In a JSON file, they would be set like this:

```
{ "Application": "exampleapp", "Use": "Production", "Owner": 
"exampleorg"}
```

In a TXT file, the same facts would be set like this:

```
Application=exampleapp
Use=Production
Owner=exampleorg
```

For Windows, the line endings used in these files must be either LF (Line feed, Unicode character 000A) or CRLF (Carriage return and line feed, Unicode characters 000D and 000A) and the file encoding used for the files must be either ANSI or UTF8 without BOM.

The examples we've looked at so far are all known as **flat facts**. However, structured facts can be returned by creating an array format. For example, in YAML, we could allow two owners by adding arrays and nested arrays to YAML. In this example, let's assume there are multiple applications, and each application can have joint ownership:

```
---
Application :
  exampleapp
Use : Production
Owner
- exampleorg
- anotherorg
- anotherapp
Use : Production
Owner : exampleorg
```

This would allow us to call `facter application.exampleapp.owner` to retrieve an array of owners or to call `facter application.anotherapp` to receive the use and owner key pairs.

Note that static external facts will always return a string type in their output.

Executable external facts

Executable external facts differ on Windows and UNIX, but both are runnable scripts that output key pairs or arrays to return facts or structured facts.

On Windows, the following file types can be used:

- Binary executables (`.com` and `.exe` files)
- Batch scripts (`.bat` and `.cmd` files)
- PowerShell scripts (`.ps1` files)

On UNIX platforms, any executable file with a valid shebang (`#!`) statement can be run. If the shebang statement is missing, the execution of the script will fail.

For both platforms, the scripts should return text. This will be read as key pairs or as YAML or JSON, which can be parsed into a structured fact.

For example, a Unix bash script that returns the PID of the `exampleapp` process as a fact, along with a fact for `exampleapp_cpu_use` and `example_memory_use`, may look like this:

```
#!/bin/bash
echo "exampleapp_pid = ${pidof exampleapp}"
echo "exampleapp_cpu_use = ${ps -C exampleapp} %cpu"
echo "exampleapp_memory_use = ${ps -C exampleapp} %mem"
```

For Windows, a PowerShell script to return the same facts would look like this:

```
Write-Output "exampleapp_pid=$((Get-Process explorer).id)"
Write-Output "exampleapp_cpu=$(Get-Process explorer).cpu)"
Write-Output "exampleapp_mem=$(Get-Process explorer).pm)"
```

> **Note**
> To find issues with external facts, you can run `facter --debug`. This will highlight if the fact is visible to Facter and if any output is not being parsed and getting ignored.

Custom facts

Custom facts are sections of Ruby code that can be used to set facts and expand on the core Facter facts. The main advantage of using custom facts over external facts surrounds the built-in mechanisms that are available. In this section, you will learn how the use of custom facts allows you to access the value of other facts within custom facts, how you can have multiple weighted resolutions, and how to use `confine` to ensure only certain nodes will attempt to run the fact.

The main disadvantage of using custom facts is that they need to be written in Ruby, which is a learning curve. It's beyond the scope of this book to dive deeper into the details of Ruby, but its basic structure and some of its core libraries, which work well on Windows and UNIX-based systems, will be shown to give enough you a head start so that you can research them further.

Like external facts, custom facts are normally distributed by Puppet modules. However, when performing local testing, there are three ways to direct Facter to locations where we store our facts locally while testing:

- The Ruby library load path
- Using the `--custom-dir` option on the Facter command (note that this can be flagged multiple times)
- Setting the `FACTERLIB` environment variable

The Ruby library load path can be checked by running `ruby -e 'puts $LOAD_PATH'`. Remember to make sure that the Ruby binary being used is the Puppet-provided version at `C:\Program Files\Puppet Labs\Puppet\puppet\bin\ruby.exe` on Windows or `/opt/puppetlabs/puppet/bin` on UNIX-based systems.

Custom facts declare themselves using `Facter.add('<fact_name>')` and use a `setcode` statement to run a code block to resolve the fact. This is the way a fact's value is determined. As a simple example, it is possible to run a UNIX shell or Windows terminal command directly by surrounding the command with backticks (`` ` ``):

```
Facter.add('exampleapp_version') do
  setcode do
    `exampleapp -version`
  end
end
```

Since there is only one command, this could also be written with a single `setcode` line:

```
Facter.add('exampleapp_version') do
  setcode `exampleapp --version`
end
```

Both would set the `exampleapp_version` fact to the output of the `exampleapp -version` command.

If your facts were more complicated and you needed to run multiple commands or manipulate the output, the command could be run using a Ruby class.

In the following example, the `Facter::Core::Execution.execute` Ruby class will run a command called `exampleapp`, with a flag of `version`, and then pipe the output of the command to `awk` to print the second returned value:

```
Facter::Core::Execution.execute('exampleapp -version' | awk '{print
$2}' )
```

PowerShell commands can be executed using the `powershell` command, like so:

```
Facter::Core::Execution.execute('powershell (Get-WindowsCapability
-Online -Name "Microsoft.Windows.PowerShell.ISE~~~~0.0.1.0").state')
```

It can be tempting to run everything as terminal commands for familiarity, but care must be taken as not everything that can be used in a terminal will work. For example, bash-style `if` statements will not work and should be written in Ruby code.

It can be useful to call the value of another fact into a variable. The following code will put the `os arch` fact into the `arch` variable:

```
arch = Facter.value('os.arch')
```

Confining custom facts

One of the main advantages of custom facts is the ability to confine the nodes they will run on. This can be achieved with the `confine` statement and by selecting the facts and values to match for the fact to run. The `confine` functions syntax follows this format:

```
confine <fact_name>: '<fact_value>'
```

The fact defined after the `confine` function will only run if the conditions are met. For example, you can confine a fact to only run on nodes with Windows-based kernels:

```
confine kernel: 'Windows'
```

An array can also be used, where matching any of the values will allow the fact to run. For example, we can check if the kernel is from Linux or Solaris:

```
confine kernel: ['Linux', 'Solaris']
```

For structured facts, the `Facter.value` method can be used to access it. For example, to test that the `os.release.major` fact is equal to `10`, the following code can be used, where `=>` is used instead of a colon (`:`) to match the fact to its value:

```
confine Facter.value(:os)['release']['major'] => '10'
```

In addition to facts, Ruby commands and library commands can be used to confine facts. For example, `confine` can be used with `Facter::Core::Execution.where` or `Facter::Core::Execution.which` to confirm a command exists in the path for Windows or Linux, respectively. Additionally, Ruby libraries such as File can be used to check this.

For example, to confine a fact if the `git` command was found in the Windows path, the following code can be run:

```
confine { Facter::Core::Execution.where('git') }
```

The following code would confine a fact to only run if `/opt/app/exampleapp` existed as a file or as a directory:

```
confine { File.exist? '/opt/app/exampleapp' }
```

To write a single fact that can cover multiple implementations and confine with granularity, we can use both resolutions (`Facter.add` statements) and multiple confine blocks. The following example shows a simple example of setting the Facter value of `whoami` to either `I am windows 10` if the kernel fact is Windows and `os.release.major` is `10` or to the `I am Sparc` string if the kernel is `sparc`:

```
Facter.add('whoami') do
  setcode do
    confine kernel: 'Windows'
    confine Facter.value(:os)['release']['major'] => '10'
    'I am windows 10'
  end
end
Facter.add('whoami') do
  setcode do
    confine kernel: 'Sparc'
    'I am Sparc'
  end
end
```

Another method of confining facts is using features. A feature is a section of Ruby code that's added to `lib/puppet/feature` in a module. For example, the `exampleapp` module could contain an `exampleapp.rb` feature that checks whether `exampleapp` was installed on either Windows or Linux:

```
require 'puppet/util/feature'
Puppet.features.add(:example_app)
do
windows= `powershell '(Get-Command exampleapp).source'`.strip
linux = `sh -c 'command -v exampleapp'`.strip
windows.empty? && linux.empty? ? false : true end
```

A custom fact could then use a `confine` statement so that only nodes with the `exampleapp` command available would run the fact:

```
Facter.add('exampleapp') do
setcode do
confine { Puppet.features.example_app? }
```

This removes the need to create additional facts and gather and process information not needed except for evaluating confinement.

> **Note**
>
> It is important to perform all logical code inside the `setcode` and `confine` blocks; otherwise, when the facts are loaded, it will run this code, rather than when the fact is queried for resolution. The order in which facts will be loaded is not predictable, so if code is required by the fact but it is outside of the blocks, it can result in ordering errors.

Weighted resolutions

Another approach to writing custom facts is to have multiple resolutions while knowing that some may return null values, but we want to work through various options. When reviewing resolutions, Facter eliminates any that are not confined. Then, it looks at the weight of each resolution. By default, this is set to `0` but it can be set using the `has_weight` function. If two resolutions have the same weight, Facter will use whichever was listed in the code first.

For example, to set the `exampleapp_version` fact with multiple resolution options, in the first resolution, it will run the command with the `version` flag weighted at `100` and then try to look for the version in the config file weighted at `50`:

```
Facter.add('exampleapp_version') do
has_weight 100
setcode do
`exampleapp --version`
```

```
end
Facter.add('exampleapp_version') do
has_weight 50
setcode do
`grep version /etc/exampleapp/exampleapp.conf | awk '{print $2}'`
end
```

This allows the command to fail so that it can be backed up with a second source.

> **Note**
>
> External facts have a weight of `1000`. So, to prevent an external fact from being able to overwrite your custom fact resolutions, you can set a value higher than `1000` on the resolution weight.

Rescue blocks

By default, Facter will error and fail to return any value if any resolution fails with an error. Using rescue blocks can allow default values to be returned as a result of failures and to opt to print warnings. This works in conjunction with weighted resolutions, where it's common to expect failures in resolutions.

A simple rescue block that returns `nil` on the failure of resolution after running the `exampleapp -version` command and logging a failure would look like this:

```
setcode do
`exampleapp --version`
rescue
  nil
  Facter.warn("exampleapp command failed")
end
```

Using `Facter.warn` ensures this message is printed to STDERR when used via the `Facter` command. When used during the Puppet catalog application, it will ensure it is printed in Puppet's logs. Returning `nil` would ensure other resolutions can be used if they return non-nil values.

Timeouts

As part of Facter 4, which was made available in Puppet 7 and above, it is now possible to add a timeout to a resolution. This can be done by adding a comma after the fact's name as part of the `Facter.add` resolution statement and using the `{timeout: <value in seconds>}` syntax, where the value in seconds can be an integer or a float. For example, to ensure a 0.2-second timeout on the `exmpleapp_version` fact, the code can be set like this:

```
Facter.add('exampleapp_version', {timeout: 0.2}) do
```

Although this is only a feature in Facter 4 and Puppet 7 and 8, in Facter 3 and Puppet 5 and higher, it is possible to put timeouts on the execution command by directly setting the `options` variable on the `execute` function. For example, the same 0.2-second timeout could be applied to the execution of the `exampleapp -version` command rather than the whole resolution by modifying the `execute` command:

```
Facter::Core::Execution.execute('exampleapp --version', options =
{:timeout => 0.2})
```

Aggregate and structured facts

Aggregate facts allow the resolutions of a fact to be broken up into chunks. These chunks can then be merged. Merging arrays or hashes creates a structured fact or performs other functions, such as adding the values of facts together.

An aggregate fact still has a `Facter.add` declaration, but within `Facter.add`, it sets the type variable to `aggregate`. Then, instead of using `setcode` sections, it uses `chunk` sections for the resolutions. By default, each `chunk` will be merged unless an aggregate block is declared to perform another function.

For example, the following code would create a structured fact called `exampleapp`. It would have `exampleapp.version` and `exampleapp.fullpath` containing the output of the commands in the chunks:

```
Facter.add(:exampleapp, :type => :aggregate) do
  Chunk(:version) do
`exampleapp -version`
  end
  Chunk(:fullpath) do
`which exampleapp`
  end
end
```

To use an aggregate block and sum facts together, you can use the following code, which makes a fact called `exampleapp_memory_usage` that takes a chunk using a fact containing the total memory used by `exampleapp` and adds it to the memory used by `exampleapp2` to give us the total memory usage:

```
Facter.add(: exampleapp_memory_usuage, :type => :aggregate) do
  chunk(:exampleapp1_usage) do
    Facter.value(:exampleapp1_usage)
  end
  chunk(:exampleapp2_usage) do
    Facter.value(:exampleapp2_usage))
  end
  aggregate do |chunks|
```

```
    total = 0
    chunks.each_count do |value|
      total += value
    end
    total
  end
end
```

A new method to return structured facts is available in Puppet 7 and 8 with Facter 4. This follows the use of dot notation in the fact's name, which allows a definition to assign different levels of a structured fact. For example, to set the `exampleapp` fact with a nested level of `exampleapp.version` and `exampleapp.pid`, you can use the following code:

```
Facter.add('exampleapp.version') do
setcode do
`exampleapp --version`
end
Facter.add('exampleapp.pid') do
setcode do
`pidof exampleapp`
end
```

This has a core advantage over using an aggregate. Unlike an aggregate, a failure of one part of the declaration will only affect that declaration; the rest will be assigned.

> **Note**
>
> This section has tried to give you enough information to get started with custom facts. In Puppet's documentation for custom facts and module code, you will find alternative syntax for many of the features we've discussed. Since it is in Ruby code, there is a greater variation of what can be declared. This book has chosen what it considers the best style and practice to follow to keep things simple and avoid listing too many options.
>
> Some modules that can be useful to follow examples further are available on GitHub:
>
> `https://github.com/puppetlabs/puppetlabs-pe_status_check/blob/main/lib/facter/`
>
> `https://github.com/puppetlabs/puppetlabs-stdlib/tree/main/lib/facter`
>
> `https://github.com/puppetlabs/puppetlabs-lvm/tree/master/lib/facter`
>
> `https://github.com/puppetlabs/puppetlabs-java/tree/main/lib/facter`

Lab

For this lab, we will create a static external fact and a custom fact and distribute them with `bolt upload` before running the facts and viewing them on the console to see if they have become visible.

For the static external fact, create a structure that sets `packtlab.use` equal to `lab` and `packlab.student` equal to your name.

For the custom fact, a `tmp_count` fact will be created, which will count the number of files in the `/tmp` directory on Linux and `C:\Users\admin\AppData\Local\Temp` on Windows. For Linux, the first resolution with a higher weighting should be `'find /tmp -type f | wc -l'`, while the second with a lower weighting should be `ls /tmp | wc -l`. For Windows, the first higher-weighted resolution should be the `(ls $env:Temp | Measure-Object -line).Lines` PowerShell command and the lower weighted resolution – that is, `(Get-ChildItem $env:Temp | Measure-Object).Count`.

All resolutions should return `undef` in the result of an error and should time out after 10 seconds.

Note that it can be useful to look at your clients' current facts on the web console so that you know how to confine them.

For each of the facts, use the `bolt` command as follows to upload them to the correct locations:

```
bolt file upload path_of_your_fact /path/to/destination --targets
windows_server_fqdn linux_sever_fqdn

bolt task run facts --targets windows_server_fqdn linux_sever_fqdn
```

Go to the web console and view the facts in your nodes to confirm they are on your clients.

You can find the example solutions at `https://github.com/PacktPublishing/Puppet-8-for-DevOps-Engineers/blob/main/ch05/tmp_count.rb` and `https://github.com/PacktPublishing/Puppet-8-for-DevOps-Engineers/blob/main/ch05/packlab.yaml`.

> **Note**
>
> When testing custom or external facts, they can be set manually with environment variables by setting FACTER_<fact_name> in UNIX environments using `export FACTER_exampleapp ="test"` or in Windows environments by using `env FACTER_exampleapp="test"` – this would hard-set the `exampleapp` fact. This method only works with custom or external facts and not core facts.

Functions

Functions are sections of Ruby code that can be run during catalog compilation and allow you to modify the catalog or calculate and return values. Puppet has many built-in functions and more can be supplied via modules from Puppet Forge, such as `https://forge.puppet.com/modules/puppetlabs/stdlib`, or custom-written functions added to modules. This book will not cover writing functions, but a complete guide can be found at `https://puppet.com/docs/puppet/latest/writing_custom_functions.html`.

In this section, we will cover the three different types of functions: statement, prefix, and chained. A selection of the core Puppet functions will be shown, grouped by purpose to demonstrate the most used and useful functions.

> **Note**
>
> A lot of functions were moved from sources such as the `stdlib` module into the core Puppet function. The full list can be reviewed at `https://puppet.com/docs/puppet/6/release_notes_puppet.html#release_notes_puppet_x-0-0`.

Statement functions

Statement functions are Puppet language-provided functions used only for their side effects, which always return `undef` values. Statement functions can omit brackets, unlike the other functions we will review in this section. You cannot add custom or Forge-provided statement functions.

Catalog statements

Catalog statements affect the content of the catalog, allowing classes to be included, dependencies and containment to affect the order of the catalog, and tags to be applied. The following shows an example syntax of catalog statements:

```
include <class name>
require <class name>
contain <class name>
tag <tag name> , *<tag name>
```

The use of `include` and `tag` was discussed in *Chapter 3*, but we didn't look at the `tag` function in much detail. The `tag` function is used within a class to mark that class, and all contained objects with the tag or list of tags.

In *Chapter 6*, we will cover the full use of `require` and `contain`.

Logging statements

Logging statements allow for string messages to be sent to log output on the Puppet server. In *Chapter 10*, server and agent logging will be reviewed in full as logging locations depend on the configuration and whether Puppet enterprise or open source is used. The syntax for a logging statement is simply `<logging level>(<string >)`.

The following log levels can be applied:

- `debug`
- `info`
- `notice`
- `warning`
- `err`
- `fail`

To log a warning message of `'code unexpected'`, the Puppet code would be as follows:

```
warning('code unexpected')
```

The string message can include variables if they are double-quoted for interpolation. So, to produce an error message of `'pa-risc is unsupported'` on a pa-risc architecture system, the Facter `os.arch` fact can be used within the error function string:

```
error("${facts['os']['arch']} is unsupported")
```

This differs from the examples this book has used up till now, particularly with the `notify` resource used in the previous chapter's examples. The `notify` resource returns to the client's logging while the log-level functions will log to the Puppet server. As `notify` is a resource and not a function, it will result in the report showing that a resource is changed every time a `notify` resource is called.

`fail` differs from the other levels as calling it as a function will terminate the compilation and result in no catalog being sent to the agent.

Prefix and chained functions

Puppet functions can be called in two ways and for many functions, either can be applicable.

Prefix functions are called by writing the name of the function and then providing a list of arguments in brackets:

```
function_name(argument, *argument)
```

Chained functions are created from an argument, a full stop (`.`), then the function name with brackets, and any further arguments in those brackets:

```
argument.function_name(argument, *argument)
```

A selection of built-in functions

There are many functions available in core Puppet, and this section will group different functions to show examples of how they can be used or refer to where in this book we will cover them in more detail. The intention of this chapter is not to give the full syntax of every function but to expose a breadth of functions. You can refer to the full functions list at `https://puppet.com/docs/puppet/latest/function.html`. Make sure you select the correct version of the documentation for the Puppet environment you are working with.

Comparison and sizing

The following functions allow you to compare and measure the size of variables. They provide additional capability beyond what can directly be done with data types.

The `length` and `size` functions are effectively the same and can both be used as prefixes or chained functions on arrays (number of elements), hashes (number of key-value pairs), strings (characters), or binaries (bytes) to confirm the relative size/length of the variable. For example, the following command would return `4` as the length of the string four and `5` as the size of the array:

```
Stringwithfour = 'four'.length()
Array_of_five = Size([8,4,5,7,0])
```

`match` is used as a chained function on a string or an array of strings with a regular expression to match patterns. It returns an array containing the string that has been matched in index 0, followed by the pattern(s) that matched. If there are no patterns in the following example, where there's a string that must start with a lowercase letter to start, then the numbers will be `6` to `8` in length. The variable matches `a123456` and returns an array containing `[ 'a123456', 'a' , '123456' ]`:

```
$matches = "a123456".match(/([a-z]{1})([1-9]{6,8})/)
```

If we tried this same regular expression on a non-matching string, `1a23456`, `undef` will be returned:

```
$nomatch = "1a23456".match(/([a-z]{1})([1-9]{6,8})/)
```

Using an array of strings (`'a123456'`, `'b1254678'`, and `'1a23456'`) with the same regular expression results in the `multi_match` variable containing an array of arrays. This is the output if `match` had been used on each string individually:

```
$multi_match = ['a123456','b1254678','1a23456'].match(/([a-z]{1})([1-
9]{6,8})/)
```

This means `multi_match` will contain `[['a123456','a','123456'],['b1254678', 'b','1254678'],undef]`.

`max` and `min` are used as prefix functions. They take an array of strings or numeric values and return the largest and smallest values in each case. Before Puppet 6.0, there was guidance as to how it would convert and handle mixed types used in these functions. However, now that it's deprecated, it is strongly advised that you ensure comparisons are like for like. In the following example, the variable highest number would contain `88`, while the lowest letter would contain `'a'`:

```
$highest_number = max( [5,3,88,46] )
$lowest_letter = ['d','b','a'].min()
```

`empty` is used as a prefix or chained function to confirm if an array or hash contains no elements or if a string or binary contains no length. In the following examples, the `empty_array` and `empty` strings would contain `true`, while the `non_empty_string` variable would contain `false`:

```
$empty_array = [].empty
$empty_string =empty('')
$nonempty_string='not_empty'.empty()
```

`compare` is used as a prefix function that compares two values and returns `-1`, `0`, or `1` if the first value is less than, equal to, or greater than the second, respectively. The two values must be of the same type and can be numeric, strings, timespans, timestamps, or semvars. For two strings, a third argument (a Boolean) can be used to check whether the comparison should ignore casing.

For example, the `numeric_compare` variable would contain `-1`, while the `string_compare` variable would contain `1` as capitals would be greater than lowercase letters and `A` would come before `b`. If the Boolean were set to `true`, it would return `1`:

```
$numeric_compare = compare(5 , 6)
$string_compare = compare('A', 'b', false)
```

Change case

The following functions change the case of strings or arrays/hashes of strings. In the case of integers, they remain unchanged and will contain other incomputable data type errors.

`capitalize`, `camelCase`, `downcase`, and `upcase` are all used as prefixes or chained functions to change the capitalization of a string or a set of strings on an iterable, such as an array. `downcase` and `upcase` can also be used on an array. All can be used on a numeric but will simply return the numeric unaffected.

CamelCase removes any underscores (_) that were used when applied. `camelCase` and `capitalize` are not recursive on an array but `upcase` and `downcase` are.

If `downcase` or `upcase` changes keys in an array while being used recursively and this creates duplicates, it will overwrite the key, using the last key-value pair that was updated in its place. To show some examples, the `upper_case` variable will contain a string called `UPANDDOWN` upon making the whole string upper case:

```
$upper_case = 'UpAnDdOwN'.upcase()
```

The `capitals` variable will contain an array called `['Up, Mix']` after capitalizing each string in the array:

```
$capitals =capitalize(['up','miX'])
```

The `downcase` variable will contain a hash of `{'lower' => 'case2'}` after downcasing both keys and overwriting the first:

```
$downcase = {'lower' = > 'case', 'Lower => 'Case2}.downcase()
```

The `camel` variable will contain `Word1Word2Word3` after removing the underscores and setting the capitalization to `camelCase`:

```
$camel = camelCase('word1_word2_word3')
```

If you're using international characters, you need to review how the Ruby system locale handles these characters as it is used to handle changes in casing.

String manipulation

The `lstrip`, `rstrip`, and `strip` functions allow spacing to be removed from strings. They are all prefixes or chained functions that are used to remove spaces from a string. `lstrip` removes leading spacing, `rstrip` removes trailing spacing, and `strip` removes both leading and trailing white spacing such as space, tab, newline, and return but not hard space. They can be used on a string or an iterable but not recursively. If used on numerics, they will return numeric unadjusted types but will result in an error on any other unsupported type.

The following example, which uses all three functions, will result in the `left` variable containing `'first second '`, the `right` variable containing `' first second'`, and the `all` variable containing `'firstsecond'`:

```
$spaces = " first second "
$left = $spaces.lstrip()
$right = rstrip($spaces)
$all = $spaces.strip()
```

Lambdas

These functions are not lambdas themselves but are most useful when used with lambdas since they allow arrays or hashes variables to be iterated over or transformed and passed to lambdas, which are sections of Puppet code. The following functions are used on variables to define their behavior: `all`, `any`, `break`, `each`, `filter`, `index`, `lest`, `map`, `next`, `return`, `reduce`, `reverse_each`, `step`, `then`, `tree_each`, `unique`, and `with`.

The syntax and behaviors of these functions will be covered in full in *Chapter 6*, but to show an example, here, we are using the `each` function and a hash containing user name keys and numbers representing their user ID. The `each` function can take each key pair as an array and allow user resources to be created with the assigned IDs:

```
$usersids = {'admin' => 1, 'operator' => 2, 'viewer' => 3}
$userids.each |$users| {
  user { $users[0]:
    id  => $users[1]
  }
}
```

> **Note**
>
> Many functions can use lambdas for error handling, which allows you to loop through the error sections, messages, and issue codes and allows for more detailed messages or actions to be taken. This will be covered in *Chapter 7*.

Templating

Templates allow you to create complicated text with simple inputs for substitution. In *Chapter 7*, we will cover templates in full, but the `template` and `epp` functions allow the ERB and EPP formats for templates to be used via the `content` attribute of the `file` resource. An example of using the ERB format and informing the `content` attribute can be found in the `exampleapp` module:

```
file { '/etc/exampleapp.conf':
  ensure  => file,
  content => template('exampleapp/exampleapp.conf.erb'),
}
```

The structure of modules and how to store template files will be covered in *Chapter 8*.

Alternatively, to use a string containing a template format and pass the value, `inline_template` and `epp_inline` can be used. For example, to use an EPP style template where it is presumed `$exampleapp_conf_template` contains a string in EPP template format, `inline_epp` will substitute the port and debug the variable values of `exampleapp_port` and `exampleapp_debugging_enabled`:

```
file { '/etc/ntp.conf':
  ensure  => file,
  content => inline_epp($exampleapp_conf_template, {'port' =>
$exampleapp_port, 'debugging' => $exampleapp_debugging_enabled}),
}
```

Hash/array

The following functions are used to either access and manipulate hash and array data beyond the normal operators available, which were discussed in *Chapter 4*, or to manipulate variables into hashes and arrays.

The `dig` function is used to search through a complex data structure by providing various keys or indices. It is particularly useful in situations where the structure is not well defined. For instance, suppose we have a data structure called `exampleapp_proc`, and we want to access the state of the process with ID `124`. If we tried to access it using a hash index such as `exampleapp_proc['exampleapp_pids']['124']['state']`, but the `124` key was not present in the hash, we would get an error and the catalog run would fail. However, by using the `dig` function instead, the notice will be undefined:

```
$exampleapp_proc = { exmpleapp_pids => { 123 => { state => running ,
user => root } }}
notice exampleapp_proc.dig('exampleapp_pids','124','state')
```

The `getvar` function is used to return parts of a structured variable using dot notation. If the variable does not exist, it will return `undef` instead of throwing an error, unlike direct access to structured variables. You can also set a default value if the value is not found; otherwise, it will return `undef`.

The first command uses `getvar` to access the `os.release.full` fact, while the second command sets the return value to `'not_found'` if the structured fact is not found:

```
getvar('facts.os.release.full')
getvar('facts.os.release.full','not_found')
```

The `join` function is used to convert an array into a string of elements using a specified delimiter. For instance, if you have an array of data center locations such as `$dc_locations = ['london', 'falkirk', 'portland', 'belfast']`, you can use the `join` function to print a colon-separated string of those locations; for example, `notice ( join($dc_locations, ':'))` This will produce the `"london:falkirk:portland:belfast"` string in the notice:

```
$dc_locations = [ 'london','falkirk','portland','belfast']
notice ( join($dc_locations, ':'))
```

However, if you use `join` on an array that contains a nested array, it will flatten the array, but it won't affect hashes or arrays within hashes. For example, `join([{'London' => ['bromley', 'brentford']}, 'Berlin', 'Falkirk', 'Grangemouth'], '@@')` would print `[ { London => [ 'bromley', 'brentford' ] }@@Berlin@@Falkirk@@Grangemouth ]` because the first element of the array is a hash and it won't be flattened despite it containing a hash:

```
$dr_locations = [ { 'London' => [
'bromley','brentford']},'Berlin',['Falkirk','Grangemouth']]
notice ( join( $dr_locations, '@@'))
```

The `keys` and `values` functions take a hash and return an array of the keys in the hash it can be run as a prefix or chained function. For example, to print the list of keys of the `offices` variable, the first two `notice` functions would print an array of `['Germany','Holland']`, while the next two would print an array of `['Berlin',Amsterdam']`:

```
$offices = {'Germany' => 'Berlin', 'Holland' => 'Amsterdam'}
notice(keys($offices))
notice($offices.keys())
notice(values($offices))
notice($offices.values())
```

These keys or values will be in the same order as they were declared in the hash. If the hash is empty, it will return an empty array.

The `split` function takes a string and, using a pattern to represent a field separator, can break a string into an array of elements. This pattern can be a string, regexp pattern, or regexp type. The following example shows how to split using each pattern method and pick different separators or multiple separators:

```
$example_split = 'north@south.east@west'
$split_on_at = split($example_split, /@/)
$split_on_fullstop = split($example_split, '[.]')
$split_on_both = split($example_split, Regexp['[.@]'])
```

The `split_on_at` variable would contain an array of `['north','south.east','west']`, `split_on_fullstop` would contain an array of `['north@south ', 'east@west']` and `split_on_both` would contain an array of `['north','south,'east','west']`.

The `sort` function takes an array and sorts the array numerically or by lexicographical order. It is not possible to mix these orderings and have numeric and lexigraphic values as this will result in an error and no conversion. Comparing characters is based on system locale and is case-sensitive unless `compare` and lambdas are used.

In its simplest form, `sort` will sort numbers and strings in ascending order – for example, we can take an unordered array of numbers and an unordered array of strings and use `sort` as a prefix or a chained function. In this example, the code will result in ordered numbers containing `[0,1,2,3,4,5,7,8,9]` and ordered strings containing `['a','b','c','d']`:

```
$unordered_numbers = [7,9,8,0,2,4,3,1,5]
$unordered_strings = ['d','c','b','a']
$ordered_numbers = $unordered_numbers.sort()
$ordered_strings = sort($unordered_strings)
```

To specify the order explicitly, you can use the `compare` function to order the variables, highlighting if they should be ascending or descending. In the following example, the integers will be ordered `[1950,1980,1984,1985]` in the ascending variable and `[1985,1984,1980,1950]` in the descending variable:

```
$ascending =(sort([1984,1950,1985,1980]) |$a,$b| { compare($a, $b) })
$descending =(sort([1984,1950,1985,1980]) |$a,$b| { compare($b, $a) })
```

As we learned when we discussed `compare` in the *Comparison and sizing* section, a Boolean can be used on `compare` to order by capitals or not.

> **Note**
>
> Instead of using the `compare` function, other functions from the *Comparison and sizing* section such as `max` or `min` can be used.

Data handling

There are several data-related functions for Hiera and encrypted EYAML. These will be covered in full in *Chapter 9*, but for reference, they are `eyaml_look_up_key`, `lookup`, and `yaml_data`. The function documentation states that a few `hiera_<type>` functions were depreciated for the `lookup` function.

The `unwrap` function was previously covered in *Chapter 3*, whereby the function was used to make sensitive data types visible/accessible in Puppet code, as necessary.

stdlib module functions

Modules will be discussed in full in *Chapter 8*, but the `stdlib` module (`https://forge.puppet.com/modules/puppetlabs/stdlib`) is so widely used that it is worth highlighting some of the functions that are available from the module as virtually every install of Puppet will make them available.

It is important to be aware that the functions in `stdlib` allow advanced behaviors that are not always best practice approaches to Puppet code, such as being able to read the contents of a YAML file into a string and using the `ensure_package` function, which is used to allow for multiple declarations of a package resource. They can provide useful workarounds in complex situations or when code is managed in multiple teams' political situations.

> **Note**
>
> Many functions have been made redundant by file type conversion, which was made available in Puppet 5, as well as other new features, but those have been left for compatibility purposes.

Array and strings

The following functions interact with strings and arrays by combining, manipulating, and producing new arrays in several ways.

The `intersection` function is a chained function that, when provided with two arrays, will produce a single array of values contained in both. For example, the following code will put the `['both']` array into the `chained_array` variable:

```
$chained_array = intersection(['first','both']['second','both])
```

The `union` function is a chained function that, when provided with two arrays, will produce a single array of unique values. In the following example, the `union_array` variable will contain the `['first','second']` array:

```
$union_array = union(['first','both'],['second','both'])
```

The `range` function is a chained function that can be provided a start, end, or step interval (if not provided, this defaults to `1`). The start and end can be strings or numerics, while the optional step should be an integer.

For example, the `onetoten` variable would contain an array of `[1,2,3,4,5,6,7,8,9,10]`, the `etog` variable would contain `['E','F','G']`, and `good_trek` would contain `['StarTrek2','startrek4',startrek6','starttrek8']`:

```
$onetoten = range(1,10)
$etog = range('E','G')
$good_trek = range('StarTrek2', 'StarTrek8', 2)
```

The `start_with` and `end_with` functions are chained functions that allow you to check if a string starts or ends with a provided string or list of strings, attempting to match any string in the list. It will return `true` or `false`, depending on the match. In the following example, `truestart` will contain `true` as `server` matches the start of `server1234`, `falseend` will contain `false` as `wales` does not end in `land`, and `trueoptions` will contain `true` as `aws104` starts with `aws` and could match other strings starting with `gcp` or `az`:

```
$truestart = 'server1234'.startswith('server')
$falseend = 'wales'.endswith('land')
$trueoptions = 'aws104'.startswith['gcp','az','aws']
```

File information

The `basename`, `dirname`, and `extname` functions can be used either as separate functions or chained together to extract the filename, directory, or extension from a file path. Here's an example:

```
$full_path = 'C:\Users\david\fact.ps1'
$file_name = basename($full_path)
$dir_name = dirname($full_path)
$ext =  $full_path.extname
```

Note that `extname` only works with filenames in the format of `filename.extension`. If the string does not contain a dot (`.`), or if the dot appears at the beginning or end of the string, it will simply return an empty string.

Lab

Having covered a wide variety of functions, let's practice using a handful of them. Let's create a class called `example_functions` that takes the parameters of the user prefix as a string and several users as an integer. This class should take two parameters: a user prefix as a string, and several users as an integer. Ensure that the prefix is in lowercase. An array of usernames should be created starting from `0` up to the number of users specified. This array should then be passed to a user resource to create the users.

Define your class with the `user` string and the number `5`.

The code should also log a warning message containing text with the contents of the `os.windows.product_name` fact or `linux` if you're not using a Windows machine.

Finally, the code should take the `fact` path and ensure every directory is audited. Hint: you will want to separate this path into an array and pass it to a file resource. `windows` and `linux` use different separators for path objects – that is, `;` and `:`. The following `if` statement should help:

```
if $facts['os.family'] == 'windows' {
}else{
}
```

You should use bolt to make stdlib available locally on our clients:

```
bolt command run "puppet module install stdlib" -t windowsclient
linuxclient
```

Then, apply the puppet class via bolt with the following command:

```
bolt puppet apply example_functions.pp -t windowsclient linuxclient
```

You can find some example solutions at https://github.com/PacktPublishing/Puppet-8-for-DevOps-Engineers/blob/main/ch05/example_functions.pp.

Deferred functions

A Deferred function (also known as an agent side function) is a function with the Deferred type applied to it. This causes the function to run locally on a client when the catalog is applied, rather than on a Puppet server during compilation. The catalog for a deferred function contains what to run on the client rather than the output of the function. The deferred type was introduced in Puppet 6.0 and is available in all later versions.

This is typically used when the compilation server can't access a necessary source in a function – for example, when retrieving a secret from a HashiCorp Vault server, where security is set up to only allow the client to access a secret.

The syntax for applying Deferred is as follows:

```
Deferred( name of function, [arguments])
```

The following is an example of retrieving a secret from vault. This can be used within a user resource for exampleapp to set the password from a Vault path of exampleapp/password:

```
user { 'exampleapp':
password => Deferred('vault_lookup::lookup', ["exampleapp/password"])
}
```

This function is from the vault_lookup module (https://forge.puppet.com/modules/puppet/vault_lookup) and requires an underlying vault client setup to be available, as per the instructions within the module and the guide from Hashicorp: https://www.hashicorp.com/resources/agent-side-lookups-with-hashicorp-vault-puppet-6.

It is important to understand the difference in using functions with `Deferred`. You cannot use a `Deferred` function to pass a variable to a string. This would result in the catalog creating a stringified version of the object. In the following example, which involves looking up a key value from `vault` called `exampleapp/message`, the first `notify` will return a string containing a stringified translation of the function name in the catalog, while the second `notify` will return the value of the vault lookup itself:

```
$deferred =>  Deferred('vault_lookup::lookup', ["exampleapp/message"])
notify {'this will return the object name':
  message => "Secret message is ${deferred"
}
notify {'this will return the message':
message => $deferred
}
```

This reflects the catalog compilation calculating the string value at compilation time. This mismatch can happen in other places, such as templates, but can be overcome by ensuring any deferred values are only used in isolation or within other deferred functions. In *Chapter 7*, you will learn how to use a deferred template.

A function can only be deferred if it uses core data types because the client only has core data types made available to it at runtime via plugin sync. In *Chapter 10*, you will learn how plugin sync works with the client.

Also, note that it is down to the implementation of the function itself whether it returns a sensitive value, and how it fails. In the case of the `vault_lookup` function, there is no graceful failure; it will return an error, resulting in an errored catalog run.

> **Note**
>
> As of Puppet 7.17.0 and Puppet 8, deferred functions can now be called on demand instead of being preprocessed. Using this method, the catalog can provide inputs to the deferred function. If the deferred function fails, then only the affected resource will fail, while all other resources will still be applied. Puppet 8 enables this behavior by default, to enable this behavior in Puppet 7, set `Puppet[:preprocess_deferred] = false` or use `--no-preprocess_deferred`.

All these behaviors apply to a local `puppet apply run` since a `puppet apply run` will generate a catalog and apply it locally.

Summary

In this chapter, you learned how the Facter tool provides system profiling information with its facts and how this can be expanded using external facts and custom facts. We warned you that there is an infrastructure cost to gathering facts and that the scale it will work with should be balanced. We stated that external facts can be simple flat files of static data or executable scripts, as allowed by the operating system. Custom facts, although written in Ruby, were shown to have several advantages over external facts. Being able to confine the custom fact to only run on certain systems allows you to choose different resolutions with a weight as to which should be selected and timeouts at the resolution level in Puppet 7 and above or the execution level in Puppet 6 and below.

Next, we reviewed functions and highlighted the vast range of tasks functions can do to manipulate the catalog or return calculated values in Puppet code. Here, we discussed catalog statements, which are used to include classes in the catalog, and logging statements, which are used to set logging messages. The other two types of functions, prefix and chained, were highlighted, along with their syntax. Then, a selection of core functions was shown, along with various categories that expose the available functions.

Then, we discussed a small selection of functions from the `stdlib` module to highlight what can be provided. Note that some of the `stdlib` functions have been deprecated and are only there for backward compatibility or to be used for edge cases, which is not a best practice.

Finally, we discussed deferred functions, which allow functions to run during the application of the catalog on a client. We highlighted the advantage of this for services that may only be available to the client, such as making API calls to secure services, or may be undesirable to be run on Puppet infrastructure shared with other services.

In the next chapter, you will learn how relationships and dependencies work between resources and classes. We will look at how scope and containment affect resources, variables, and classes and how to structure code and necessary dependencies without encountering common pitfalls and dependency hell.

Part 2 – Structuring, Ordering, and Managing Data in the Puppet Language

This part will look at the more advanced Puppet language features. We will show how to manage dependencies and flow within code using iteration and conditions. We will then see how to use best practices to structure Puppet into modules using roles and profile patterns. Puppet Forge will be shown to be a useful source of pre-built modules and we will look at how to understand and review the source and content of those modules. We will then look at how to manage data with Puppet using Hiera and understand the best practices in terms of when to use separate data sources and variables.

This part has the following chapters:

- *Chapter 6, Relationships, Ordering, and Scope*
- *Chapter 7, Templating, Iterating, and Conditionals*
- *Chapter 8, Developing and Managing Modules*
- *Chapter 9, Handling Data with Puppet*

6

Relationships, Ordering, and Scope

In this chapter, we will be discussing relationships, ordering, and scope in Puppet. These topics are often considered complicated because Puppet's approach differs greatly from traditional languages. However, we will show you how to manage these aspects effectively and avoid unnecessary complexities.

We will start by discussing Puppet's approach to relationships and ordering. By default, Puppet treats resources as independent and can apply them in any order in the catalog. However, where ordering is necessary, we will show you how to use metaparameters such as `before`, `after`, `notify`, and `subscribe` to enforce ordering and create relationships between resources.

After that, we will cover the concept of containment. We will explain that including classes are not contained within their calling classes, so relationships/dependencies made between classes do not automatically create relationships and dependencies with the resources in those classes. To address this, we will introduce the `contain` function, which allows you to contain the resources within a class and create those relationships.

Finally, we will discuss scopes and how variables and resource defaults can have different visibility depending on where they are in the code and their relative scope. We will then provide best practices and pitfalls to ensure that you take the simplest path and avoid unnecessary complexity.

Overall, this chapter will equip you with the knowledge and skills to effectively manage relationships, ordering, and scope in Puppet.

In this chapter, we're going to cover the following main topics:

- Relationships and ordering
- Containment
- Scope
- Best practices and pitfalls

Technical requirements

All examples and labs in this chapter can be run within your own local dev environment.

Relationships and ordering

By default, Puppet treats all resources as independent of each other, which means they can be applied in any order. This is different from traditional declarative code, which runs line by line and executes in the order it is written. One of the main advantages of Puppet's approach is that if a single part of the code fails, Puppet will continue to apply all other resources. This eliminates the need to stop or have substantial failure handling in place to continue code. As a result, Puppet can bring a client server as close to the desired state as possible, even if some resources fail.

It's clear that some resources will be dependent on each other, such as a configuration file that can only exist after a package has been installed. Puppet provides metaparameters to create these dependencies:

- `before`: The resource should be applied before the named resource(s).
- `require`: The resource should be applied after the named resource(s).
- `notify`: The resource should be applied before the named resource(s). The named resource refreshes if the resource changes.
- `subscribe`: The resource should be applied after the named resource(s). The resource refreshes if the named resource changes.

The `before` and `require` metaparameters can be used to enforce a dependency relationship. However, it's important to note that a relationship only needs to be applied in one direction. Therefore, there is no need to use both `before` and `require` on either side of a dependency.

For example, to indicate that the `httpd` package should be installed before managing a file, either `before` or `require` can be used, as shown:

```
package { 'httpd':
  ensure => latest,
  before => File['/etc/httpd.conf'],
}

file { '/etc/httpd.conf':
  ensure => file,
  require => Package['httpd'],
}
```

A dependency chart, also known as a **Directed Acyclic Graph** (**DAG**), can be created using the Puppet validator at `https://validate.puppet.com/` by selecting the **show relationships** option. Alternatively, the DAG can be generated by selecting **open node graphs to the side** in the **Puppet Development Kit** (**PDK**), or by using the `--graph` option with the `puppet` command to produce a dot file that can be used to create a graphic in an appropriate program.

In *Figure 6.1*, the `require` on the file resource has been removed to produce a DAG for the example code:

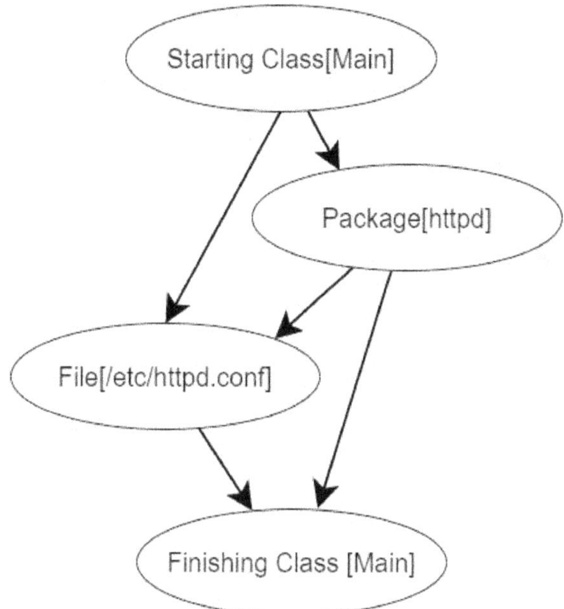

Figure 6.1 – A DAG of resource dependencies

If both the `before` and `require` metaparameters were present, an extra arrow would be visible in the DAG, but it would have no effect on the compilation or resources applied. It's worth noting that the starting and finishing classes, named `Main`, in the example code reflect that the code is not contained within a class and the code is at a global level. This will be discussed further in the *Scope* section.

In a DAG, loops are not expected, so the flow of dependencies should only go downward. If a third resource, such as a service, is added that should be enforced before the `httpd` package after the `/etc/httpd.conf` file, the DAG would look like this:

```
service { 'httpd':
  ensure  => running,
  before  => Package['httpd'],
  require => File['/etc/httpd.conf'],
}
```

This would result in a dependency cycle, as illustrated in *Figure 6.2*. When compiled, the code would produce an error, as there would be no way to determine the order in which to apply the resources.

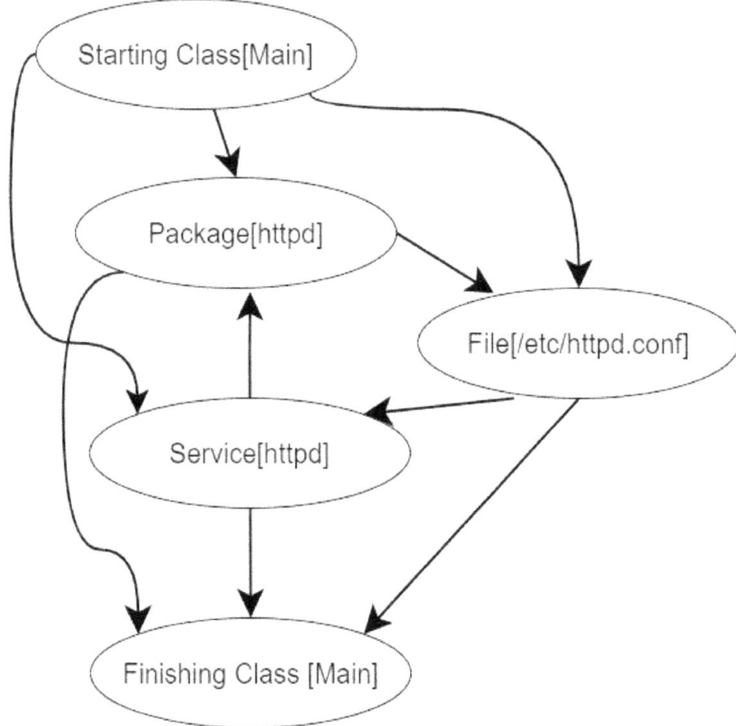

Figure 6.2 – A DAG showing a dependency cycle

It is also possible to represent multiple dependencies with an array, which can be an array of names either of the same type or of different types. For example, if a package were required by two files and two services for `exampleapp`, it could be represented like this:

```
package { 'exampleapp':
  ensure => latest,
  before => [File['/opt/exampleapp.content','/var/exampleapp.
variables],Service['exampleapp','exampleapp2']]
}
```

Sometimes, it can be easier to have all the resource dependencies on one side rather than on each individual resource.

As was mentioned in *Chapter 3*, some Puppet types have automatic rules for creating dependencies, which can be found in the documentation for the Puppet type under `Autorequires` either online or using the Puppet `describe` command. For example, the user type autorequires any group under Puppet's control, that a user resource has as its primary or secondary group.

As well as an ordering concept, Puppet has the `refresh` attribute, so if a resource has a dependency on another resource, it will refresh itself. This is useful in situations such as when a configuration file is updating and the service should restart to reread the configuration file.

The `notify` and `subscribe` metaparameters create the same dependency as `before` and `require` but add the `refresh` attribute to the dependent resource. Of the built-in Puppet types, `service exec` and `package` can be refreshed. If a `notify` or `subscribe` metaparameter is used with a resource type incapable of refreshing, it will just enforce the dependency and do nothing on a refresh event.

> **Note**
> The `notify` metaparameter should not be confused with the `notify` resource type used to send messages to the agents log.

For example, a `service` resource could use `subscribe` or `notify` from the `file` resource so that the service would be dependent on the file being created. It would also receive a refresh event if the file was updated and restart the service assuming the provider had the capability. As shown in the following code, again we show both sides of the dependency, although only one relationship attribute should be given:

```
service { 'httpd':
  ensure => running,
  subscribe => File['/etc/httpd.conf'],
}
file { '/etc/httpd.conf':
  ensure => file,
  notify => Service['httpd'],
}
```

In a DAG diagram, this would be identical to using `before` and `require`, and it can use the same resource reference or arrays of resource references.

The default behaviors and parameters of a refresh event for each type are shown in *Table 6.1*. Here, we see that by default, a service will use the provider's `restart` variable if it is provided. Otherwise, `hasrestart` can define an `init` script or `restart` can define a custom restart script. If no `init` script is provided, the service name will be searched for in the process tree, but it is strongly advised to provide clear service management scripts.

For the package type, the default behavior is to ignore the `restart` event, but the parameter can be set to reinstall the package as the result of a `refresh` event.

`Exec` will rerun its command on a refresh but can be changed to run a different `refresh` command or to only run as a result of a `refresh` event.

Type	Default Behavior	Parameter(s)
`Service`	Restart the service if the provider has a restart feature; otherwise, stop and start	`hasrestart` `restart`
`Package`	Ignore refresh event	`reinstall_on_refresh`
`Exec`	Rerun the command	`refresh` `refresh only`

Table 6.1– Puppet native type refresh options

Metaparameter dependencies can produce three types of errors. The first is missing dependencies, where the resource is not found in the compiled catalog. This should normally be investigated for typos or logic, meaning the resource is not included. The second type of error is failed dependencies, where an issue with a resource means none of its dependencies can be applied. Troubleshooting this resource and rerunning Puppet should then allow all dependent resources to be applied. The third type of error is the dependency cycle, which we discussed and showed in *figure 6.2*, where producing a DAG can help identify where the loop is and fix the dependency logic.

Despite having said so far that resources have no order beyond dependencies, this is not quite true since Puppet runs in what is known as **manifest order**. So, an individual manifest file will be applied in the order it is written unless dependencies change that. Although this could allow you not to use dependencies, the main purpose is to prevent random compilations causing code to behave differently on different servers as it could do if read in at random.

> **Note**
>
> Puppet went through a strange philosophical/purity argument around ordering in earlier versions. It was viewed as necessary to break the bad habits of developers assuming ordering would be like in other languages, line by line. So, Puppet initially chose a random order. This was chaotic and resulted in code that might work in your lab but ran in a different order in production and broke.

A variation of the dependency metaparameters is chaining arrows, where `before` and `require` are represented by `->` and `<-` and `notify` and `subscribe` by `~>` and `<~`. They are generally used to show relationships between classes, such as to represent a module pattern, which will be seen in *Chapter 8*. For example, if we wanted an `install` class to apply before a `config` class and then for a `service` class to be applied and refreshed if the `config` class was updated, it could be represented as follows:

```
include examplemodule::install, examplemodule::config,
examplemodule::service
Class['examplemodule::install']
-> Class['examplemodule::config']
~> Class['examplemodule::service']
```

The `include` function is necessary, as discussed in *Chapter 3*, to ensure the classes are added to a catalog.

For style purposes, only the right-facing arrows are recommended to be used, to make it consistent while reading. While dependency parameters can be used in classes and resource declarations and chaining arrows on other resource types, it is not recommended to do so to make it clearer to read.

In simpler cases, the required function can also be used from within a class to create a dependency on other classes. However, no `refresh` or `before` equivalent exists, so for styling and consistency, it's generally easier to use ordering arrows. A simple example, using the `require` function to represent that the `install` class should applied before the `config` class, would be as follows:

```
class examplemodule::config {
  require examplemodule::install
}
```

What we have just discussed in terms of the approach to class dependency is not quite as simple as it may seem because Puppet classes do not actually contain other classes. A class will include other classes by default, so the dependencies do not cover them. We will now look at what this containment problem means and how to handle it.

Containment

Containment in Puppet means that included classes are not contained in the same way as resources in a class; so, when setting up a dependency to a class that includes another class via the `include` function or a `class` resource, the dependency will only cover the resources. For example, say we created a requirement for `class1` to be applied before `class2` and `class2` contained a `package` resource and an `include` call to `class3`, as shown in the following code:

```
include examplemodule::class1, examplemodule::class2
Class['examplemodule::class1'] -> Class['examplemodule::class2']
class examplemodule::class2 {
```

```
  include examplemodule::class3
  package{'PDS':}
}
```

So, while there might be an assumption that this would ensure `class1` was before `class3` but `class1` was before `class2`, it doesn't, as can be seen in the DAG diagram in *Figure 6.3*.

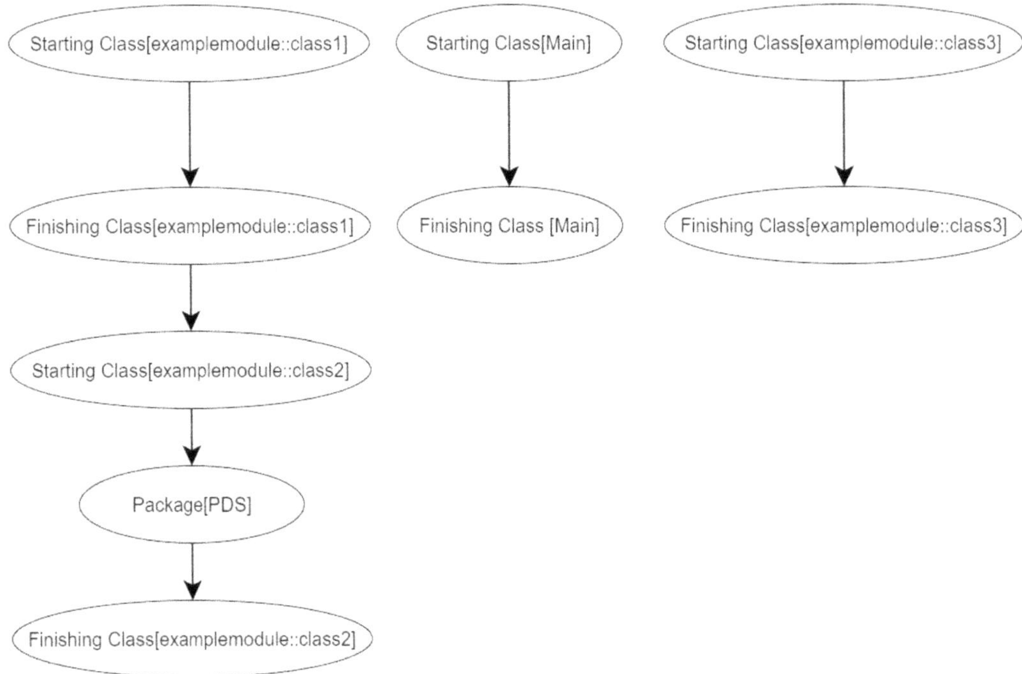

Figure 6.3 – DAG showing lack of containment

Remembering back to *Chapter 3*, where the `include` function was introduced, this containment is not automatic because we may want to include this class in different places for different situations and for it only to appear once in the catalog without dependency or containment issues.

To contain a class, the `contain` function is used. Change the `include` line to `contain examplemodule::class3`, which will change the DAG diagram to contain `examplemodule::class3`, as We can see in *Figure 6.4*.

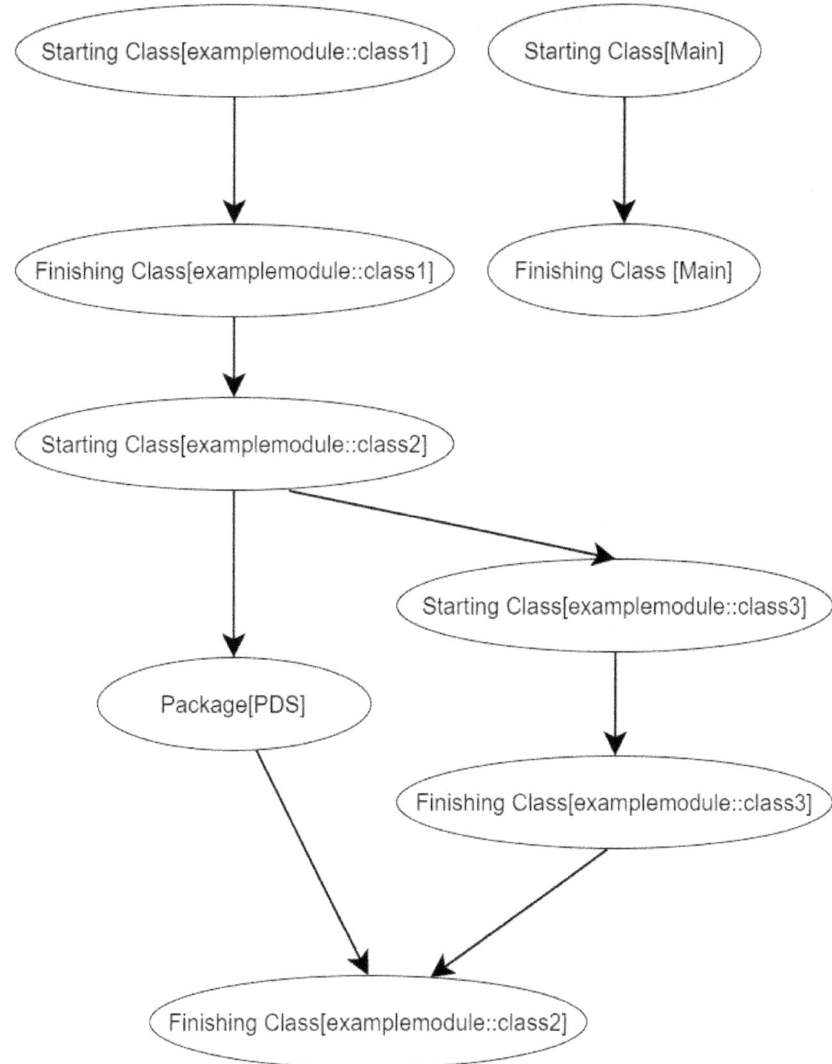

Figure 6.4 – DAG showing the use of the contain function

If a `class` resource is used alongside a `contain` statement, it must appear in manifest order after the `class` resource. Failure to do so will cause the `class` resource to interpret the `contain` statement as an attempt to declare a duplicate resource, resulting in an error. For example, if the following code is used, the attribute can be passed successfully:

```
class {'examplemodule::class3':
  attribute1 => 'value1''
}
contain examplemodule::class3
```

The immediate question to this containment problem might be why not just use `contain` for everything? That comes down to the needless and confusing dependencies it could create. If our original example were updated to use `contain` instead of `class` and we had another class, `anothermodule:class`, which required `examplemodule:class3` to be in the catalog, we could add the following code:

```
class anothermodule::class {
  contain examplemodule::class3
  package{'PTOP':}
}
```

Then, the DAG would look like *Figure 6.5*. It can be immediately seen that we have created needless dependencies just with a small number of classes.

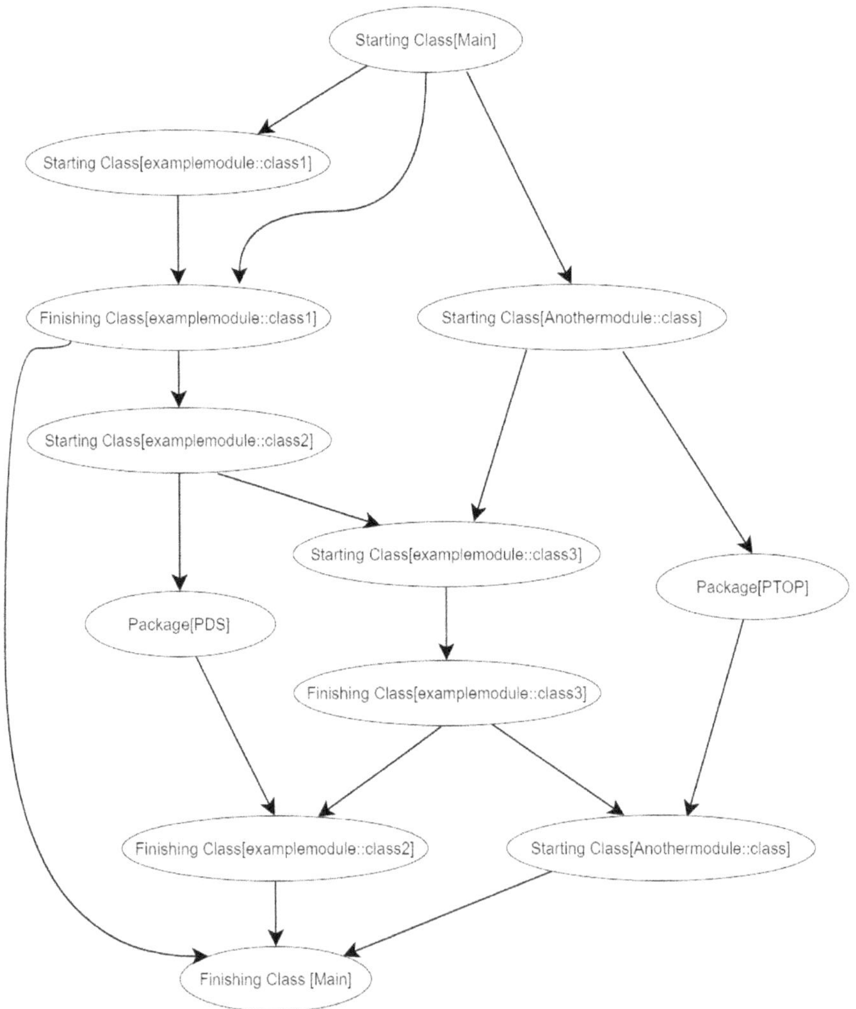

Figure 6.5 – DAG showing cycle caused by the overuse of contain

Worse, it can be easy to create a cyclical dependency. If, for example, the `security::default` class were to be included in all application classes, a cyclical dependency could be created by a `require` function being used between a class, `application2`, requiring the `application1` class, as shown in the following code:

```
class application1 {
  contain security::default
}
class application2 {
  contain security::default
  require application1
}
```

This would produce the DAG shown in *Figure 6.6*. If only includes were used, we would have avoided the need for an unnecessary relationship from the application classes to `security::default`:

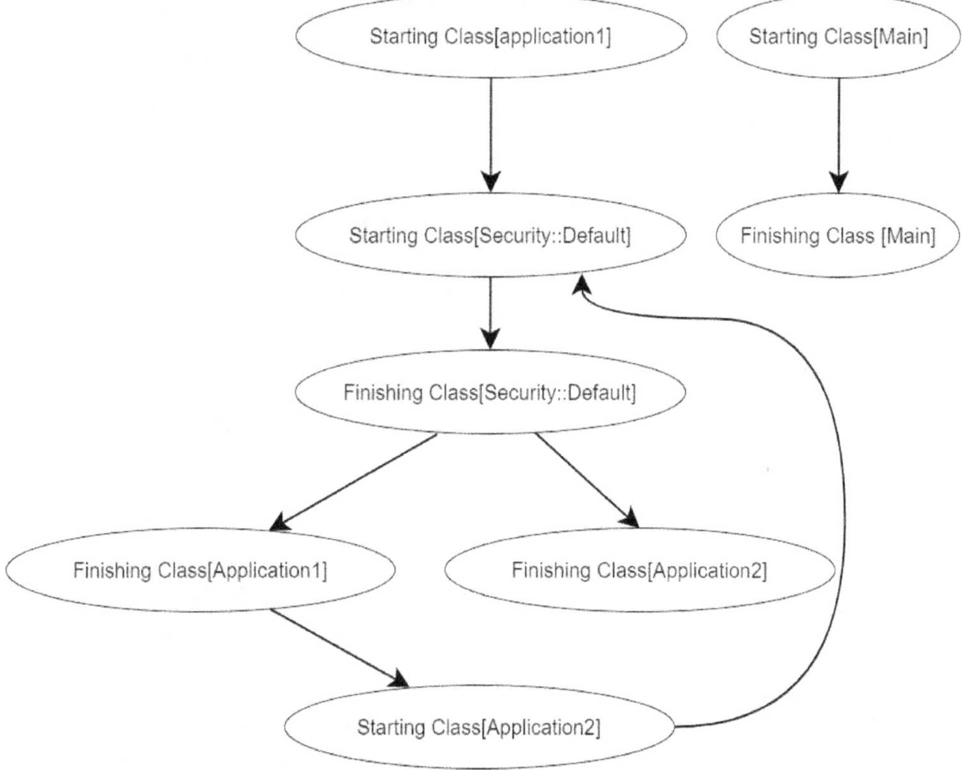

Figure 6.6 – DAG showing how the over-use of contain causes cyclic dependencies

In the *Best practices and pitfalls* section, we will further discuss how to avoid worrying about containment using consistent patterns.

Before the `contain` function was introduced in Puppet 3.4, there was another approach, which you may see in heritage code: using `anchor` resources. This can either be done with a specific anchor resource provided by the `stdlib` module or any other pair or resources in the class. To ensure that the current class contains `examplemodule::class3`, the code using the `anchor` resource directly would look as follows:

```
anchor {['start', 'stop']: }
include examplemodule::class3
Anchor['start'] -> Class[' examplemodule::class3'] -> Anchor['stop']
```

Alternatively, if the two package resources, `pdk` and `cowsay`, were in this class, they could be borrowed to create a relationship and contain the class:

```
Package['pds'] -> Class[' examplemodule::class3'] -> Package['cowsay']
```

The issue with this pattern is that it clutters up the DAG with extra anchor resources or unnecessary relationships, which can be confusing. Therefore, if you find that anchors are being used, it is recommended that you modernize your approach by using the `contain` keyword instead.

Having discussed dependencies and the containment of resources and classes, we will now see how variables and resource defaults are scoped across the Puppet language.

Scope

In Puppet terms, scope reflects the location in the code where variables can be directly accessed without the use of namespaces and where resource defaults can be affected.

There are three levels of scope:

- **Top scope**: Any code outside of a class, type, or node definition. Any variable or resource declaration in the top scope will be accessible to be read or used anywhere.
- **Node scope**: Any code defined in a node definition. Any variables and resource defaults in the node scope will be visible to nodes that match the node definition at the node and local scope levels.
- **Local scope**: Any code defined in a class, defined type, or lambda. So, any variables and resource defaults defined will only be visible within that specific resource.

Both **External Node Classifiers** (**ENCs**) and node definitions will be discussed in *Chapter 11*. All we need to understand for this section is that an ENC is an executable script that returns variables and classes to be applied to a host. This script can inject custom logic and data by performing various actions, such as performing a database lookup or using AWS Lambda. It can also be used to access third-party sources, such as **Configuration Management Databases** (**CMDBs**). The variables returned are at the top-scope level, while the classes are at the node-scope level. This allows the provided variables to be visible anywhere, but only for the classes declared to have access to node-scope variables. In contrast, a node definition is a section of code that is applied to matching nodes.

Classes have what is known as a named scope, where the name of the class is used in the namespace. For example, a variable created in `exampleclass` called `test` can be accessed from anywhere via `exampleclass::test`. Variables created in the global scope, such as `site.pp`, can be accessed from an empty namespace by calling `::variablename`. However, accessing data like this is generally not recommended. In *Chapter 9*, we will show how to centrally manage data.

Node-scope definitions and local-scope definitions on lambdas and defined types are anonymous and can only be accessed directly by name from where they are visible. It is also possible to override higher scope variables by declaring a variable in the current scope, such as a class overriding a global variable with its own.

To show this, consider the following code:

```
$top='toptest'
$test='testing123'
notify "Top = ${top} node = ${node} local = ${local} test =
${testing}"
notify "Access directly ${example::local}"
node default {
  include example
  $test='hello world'
  $node='nodetest'
  notify "Top = ${top} node = ${node} local = ${local} test =
${testing}"
  notify "Access directly ${example::local}"
}
class example {
  test='an example'
  $local ='localtest'
  notify "Top = ${top} node = ${node} local = ${local} test =
${testing}"
}
```

The first `notify` would fail to find the local or node variable since it is in the global scope, and `testing` would be set to `testing123`. The second `notify` would directly access the local namespace, `example`, and print `localtest`. The third `notify` would be unable to access the local variable and would print `hello world`. The fourth `notify` would again access the local scope via the namespace. The final `notify` would be able to access all the variables, and `local` would be set to `localtest`. This example shows the flow of variables between scopes.

Resource titles and resource references are unaffected by the scope and can be declared in any scope. For example, a resource can declare a dependency for any resource in the catalog. However, it is not good practice to rely on accessing external variables like this.

Best practices and pitfalls

In earlier versions of Puppet, scope, dependencies, and containment were some of the most challenging issues, which led to significant problems for newer developers. One major solution that largely addressed these issues was the widespread adoption of the roles and profiles method, which will be covered in full detail in *Chapter 8*. Hiera data will be covered in *Chapter 9*.

The roles and profiles method involves grouping single-use component modules that perform one independent function well. For instance, a component module could install and configure Oracle. The module structure would contain a group of manifests with specific purposes, such as installing packages or managing services. This simplifies module organization and allows for the easier ordering of classes. For example, the `install` class can be applied before the `service` class.

Component modules should function independently of one another and have no direct dependencies across modules. The profile layer groups modules together to create technology stacks and can put modules in order if necessary. Roles abstract another layer up to create business solutions using these technology stacks and can order the profiles. In this structure, any global or node data should come from Hiera instead of being set in node or global scopes, which reduces code complexity. It may feel counterintuitive to developers to avoid setting global variables in code, but following this practice is recommended.

As mentioned in *Chapter 3*, it is advisable to avoid using resource collectors/exported resources. However, it is worth noting that they can be used as part of chaining arrows. Using them can be risky as it may result in unpredictable outcomes and create significant dependency cycles that are difficult to map until runtime. Dependencies should always be created as required, and you should not rely on manifest ordering for this purpose. Omitting these dependencies could significantly reduce code maintainability and create complications during future refactoring.

Use chaining arrows for class dependencies and contain them only where necessary, as in the roles and profiles method. Avoid enforcing resource defaults globally, such as in `site.pp` or node definitions, as this approach makes the code unpredictable for everyone, especially when working with multiple application teams who may not be aware of these defaults in their own code. In summary, avoid attempting overly complex or familiar approaches from other languages and instead follow established roles/profiles and Hiera patterns. Review roles/profiles and Hiera patterns carefully and refactor any code that does not follow these guidelines.

Lab – overview of relationships, ordering, and scope

In this lab, we will provide some code to review and run to ensure the concepts discussed are understood. All the code can be found at `https://github.com/PacktPublishing/Puppet-8-for-DevOps-Engineers/tree/main/ch06`.

The code at `https://github.com/PacktPublishing/Puppet-8-for-DevOps-Engineers/blob/main/ch06/lab6_1.pp` currently has no dependencies. To meet the following requirements, the code needs to be adjusted accordingly:

- The `install` class and all its resources should run before `config` and `service`
- The `config` class and all its resources should run before `service`
- The `service` class and all its resources should be refreshed if any resource in the `config` class is updated
- The `httpd` package should be installed before the `exampleapp` package
- The `exampleuser` user should be created after the `examplegroup` group
- The `/etc/exampleapp/` directory should be created after the `exampleuser` user and the `examplegroup` group
- The `/etc/exampleapp/exampleapp.conf` file should be created after the `exampleuser` user, the `examplegroup` group, and the `/etc/exampleapp` directory
- The `httpd` service should start before the `exampleapp` service, and the `exampleapp` service should refresh if the `httpd` service restarts

It is recommended to use `https://validate.puppet.com/` to check your Puppet code, as you should not rely on manifest ordering alone. Additionally, it's important to remember that some resources have auto-requirement behavior. A sample example is provided at `https://github.com/PacktPublishing/Puppet-8-for-DevOps-Engineers/blob/main/ch06/lab6_2.pp`. Examine the code and see what the `notify` function would print.

Summary

In this chapter, we discussed how resources are assumed to be applied in any order by default, and how metaparameters such as `before`, `require`, `notify`, and `subscribe` can be used to define any required order. We learned that DAGs can be used to visualize dependencies between resources, and that dependency cycles should be avoided to ensure the catalog can be applied successfully. We also discussed how certain resources automatically apply dependencies, such as a user requiring its primary group. The `notify` and `subscribe` metaparameters were explained, and their use of `refresh` was highlighted as particularly useful for resources such as `exec`, `package`, and `service`. This allows for these resources to be restarted, reinstalled, or rerun when necessary, such as when a configuration file changes. Additionally, we acknowledged that although resources should be assumed to have no order, they are in fact applied in the order they are written in a manifest to ensure consistency across environments. We also discussed the three types of errors that can occur: cyclic, missing dependencies, and dependent resource failures.

After that, we discussed chaining arrows as a variation of metaparameters, allowing them to be used between classes. We emphasized that only right-facing arrows should be used to comply with the style guide. While metaparameters can be used on classes and chaining arrows on resources for consistency and styling, we recommend avoiding this practice. Instead, we showed that the `require` function can be used within a class that is dependent on another class for relatively simple class dependencies.

We then discussed the issue of containment, which arises when including classes within other classes does not create resource dependencies. This was achieved using the `contain` function instead of an `include` function within a class causing this class to contain the other classes' resources and creating the dependencies. We discussed how this may bring the temptation to contain all classes, but we demonstrated that this would create needless or cyclic dependencies. The older anchor pattern was shown since heritage code could still contain this. We highlighted that the `anchor` function is no longer a recommended approach and, where found, it should be modernized to use the `contain` function.

The scope was shown to affect variables and defaults for resources, where the global scope is anything set outside of the class, type, or node definition. The node scope is anything in a node definition and the local scope is anything in a class, type, or lambda.

Finally, for best practice, it is recommended to follow the roles and profiles method to ensure consistency in dependencies and ordering. It is also recommended to use Hiera instead of complex variable usage and to avoid setting resource defaults in the global scope, such as `site.pp` or node definitions. It is important to never rely on manifest ordering, use explicit dependencies to ensure consistency.

The next chapter will explore templates, iteration, and conditional statements in Puppet. It will demonstrate how Puppet can generate file content by leveraging variables, conditions, and text manipulation functions. Additionally, it will explain how iterative functions and lambda code sections enable Puppet to loop through and manipulate collections of data. Finally, the chapter will cover the usage of conditional statements in Puppet to create different configurations based on conditional logic.

7

Templating, Iterating, and Conditionals

This chapter will cover advanced structures for the Puppet language, including templates that enable the insertion of variables into templated files. The two formats available in Puppet, **Embedded Ruby** (**ERB**) templates, which are based on native Ruby templating, and **Embedded Puppet** (**EPP**) templates, which are modern Puppet language-based templates, will be discussed, highlighting the differences between the two and the core advantages of using EPP over ERB.

Additionally, the chapter will delve into the use of iteration and loops in Puppet, showing how iterative functions are used with sections of code known as lambdas in Puppet instead of more traditional `loop` keywords of other languages. Finally, the chapter will examine the different types of conditional statements available in Puppet, including `if`, `case`, and `unless` statements, which are typical of any programming language, and the Puppet-specific selector, which allows a value on a key or variable to be chosen based on a fact or variable. The chapter will also examine the use of regular expressions within conditionals in detail.

In this chapter, we're going to cover the following main topics:

- Templating formats in Puppet – EPP and ERB
- Iteration and loops
- Conditional statements

Technical requirement

All code in this section can be tested on the local development server.

Templating formats in Puppet – EPP and ERB

Templating in Puppet allows for the generation of content in a standard format, by substituting variables and using conditional logic to customize the content. Puppet supports two templating formats: ERB, which is a native Ruby templating format (`https://github.com/ruby/erb`) and has been available in all versions of Puppet; and EPP templates, which are based on the Puppet language, were introduced in Puppet 4, and are available in later versions of Puppet 3 with the future parser enabled.

Templates provide greater flexibility than strings but are less flexible than using resources such as `file_line`, `augueas`, or `concat` for controlling individual or groups of settings. Therefore, a balance of complexity needs to be struck when deciding whether to use templates or resources.

For relatively short `heredoc` files or simple strings, templates with variable interpolation may be sufficient. However, for more complex files and particular files where multiple modules may be managing different settings or where manual edits may be accepted, using resources for each setting or section would be less complex and more manageable.

In older code, it is possible to find that templates were overused, which can reflect the lack of availability of resource types such as `file_line` in previous versions of Puppet. It is important to review what state was being attempted to be achieved and ensure that by using templates to control all the content settings, a whole file is not being unnecessarily enforced, which may contain settings that have become redundant as the underlying application associated with the configuration file has updated and changed its configuration settings.

While there is no reason to use ERB for new code, many forge modules and legacy code bases may contain ERB, and thus both formats will be covered in this section to provide understanding. After showing the syntax of both formats, the advantages of using EPP and reasons to convert ERB to EPP will be discussed.

Templates can be generated either by using content in template files or via a string, which is known as an inline template. For template files, ERB uses the `template` function and EPP uses the `epp` function. For inline templates, EPP uses the `inline_epp` function and ERB uses the `inline_template` function.

EPP templates

An EPP template file is a text file that contains a mixture of text and Puppet language expressions surrounded by tags. These tags indicate how the Puppet expression should be evaluated and can modify the text in the template, allowing for the creation of a file based on Puppet language features such as variable interpolation, logic statements, and functions.

Table 7.1 shows the available tag types that can be used:

<table>
<tr><th>Tag Name</th><th>Starting tag (with trimming)</th><th>Ending tag (with trimming)</th><th>Purpose</th></tr>
<tr><td>Parameter</td><td><code><% |</code> (<code><%- |</code>)</td><td><code>|%></code> (<code>| -%></code>)</td><td>Declare parameters accepted by the template</td></tr>
<tr><td>Non-printing expression</td><td><code><%</code> (<code><%-</code>)</td><td><code>%></code> (<code><%-</code>)</td><td>Evaluate the Puppet code but don't print</td></tr>
<tr><td>Expression printing</td><td><code><%=</code></td><td><code>%></code> (<code>-%></code>)</td><td>Evaluate the code to a value to print</td></tr>
<tr><td>Comment</td><td><code><%#</code> (<code><%#-</code>)</td><td><code>%></code> (<code>-%></code>)</td><td>Allow the addition of comment lines just for the template file itself</td></tr>
</table>

Table 7.1 – EPP template tags

When a template is evaluated, it switches between text mode and Puppet mode as it encounters start tags, and returns to text mode as it reaches end tags. In text mode, it outputs the text as content, and when it finds a tag, the Puppet code between the start and end tags is evaluated, depending on the kind of start tag.

As noted in *Table 7.1*, some of the tags can be modified with a hyphen (`-`) to trim spaces and new lines as appropriate.

The parameter tag is optional and, if used, must be the first content in a template file, except for a comment tag, which must use a closing hyphen. It produces behavior similar to how parameters can be declared at the start of Puppet classes, as was shown in *Chapter 8*. The parameters follow the same pattern as a class, so they can optionally include a type at the beginning. They must then have a dollar (`$`) symbol followed by a variable name, optionally followed by an equal (=) symbol and a default value, and finally, they must end with a comma.

For example, to have an options parameter containing a string set to an empty string by default, an `application_mode` parameter, which can contain full, partial, or none strings and defaults to node, and a `cluster_enabled` parameter, which is a Boolean, the following code would start our template:

```
<%- |
String $options = '',
Enum[full,partial,none] $application_mode = 'none',
Boolean $cluster_enabled,
|-%>
```

When parameters are passed to an EPP template, they become local scope and can just be called directly by name, but variables from the calling class must be called by full namespace name; this is similar to a defined type. Any parameters without a default value, such as `cluster_enabled` in the preceding example, are mandatory and must be passed in.

> **Note**
>
> It is recommended to always use hyphens with parameters to avoid any accidental white space at the start of a template.

If parameters are not used, class variables can be directly accessed using the class scope, such as `$example_module::example_param`.

Parameters allow a template to be more flexible if it is to be used in several different places, ensuring the data is more defined and locked down to requirements and making it clear at a glance what data is consumed. It may become a better option to just use variables when you need to use a lot of variables and parameters would just not scale. Passing parameters not defined in the parameter list will result in a syntax error, although if no parameters are used, any parameters can be passed to the template. Later in this section, it will be seen how to pass a hash to the `epp_template` function when referencing it.

The `comment` tag in Puppet templates allows for comments to be added within the template file itself. These comments will not appear in the output when the template is evaluated and its content is generated. Here's an example of what a comment would look like in a Puppet template:

```
<%#- An example comment. -%>
```

> **Note**
>
> The `<%#-` hyphen trimming feature is available from Puppet 6.0.0 onward. Before this, the trimming behavior was assumed.

An expression printing tag puts the returned value of a Puppet expression into the output. This can be a variable or fact, the output of a function, or the output for operators. The final output is a string and will be automatically converted if necessary. At its simplest level, this can be used to print a fact as a value. For example, the following line will read `application = exampleapp` if the `application_name` fact contained the `exampleapp` value:

```
application = <%= $facts[application_name] -%>
```

This example is also the first time variables have been shown in this context, but they are accessed in the same way as variables in regular Puppet code.

Non-printing tags contain iterative and conditional logic. This is different from other tags because the effect of the tag can span multiple lines until another tag closes the iterator or conditional logic. For example, the `if` statement (which will be covered in the *Conditional statement* section) opens with a curly brace, `{`, and closes with a curly brace, `}`. In the following example, we can ensure that if the application name returns `undef` from the `getvar` function, it will not output `application =` as it would have in our previous example. Instead, it will ignore the line if the variable is not defined:

```
<% if getvar(facts.application_name) { -%>
 application = <%= $facts[application_name] -%>
<% } -%>
```

Multiple levels of non-printing tags can be used to create nested `if` or `case` statements as appropriate.

There are some syntax mistakes to be careful with. If a non-printing expression tag contains a comment, it will essentially comment to the end of the line and require the close tag on the next line, as per this example:

```
<%-# I don't finish commenting here  -%> but on the next line
-%>
```

This mistake could clearly happen with a mistyped `comment` tag so care must be taken and any tags, not just a comment closing tag, would be ignored till the new line.

To include literal `<%` or `%>` characters in the template output without having them evaluated as EPP tags, you can use an additional `%` character to escape them. For instance, to output `<% Puppet expression example %>` as text, you would write `<%% Puppet expression example %%%>`. Note that the escape only applies to the first `<%` or `%>` encountered, so if you need to escape only one of them in a line, you can use the escape once and then the other symbol normally.

EPP templates can be validated using the puppet epp validate <template_name.epp> command, and in *Chapter 8*, it will be seen that the **Puppet Development Kit** (**PDK**) will run this command as part of its validation.

To test the rendering of templates, the `render` command can be used with a hash of values as required: puppet epp render <template_name.epp> --values '{key1 => value1, key2 => value2}'.

> **Note**
>
> The full specification for EPP templates can be viewed online at `https://github.com/puppetlabs/puppet-specifications/blob/master/language/templates.md`.

After reviewing the syntax of the EPP template file, let's see how to use the epp function in a Puppet resource. The epp function can be used with resources such as `file` by passing it to the `content` attribute. Additionally, a key-value hash can be provided to specify the parameters, as discussed in the previous section on the `parameter` tag:

```
file { '/etc/exampleapp.conf':
  ensure => file,
  content => epp('exampleapp/exampleapp.conf.epp', {'version' => '1',
'clustered' => false}),
}
```

> **Note**
>
> If you wish to try the preceding example on your developer environment, create a suitable template file on your system and change the `exampleapp` module name to the absolute path containing the template, such as `/var/tmp` or `C:\Users\David Sandilands`.

The namespace used in `epp` assumes that either it will form a module path, `<modulename/templatename.epp>`, which translates to `modulepath/modulename/templates/templatename.epp`, or it will be an absolute path on disk. In *Chapter 8*, the structure of modules will be covered in detail.

Inline templates are similar to regular templates, but instead of using a separate template file, they require a string or variable to be passed to them. They are generally used for workarounds or where using a heredoc feels easier than using a template file.

One example of a workaround is when using the Vault module, which was discussed in *Chapter 5*, to retrieve secrets using deferred functions. The Vault module returns a key-value pair, but we may only want to access the value of the password. As the value is deferred, it can't be manipulated as a string. Using the `inline_epp` function, as shown in the following example, allows us to unwrap the string during agent runtime and apply it to the file:

```
$vault_keypair = { 'password' => Deferred('vault_lookup::lookup',
["secret/examleapp", 'https://vault:8200']), }
file { '/etc/exampleapp_secret.conf':
  ensure => file,
  content => Deferred('inline_epp', ['PASSWORD=<%= $password.unwrap
%>', $vault_keypair]),
}
```

Having covered EPP templates for the management of heritage code, we will now review how ERB is different.

ERB templates

ERB templates are similar to EPP templates, but there are some differences worth noting. ERB templates are text files that contain a mixture of text and Ruby language expressions surrounded by tags. ERB uses the same tags as EPP, except it does not have parameter tags, and it is not possible to pass parameters.

In ERB, a template has a local scope and a parent scope, which is the class or defined type evaluating the template. Variables in the current scope can be accessed using the @ symbol, which is how Ruby normally accesses variables. To access variables out of scope, you can use the scope object or the older scope.lookup function, which was used before the hash format was introduced into Puppet.

To give some simple Ruby examples, you can use an if statement to check whether the exampleapp_extras variable does not contain NONE, and to output the extras <exampleapp_version> string in the template. You can also use an unless statement to check whether the exampleapp_key variable is not nil, and to output the key <exampleapp_nill> string if it has a defined value:

```
<% if @exampleapp_extras!= "NONE" %>extras<%= @exampleapp_version%><%
end %>
<% unless @exampleapp_key.nil? -%>
key <%= @exampleapp_%>
<% end -%>
```

Iteration in Ruby is similar, with the each function also available. The following example shows an array of settings from a variable being output one by one in the template content using iteration:

```
<% @array_of_settings.each do |setting| -%>
<%= val %>
<% end -%>
```

Data from Puppet variables will be translated from their Puppet type to the equivalent Ruby type (more information can be found in the official Puppet documentation at https://www.puppet.com/docs/puppet/latest/lang_template_erb.html#erb_variables-puppet-data-types-ruby). However, it is beyond the scope of this book to discuss how Ruby can transform this data.

It is also possible to call Puppet functions in ERB templates using the <%scope.function_name(<Name of function>, <Array of Arguments>)%> syntax.

For example, to use the downcase function on the example_variable variable and output the result to the template, the following code can be used:

```
<%= scope.call_function('downcase', [@example_variable]) %>
```

Validating the syntax of an ERB template can be done by running the erb command: erb -P -x -T '-' example.erb | ruby -c. As with EPP, the PDK will check for both types of templates when running validation. Unfortunately, there is no way to render an ERB template.

Using the content of an ERB template file in a file looks very similar to EPP, but as discussed, it does not have parameters and uses the `template()` function. Converting the EPP example would look like this:

```
file { '/etc/exampleapp.conf':
  ensure => file,
  content => template('exampleapp/exampleapp.conf.erb'),
}
```

It is possible to pass and evaluate multiple template files, which will be concatenated together. For example, updating content as follows would combine the two templates:

```
content => [ template('exampleapp/exampleapp.conf.erb'),
template('exampleapp/exampleapp2.conf.erb')]
```

Inline ERB just uses the `inline_template` function in the same system as inline EPP, and was often written in the past to allow Ruby code to provide a workaround for the lack of iteration/loops provided by past versions of Puppet and perform data transformation.

Now that ERB has been discussed, it is time to highlight why EPP is preferred over ERB and reasons to consider converting heritage code.

EPP and ERB comparison

After reviewing both syntax templates, it is clear that EPP has several advantages over ERB. Firstly, EPP has significantly better performance than ERB. ERB creates a scope object for all facts and top-scope variables each time a template is evaluated, while EPP only uses facts and variables relevant to the template. In environments with a large number of facts, this can have a significant impact on performance.

Furthermore, EPP provides greater security because templates can be provided with a limited scope of data to be used and validate that all data exists before use. ERB, on the other hand, has no built-in validation, and non-existent variables will simply be dropped. For example, if a variable in a class has not been evaluated before the template is used, ERB will not catch this.

EPP can also be seen as easier to use since it is in the Puppet DSL style and does not require any Ruby knowledge. This makes it easier to code, especially with the ability to use the `puppet epp render` and `validate` commands. Additionally, EPP is under more active development, and recent features, such as templates being able to automatically unwrap sensitive variables in 6.20 and later, will only be available in EPP.

Iteration and loops

Puppet's approach to iteration and loops is influenced by the fact that its variables are immutable, meaning that once they are set, they cannot be changed. This makes many normal approaches used with `loop` or `do` keywords to transform data impossible. In early versions of the language, this was worked around by passing an array to defined types, as discussed in *Chapter 3*, or using inline ERB templates with Ruby code to manipulate arrays and hashes.

However, the issue with the defined type approach was that the code doing the work was abstracted away and not visible. Furthermore, every time a different type of iteration was required, it would require its own defined type, bloating the code. Therefore, it is important to review heritage code and refactor these patterns to the approaches that will be discussed.

In modern Puppet, the approach taken is to use iterative functions that pass data from arrays and hashes to lambdas. A lambda is a function with no name, so it cannot be called anywhere else except by a function. A lambda can be attached to any function call, including custom functions. *Table 7.2* provides a full list of functions involved with iteration and lambdas. While some functions may not be considered iterators, they have similar behaviors. It should also be noted that other functions could be combined/chained into these examples, such as `unique`:

Function name	**Purpose**	**Return type**	**Parameters**
`all`	Runs through all elements until `false` is returned from the lambda or completes and returns `true`.	`true` or `false`	1 or 2
`any`	Runs through all elements until `true` is returned from the lambda or completes and returns `false`.	`true` or `false`	1 or 2
`break`	Used within a lambda and stops the iteration.	n/a	n/a
`each`, `reverse_each`, `tree_each`	Passes each element of a hash or array in turn for the lambda to process (reverse order or recursive variations).	n/a	1 or 2
`filter`	Runs through all elements and matches with lambda code, returning matching elements in an array.	Array	1 or 2
`index`	Runs through all elements and on the first match within lambda code, returns the index of the matching element.	Integer	1 or 2

`lest`	The function takes one argument; if this value is undefined it will run a lambda and return the outcome. If the argument is not undefined it will return the argument.	Any valid type	0
`map`	Runs through all elements and applies lambda code to that element. Returns an array of elements post lambda.	array	1 or 2
`next`	Used within a lambda to change the value of the next element in the iteration.	n/a	n/a
`reduce`	Runs through all elements and applies lambda code passing the outcome forward to each iteration.	array	2
`return`	Used to cause a lambda to return (cannot be used at top scope).	n/a	n/a
`slice`	Runs through a sliced size of elements such as three elements per iteration.	Array	1 or size of slice
`step`	Chained into another iterable function passing a sequence of elements incrementing at a step size from a starting element to a finishing element.	Iterable	n/a
`then`	Takes one argument and if it is not undefined, it will call a lambda with the argument. Otherwise, it will return `undefined`.	Any valid type	1
`with`	Takes one argument and unconditionally passes it to a lambda and runs with the argument. Returns the result of the lambda.	Any valid type	1

Table 7.2 – Functions for iteration and lambdas

The basic syntax structure for iterative functions using lambdas in Puppet is as follows:

```
<function acting on data> | <parameter(s)> | { lambda of Puppet code }
```

As an example, consider using the `each` function to loop around an array, with a single parameter (optionally typed), and printing the output:

```
['first', 'second', 'third'].each | String $x | { notice $x }
```

This would result in the `notice` function printing for each string in the array, similar to a `for` loop with a `print/echo` command in most languages.

The `each` function can also use two parameters, which would give the index as the first parameter and the content of that index as the second parameter. The following code would print `index 2 contains second` for the second iteration of the lambda:

```
['first', 'second', 'third'].each | $index $value | { notice "index
$index contains $value" }
```

> **Note**
>
> To test these examples on your developer desktop, simply run `puppet apply -e '<example code>'`.

To clarify, when using the `each` function with a single parameter on a hash, each key-value pair will be passed as an array to the lambda. For example, running the code `[{ key1 => 'val1', key2 => 'val2' }].each | $key_pair | { notice $key_pair }` will output two arrays, one for each key-value pair: `['key1', 'val1']` and `['key2', 'val2']`.

If two parameters are used in the lambda, the first parameter will represent the key and the second parameter will represent the value. For example, running the code `[{ key1 => 'val1', key2 => 'val2' }].each | $key, $value | { notice "$key contains $value" }` will output two strings, one for each key-value pair: `key1 contains val1` and `key2 contains val2`

It is also worth noting that other data types, such as strings, can be automatically converted into arrays where each character in the string will be treated as an element. Additionally, a range of numbers can be declared using an `Integer` type; for example, running the code `Integer[100, 150].each | Integer $number | { notice $number }` will output all integers from `100` to `150`.

Finally, iteration can be nested; for example, to handle a hash with array values, an iterative function can be used within the lambda. Running the following code will output each value in the array of the `key1` key-value pair – `'value1'` and `'value2'`:

```
[{ key1 => ['value1', 'value2'], key2 => 'val2' }].each | $key,
$value_array | {
  $value_array.each | $value | {
    notice $value
  }
}
```

Overall, this section provides an overview of the most commonly used functions in Puppet, but for more in-depth descriptions, users can refer to the official documentation at `https://www.puppet.com/docs/puppet/latest/function.html` .

Iterative loops

The main function reviewed so far is `each`, and there are several functions that perform this loop of elements or manipulate the loop. `reverse_each` simply takes the reverse order of the elements, as its name suggests. `tree_each` allows values in nested arrays/hashes to be returned with different behaviors depending on the flags provided. It is relatively complicated and niche. The `slice` function allows us to take a specific number of elements in each iteration. For example, the following code would pass arrays of three numbers at a time from the sequence to the lambda:

```
Integer[100, 151].slice(3) | Array $numbers | { notice $numbers }
```

On the last iteration, it would provide the remaining elements; in this example, an array with just `[151]`. It is also possible to use multiple parameters, but the number of parameters must be the same as the `slice` size:

```
Integer[100, 151].slice(3) | Integer $first, Integer $second, Integer
$third  | { notice $numbers }
```

The `step` function allows us to choose which elements of an iterable we wish to pass. In this code example, it would start at the first element, then the fourth, the seventh, and so on:

```
Integer[100, 150].step(3) | Integer $numbers | { notice $numbers }
```

This can be useful when chained into another iterable function. The next type of function is for matching patterns. This is a different style of iteration function where instead of just passing the elements to the lambda to perform some action, the iterative function defines how the lambda will return. For example, `all` is looking for all elements to match the check in the lambda to return `true`. If any of the lambdas return `false`, the function will return `false`. For example, the following code would print `true` because all the elements were greater than `99`:

```
Integer[100, 151].all | Integer $number | { $number > 99 }.notice()
```

The any function is the opposite of `all`, returning `false` if there are no matches and `true` if the lambda returns `true` in any iteration. The `index` function is similar to any, but instead of returning `true` or `false`, it returns the index number of the element that matches, or `undef` if there is no match. For example, the following code would print `20`, as `number` would match the 20th element:

```
Integer[100, 151].index | Integer $number | { $number == 120
}.notice()
```

All of the functions can work with two parameters on arrays or hashes, as shown in the examples for the `each` function.

Data transformation

Data transformation is another way iterators are used to iterate through elements and perform adjustments before returning them. This is one of the main reasons why the iterator to lambda pattern was developed, as Puppet is unable to reassign variables. For example, the map function iterates over each element and applies a lambda whose result is stored in an array. For example, the following code would divide each element by `1024` and return an array of `[2, 3, 1]`:

```
[2048, 3096, 1024].map | $size | { $size / 1024 } .notice()
```

The `filter` function takes each element in the iteration and applies the code in the lambda. If the lambda returns `true`, the element will be added to an array for return. Otherwise, if `false`, it will continue to the next iteration. For example, the following filter would iterate through each array, checking whether the size is greater than `0`, which would result in an output of `[[1, 2, 3], ['a', 'b', 'c']]`, with the second element's empty array being removed:

```
[[1,2,3], [], [a,b,c]].filter | $array | { $array.size > 0 }
.notice()
```

Both `filter` and `map` can handle one or two parameters, as shown with the `each` function on arrays and hashes.

The `reduce` function allows for cumulative work to take place in the lambdas. It is different from the other functions and requires two parameters: the first parameter keeps its value through the iterations, while the second is the element. Additionally, the starting value can be chosen for the first parameter by passing a value to the `reduce` function. In this example, the `total` parameter would start at `1` and, in each round, add the element to its total, resulting in `15` being returned and printed:

```
[2, 4, 8].reduce(1) | $total, $number | { $total + $number }
.notice()
```

Within a lambda, it is also possible to change the flow of iteration. The `next` function can change what the next element will be: `undef` if nothing is provided to the `next` function, or the value provided to the `next` function. The `break` function stops the iterator at that point in the code and returns to the iterative function, effectively ending the iterator at that point. The `return` function, in comparison, returns from the iterative function, so it will not complete at all and returns to the containing class, function, or defined type.

To demonstrate this change in flow, the first example using map over a series of numbers will run a `next` function, and when the element matches `101`, it will replace the next element, `102`, with `1984`, and then a `break` function will be run when the element is greater than `104`. So, with the notice at the end printing, this will return an array of `[100, 1984, 102]`:

```
Integer[100, 151].map | Integer $number | { if $number == 101 {
next(1984) } if $number > 103 { break() } $number }.notice()
```

To highlight the different behavior of replacing the `break` function with a `return` function, the following example will result in nothing being printed:

```
class example {

  Integer[100, 151].map | Integer $number | { if $number == 101 {
next(1984) } if $number > 103 { return() } $number }.notice()

}
```

This is the reason, in this example, that we have put the function within a class because `return` cannot be called at a top scope level only within a class, function, or defined type.

Nested data

The last types of functions are useful when handling nested data or generally handling undefined values. `then` is chained after a lambda and if it receives `undef` outputted from that lambda, it will return `undef`; otherwise, it will pass the value to another lambda. So the following example would use the `dig` function to attempt to access a `c` element that doesn't exist in the second hash in the array, and as `then` would receive `undef`, it would therefore return `undef`:

```
$example = {first => { second => [{a => 10, b => 20}, {d => 30, e =>
40}]}}
$example.dig(first, second, 1, c ).then |$x| { $x / 10 }.notice()
```

To clarify the previous statement, if `dig(first, second, 1, d)` is changed to `dig(first, second, 2, d)`, it would then pass `30` to the lambda, which would divide by `10` and print `3`.

`lest` is the opposite of `then` and returns the value it is defined; otherwise, it passes `undef` to the lambda, which can take an action such as setting a default value instead. This can be useful; when used alongside `then`, taking the preceding example, adding `lest` would allow a value of `undef` to be returned if it is `0`:

```
$example.dig(first, second, 1, c ).then |$x| { $x / 10 }.lest() || { 0
} .notice()
```

The `with` function is somewhat of a specialist edge, as it is used to pass values through if our lambda is able to handle `undef` or defined values.

So, having reviewed the various functions and seeing the data transformation and explorations of data possible, it is worth highlighting again how, instead of using a defined type, as was done in the past, when we need to create multiple resources. So, for example, to create a directory for as many instances as requested, the following code could be used:

```
Integer[1,$instance_number].each |Integer $id | {
  file {"/opt/app/exampleapp/instance${id}":
```

```
      ensure => directory,
    }
  }
```

Having covered how Puppet can perform loops and iteration, the conditional statements will be reviewed next.

Conditional statements

Puppet has the conditional statements you would expect of any language, with `if`, `unless`, and `case` allowing code behavior to be different depending on things such as facts or data from external sources. Puppet additionally uses selectors, which are similar to a `case` statement but return a value instead of executing code on result.

If and unless statements

The `if` statement follows a specific syntax that includes the `if` keyword followed by a condition, an opening curly brace ({), the Puppet code to execute if the condition evaluates to `true`, and a closing curly brace (}).

The following example is a simple check on a Boolean in `example_bool` to print a notice if it contains `true`:

```
if $example_bool {
  notice 'It was true'
}
```

This can be optionally extended by adding an `else` keyword after the closing curly brace (}) of the `if` statement and then using an opening curly brace ({) with Puppet code to perform if the condition is `false`. This is then closed with a closing curly brace (}). To also print when `example_bool` is `false`, the code would be updated as follows:

```
if $example_bool {
  notice 'It was true'
}else{
  notice 'It was false'
}
```

Similarly, to perform multiple `if` checks together, the `elsif` keyword can be used after the `if` statement's closing curly brace (`}`), allowing the same `if` syntax to be followed again. This can be nested and repeated as required. To provide an example, following up on the Boolean check with `elsif`, we can add a second check to see whether the value variable is greater than `2`, print a notice if it is, and an `else` statement that prints that both conditions were `false`:

```
if $example_bool {
  notice 'It was true'
}elsif $value > 2{
  notice 'It was false and value is greater than 2'
}else {
  notice 'It was false and the value was 2 or less'
}
```

The `unless` statement is simply the inverse of the `if` statement. It allows you to avoid having to negate a condition, and it can also be used in combination with an `if` statement. However, it has no equivalent to `elsif` and will cause a compilation failure if used. To demonstrate this with the previous `if` example, the `unless` statement can be used instead to check whether `example_bool` is `false` and print a notice in that case:

```
unless $example_bool {
  notice 'It was false'
}else{
  notice 'It was true'
}
```

Any non-Boolean values used in the conditions will be converted to Booleans as per the data type rules, as covered in *Chapter 4*.

The Puppet style guide recommends for the lines which follow the keywords `if` and `unless` that the code should be indented by two spaces and aligned as was shown in the examples.

Case statement

The `case` statement works by matching the value outputted from a control expression. This is commonly the content of a fact or variable, but it could also be an expression or function. The format starts with the `case` keyword, followed by a control expression, resolving to a value and enclosed within curly braces. Each following line starts with either a matching case or a comma-separated list of cases, followed by a colon, and the Puppet code enclosed within curly braces to be applied for the matching case. The `case` statement is then closed with a curly brace.

For example, to test the value of the `hardwareisa` fact and include a profile based on the type of processor architecture in use, the following code can be used. It includes a Unix profile for `sparc` or `powerpc` values, the Linux 32-bit profile for `i686` and `i386` values, the 64-bit profile for any value ending in `64`, and the default profile for any value failing to match a case:

```
case $facts[' hardwareisa'] {
  'sparc', 'powerpc': { include profile::unix}
  'i686', 'i386': { include profile::linux::32bit}
  /(*64)/: { include profile::linux::64bit}
  default: { include profile::default }
}
```

The Puppet style guide recommends always using a `default` case, which can be a failure or even just an empty curly brace.

Selectors

Selectors are similar to `case` statements, but instead of applying Puppet code for the matching case, they return a value. Selectors can be used wherever a value is expected, such as variable assignment, resource attributes, and function arguments. The Puppet style guide recommends only using selectors in variable assignments to improve readability, but it may be seen in legacy code in resource attributes as it was previously a popular pattern.

The syntax of a selector is a control expression resolving to a value, a question mark (`?`), and an opening curly brace (`{`). It then has a number of case matches, starting with a single case or the `default` keyword, a hash rocket (`=>`), the value to return, and a closing comma. The selector is then closed with a closing curly brace (`}`).

The following example shows the `apache_package_name` variable assigned based on the output of the `os.family` fact, using the `httpd` name for Red Hat, `apache2` for Debian or Ubuntu, `apache-httpd` for Windows, and defaulting to `httpd` if there is no match. A package resource could then use this name to install the relevant package:

```
$apache_package_name = $facts['os']['family'] ? {
  'RedHat' => 'httpd',
  /(Debian|Ubuntu)/ => ' apache2 ',
  'Windows' => 'apache-httpd',
  default => 'httpd',
}
package { $apache_package_name }
```

As with the `case` statement, the Puppet style guide recommends that selectors always use a `default` case, which can be a failure or even just an empty curly brace.

Capture variables

If the `case` statement used was a regular expression, then what is known as capture variables will be available within the associated code as number variables such as `$1` and `$2`, with the entire match available at `$0`.

To modify the example in the *Case statement* section, if the match was against `amd64`, this would include `profile::linux::amd64`:

```
/(*64)/: { include profile::linux::${1} }
```

Having reviewed all aspects of templates, conditionals, iterations, and loops, we will now use a lab to recap and bring these concepts together.

Lab – creating and testing templates containing loops and conditions

In this lab, we will bring together everything you have seen in this chapter so far, testing and validating some example templates and creating a template using logic and iteration:

1. Download the template files at `https://github.com/PacktPublishing/Puppet-8-for-DevOps-Engineers/tree/main/ch07/templates_to_check` and validate and parse each of them to ensure the templates produce the following (<> shows where interpolation needs to happen):

    - **“This template was run on a machine with <# number of cpus of your machine> cpus”**
    - **“The custom fact pack.lab was < set to content of fact or the string ‘not set’ > “**

> Hint
>
> You would want to use the `getvar` function to test the fact and test it by passing in a hash, setting it when parsing to test it.

    - **“The system uptime is < showing only days hours mins if they are non zero > “**
    - **“This machine is <not/is> virtual <and runs on <$virtual>”**

2. Create a template that prints the following

“This is a < os family > machine running version < os release full > ““The following directories are in the path < list each path >”

> **Hint**
>
> Use the `split` function (`https://www.puppet.com/docs/puppet/latest/function.html#split`) to separate the path fact string into an array that can be iterated, and remember that Windows paths are split by ; while Unix/Linux based paths are split by : . See the answer at `https://github.com/PacktPublishing/Puppet-8-for-DevOps-Engineers/blob/main/ch07/2_sample.epp`.

Summary

In this chapter, the use of templates in Puppet was examined. It showed Puppet's two types of templates – EPP and ERB – which worked in similar ways, using a mix of plain text and tags surrounding code to allow for Puppet/Ruby to apply logic and variables and create more complex content when evaluated. It was warned that the level of complexity should be carefully considered before using templates instead of functions such as `file_line`, and controlling resources individually. Additionally, because of the lack of functions for a single line or setting control in files, templates had been overused, and heritage code should be examined carefully to ensure that a template was the correct level of complexity.

EPP was shown to be the recommended way of producing templates since it was in the Puppet language and easier to learn for a Puppet developer. It was also more secure since it could limit its scope with parameters, and similarly, more performant since it only created a scope for the variables and facts required, unlike ERB, which, for every template used, would generate a scope for all facts. Furthermore, it was mentioned that all Puppet future development work would be for EPP, as was shown by the inclusion of automatic unwrapping of sensitive variables in EPP files, and the capability of rendering a template was only available in EPP.

EPP and ERB templates were shown to be referenced and evaluated from files via the `epp` and `template` functions, where multiple files could be combined together. It was also shown that inline templates were possible via the `inline-template` or `inline-epp` functions, where the text could be passed directly to the function instead of being stored in a file.

Iteration and loops were then shown, highlighting Puppet's previous lack of capability to do this with the immutable nature of Puppet variables making more traditional `loop` keywords impractical. Puppet was shown to instead use iterative functions on arrays and hashes, which passed values as parameters to lambda functions. These unnamed lambda functions, which could only be called by other functions, allow the creation of a scope entirely local to the lambda function and, therefore, allow variable names to be reused. The iterative function chooses how values should be passed to the lambda, such as each passing one value or key pair at a time, or `Reduce`, which allows the use of a parameter that is passed through each lambda function along with each value and key pair and can be useful to do cumulative transformations.

Puppet's conditional logic was then discussed, showing it to be similar to most other languages. `if` checks evaluate a check to a Boolean statement/comparison, which if `true`, is acted on with Puppet code. The `else` keyword allows an action to take place if the Boolean was `false`, and the `elsif` keyword allows the chaining of checks together. `unless` was shown to be the inverse of `if`, acting on the check if it was negative and allowing `else` to act if it was `true`, although it had no equivalent of `elsif`.

The `case` keyword was then discussed. We showed that it works by taking the value and matching it to run Puppet code based on the match or to a default if no value was found. The `selector` keyword was shown to be similar to the behavior of the `case` statement but is used instead of running Puppet code to assign a value. It was highlighted that, despite it being a common pattern in the past to use a selector within resources, this was no longer considered best practice. Finally, capture variables were shown as variables available to conditionals that had used regular expressions to show what the match had been.

Having now reviewed the core Puppet language, it is time to learn how to structure the manifest file and classes we have used so far. The next chapter will demonstrate how modules provide the necessary structure to hold the manifests, classes, and other configuration and implementation files that we have examined. The roles and profiles pattern will also be explored, providing an additional abstraction to represent the technology stacks and business needs of customers. Moreover, Puppet Forge will be demonstrated as a source of modules that can be consumed to reduce the need for development and to collaborate with the Puppet community to enhance the available code.

8
Developing and Managing Modules

Having reviewed many aspects of the Puppet language, it is clear that using manifest files and classes alone would not scale or provide the structure needed as a code base grows for a diverse range of servers and customer requirements. In this chapter, we will review the components required to create Puppet code at scale. We will be looking at Puppet **modules**, which allow us to bundle code and data focused on a single technology implementation, thus making it easy to share and combine with other implementations. Then, we'll explore the **roles and profiles method** to show you how profiles can group modules to create technology stacks and roles, then combine profiles to match business requirements. After, we'll cover the **Puppet Development Kit** (**PDK**), showing how it can automate the process of creating and managing modules. The directories and files templated by the PDK will be shown, highlighting its built-in validation and linting checks, as well as unit compilation checks. Next, we'll look at **Rspec**, as a method that expands on this to provide more thorough unit testing, as well as **ServerSpec**, which is used for server testing, at a high level. Then, we'll discuss the **Puppet Forge** catalog, which acts as the source of modules developed by Puppet itself, as well as vendors and members of the community. We will show you how to filter various aspects of modules to understand how they are supported, their compatibility with OS and Puppet versions, and scoring/scanning ratings so that you can choose the best module for your organization's needs.

In this chapter, we're going to cover the following main topics:

- What is a module and what is in it?
- Roles and profiles method
- The PDK and how to write and test a module
- Testing using RSpec with PDK
- Understanding Puppet Forge

Technical requirements

This chapter does not require any infrastructure to be deployed. All actions can be performed from your developer desktop.

What is a module and what is in it?

Modules provide us with a way to group code and data, making it easier to share and reuse code that is part of a specific technology implementation. Almost all of your Puppet code will be stored in modules of one kind or another. You should work out the scope of your module to create focused modules with a single clear responsibility. If you were deploying a LAMP or WAMP stack, you would not make a single module that configured all components; instead, you would break it into individual modules, including OS settings, MySQL, and Apache. This allows greater code reuse and reduces the complexity of any individual module.

A module is a directory named with similar criteria to a class, so it must begin with a lowercase letter and can only contain lowercase letters, numbers, and underscores. Unlike classes, modules cannot be nested and do not use the `::` symbol. Reserved words and class names should not be used as module names.

Modules have a directory structure that allows Puppet to know where various types of code and data will be stored and autoloaded as requested. As discussed in *Chapter 6*, a scope namespace and file service namespace for that module will be created. The core code and data are stored in the following directories:

- `data`: Contains module-based data for parameter defaults, which will be covered in *Chapter 9*.
- `examples`: Contains examples of how to declare the modules' classes and defined types.
- `files`: Contains static file content that can be placed by Puppet.
- `manifests`: Contains all the manifests of the module and directories that provide structure.
- `template`: Contains the EPP and ERB template files to be used by Puppet code.
- `tasks`: Contains tasks for procedural work. This will be covered in *Chapter 12*

Modules also have what is known as **plugins**, allowing them to distribute various custom Puppet components to the Puppet server or agent, as relevant. Some of these plugins are as follows:

- `lib/facter`: Custom facts written in **Ruby** that are used on the agent
- `lib/puppet/functions`: Custom functions written in Ruby that are used by the server
- `lib/puppet/type`: Custom resource types that are used on both the server and agent

- `lib/puppet/provider`: Custom resource providers written in Puppet that are used on both the server and agent
- `lib/augeaus/lenses`: Custom Augeas lenses that are used on the agent side
- `facts.d`: External facts or static scripts that are used on the agent
- `functions`: Customs functions written in Puppet that are used on the server

It is important to note that certain plugin types, such as resource types, are not isolated fully in environments. Environments will be discussed in detail in *Chapter 11*, where we will focus on classification and release management, but for now, note that environments allow isolated code bases to be used by nodes so that they can use different versions of code. This is due to the way Ruby loads the first resource type and makes it global, ignoring any duplicates found. Therefore, it is important to consult the Puppet documentation: `https://puppet.com/docs/puppet/latest/environment_isolation.html#environment_isolation`. You can configure environment isolation as necessary if you use modules containing plugins that cannot be isolated. Puppet Enterprise provides isolation in environments by default.

Modules can be used in different ways. While most of this chapter will focus on modules that use Puppet code for configuration, modules can be used to distribute an item or items. An example of this is the PowerShell module in Puppet Forge (`https://forge.puppet.com/modules/puppetlabs/powershell`), which is used only to distribute a new exec provider using the provider plugin directory.

Focusing on the `manifests` directory, the manifest files will have the same name as the class name they contain. The major exception to this is the main manifest, which is named `init.pp`, but has the class name of the module. This main manifest is often used as the entry point to the module. As discussed in *Chapter 6*, a module namespace is created for the module, allowing us to include the module in code by running `include <module name>`.

The classes should be self-contained and small, focusing on one aspect. A common piece of advice on how to identify a class that is too big is when it is too large to view in a single editor screen. With this in mind, one of the most common patterns when starting with modules is to use the main manifest, `init.pp`, as an entry point that takes parameters to be used across the module. After, it calls other classes that are used and sets their ordering. An example of this is using an `install` class to install resources such as packages, a `config` class to add any configuration files or users, and a `service` class to manage the service. The following code shows an example of a main manifest for this pattern:

```
class exampleapp (
  Boolean $package_managed = true,
  Integer $package_version = 3,
  Boolean $user_managed = true,
  Integer $user_id= 10,
```

```
    Boolean $service_enable = true,
    Integer $jmx_heap_size = 1024,
    Integer $thread_number = 10,
  )
  {
    contain exampleapp::install
    contain exampleapp::config
    contain exampleapp::service

    Class['exampleapp::install']
    -> Class['exampleapp::config']
    ~> Class['exampleapp::service']
  }
```

Considering the available parameters, such as the public API of the module, would allow the module to be flexible; in addition, consistency should be maintained when naming these parameters. Here, we use an approach where parameters are named based on their effect. So, for `exampleapp`, it can be seen that both the package and user take a Boolean to declare if the module is managing them as a resource. A Boolean is used on `service_enable` to decide if the service is enabled on boot, while integers are provided for `user_id` and `package_version`. Two additional integers are then used for configuring the application, setting the Java memory size, and the thread number. These parameters can be accessed using the module namespace and performing a data lookup for the variable at `modulename::variablename`. This is known as automatic parameter lookup and will be covered in detail when we review how Puppet handles data in *Chapter 9*.

> **Note**
>
> Parameters and other aspects of a module can be documented via comments in the header of the code, which can then be generated in different formats using the Puppet Strings gem. Details can be found in the `Puppet Strings style guide` and on the following web page: `https://puppet.com/docs/puppet/latest/puppet_strings_style.html`.

We adopt the approach of using containment and ordering on the classes within the module. This ensures, if requested, that the package is installed, the configuration is added, and the service is then managed or refreshed since each stage depends on the next. The `contain` keyword should not be considered a default to be used instead of `include`; it should only be used when it is this component module style and when the classes will only ever only be used within the main class. In the *Roles and profiles method* section, you will see where containment and ordering like this would be inappropriate.

From this, we can see how these subclasses use the parameters from the main manifest. For example, the `install` manifest uses an `if` statement on the `package_managed` variable; if it's `True`, it installs the version of the `exampleapp` package set by the `package_version` variable:

```
class exampleapp::install {
if $exampleapp::package_managed {
  package { 'exampleapp':
    ensure => $exampleapp::package_version
  }
 }
}
```

For the `config` manifest, we can see how the `jmx_heap_size` and `thread_number` variables can be substituted into a template by using the module namespace to access the template stored in the templates folder in the `exampleapp` module:

```
class exampleapp::config {
  file { '/etc/exampleapp/app.conf':
    ensure  => file,
    content => epp('exampleapp/app.conf.epp', {' jmx_heap_size ' =>
$exampleapp::jmx_heap_size , ' thread_number' => $exampleapp::thread_
number }),
  }
}
```

The `service` class is very similar in style to the `install` class. It uses an `if` statement; if this is `True`, it adds a service of `exampleapp` and sets the `enable` parameter based on the `service_enable` variable:

```
class exampleapp::service
  if $exampleapp::service_managed {
    service{'exampleapp':
    ensure => 'running',
    enable => $exampleapp::service_enable ,
    }
  }
}
```

> **Note**
>
> A common module pattern was to use a `params.pp` file to manage default module data. Hiera 5 can now manage module-level data in a more structured way than a manifest file, as will be shown in *Chapter 9*. The `params.pp` file is still commonly seen in code, particularly where the structure of the data is simple and there is little value in changing it to Hiera.

The `examples` directory could contain a file called `init.pp`, specifying how the class could be called:

```
class { 'exampleapp':
  package_managed => true,
  package_version => 3,
  user_managed => true,
  user_id => 10,
  service_enable => true,
  jmx_heap_size => 1024,
  thread_number => 10,
}
```

The naming of files is not important in the `examples` directory; there could be several different examples to show common selections of module attributes for different setups. For example, a module might show how it can be included with the minimum default values, but also the required attributes for a specific architecture, such as deploying in a cluster setup.

As the configuration use case becomes more complex, the other most common approach to the module structure can be seen in the Apache module in the Puppet Forge catalog: `https://forge.puppet.com/modules/puppetlabs/apache`. Instead of grouping resources based on simple `package` and `config` classes, the `apache` module breaks down the different components of the application. In this example, the main manifest for Apache performs a default installation of Apache, with a default virtual host and a document root directory, and starts the Apache service. This can be configured by using the relevant module parameters. Here, the Apache service is managed in a separate class, but the resources that would normally have been in the `package` and `config` classes are managed within the Apache main class. There are also various implementation classes, such as `vhosts`, `mod`, and `mpm`, for different `apache` configuration items. This gives the main class a clear purpose of performing the basic installation and configuring the `apache` server so that the implementation classes can focus on specific customizations. For example, the `vhosts` classes are defined types that can be defined for each virtual host the `apache` server requires.

These examples provide a structure that you can follow for your modules and can be adapted as needed. However, the key lessons to take away are that modules should focus on a single task, only manage their resources (no cross-module dependencies), and be granular and portable.

In this section, we looked at the directory and file structure of a Puppet module and two common patterns for creating modules. These modules, by themselves, configure focused individual technical implementations. In the next section, we will see how a module's structure can be used to combine modules to produce technical stacks and, by combining technical stacks and configuration, meet the business requirements customers have for servers.

Lab – reviewing the apache module

It would be impractical to print all the key parts of the `apache` module's code in this book. However, reviewing the code at `https://forge.puppet.com/modules/puppetlabs/apache` and reading the `examples` directory to see how the main class, `apache`, is combined with various classes within the module to configure different components will help you understand how this module pattern can be structured.

Roles and profiles method

In the previous section, the modules we discussed are what are known as component modules since they cover single implementations. These modules are mostly of interest to users directly involved with the technology implementation, such as Unix or Windows administrators, where understanding that specific resources have been applied is the most important aspect of the configuration. But different users are not interested in how a node is configured; instead, they are interested in what it does. An application specialist, for example, would care that, for their application, Tomcat and MySQL were installed, not how it was configured. A project manager would care that they got a server that met a business need but not what technical stacks were used. The project manager may also see each implementation as unique, but there will often be a lot of similarities, such as various technology stacks being used across multiple applications using Apache or Java with variations of settings based on location or environment.

Without providing some sort of structure for these levels of logic, trying to apply these modules to nodes would require a lot of duplication and complexity of `if` statements.

A pattern called roles and profiles uses a modular structure to achieve this. Role and profile are not keywords but just patterns to use within modules and classes. A simple approach can be to have a module called `role` and a module called `profile`, with each class in these modules representing a role or profile.

These roles represent the business need that a customer, such as a project manager, would require, while a profile would reflect a technical stack for an application.

When looking at how to apply the role and profiles method to the configuration of pre-existing applications, it's important to look from the role down to the module to avoid the natural temptation of trying to find the technical solution first without the business logic. This involves breaking things down into components and granularly thinking about what it is and not what it looks like. One trick to identifying roles is often to take hostnames, which generally contain information about the location, environment use, and application. For example, a hostname may look like `fk1ora005prd`, where `fk1` is a data center location, `005` is a numerator, and `prd` is the production environment, leaving `ora` as the Oracle application, which would match the role's name. So, roles should be business names such as `buildserver`, `proxyserver`, or `ecomwebserver`, while the profile should be the technology stack name, such as Apache, Jenkins, or nginx.

This naming isn't always perfect and sometimes, some of the terms may simply be the same ones that project managers use for ordering Oracle servers. They may be unaware of the underlying profiles of the Oracle role, which would include an Oracle profile and other relevant profiles.

In this case, a `role` class should simply call the required profiles with no variables, as shown here:

```
class role::exampleapp {
  include profile::core
  include profile::java
  include profile::apache
}
```

In contrast, a `profile` class should contain parameters to customize the module and class declarations to add the required modules:

```
class profile::java(
  Pattern[/present|installed|latest|^[.+_0-9a-zA-Z:~-]+$/] $java_
version
  String $java_distribution
) {
class { 'java':
  version      => $java_version,
  distribution =>  $java_distribution,
  }
}
```

In *Chapter 9*, which deals with Puppet and data, you'll see how Hiera can model the data that overrides the profile defaults. A server should only have one role; if it needs two roles, then it is in itself a new role, but a role will have multiple profiles. *Figure 8.1* shows a simple example of using roles, profiles, and modules and how the classes would include one another. With this setup, as we'll see later in *Chapter 11*, classifying a box is as simple as ensuring the right role is assigned to the node:

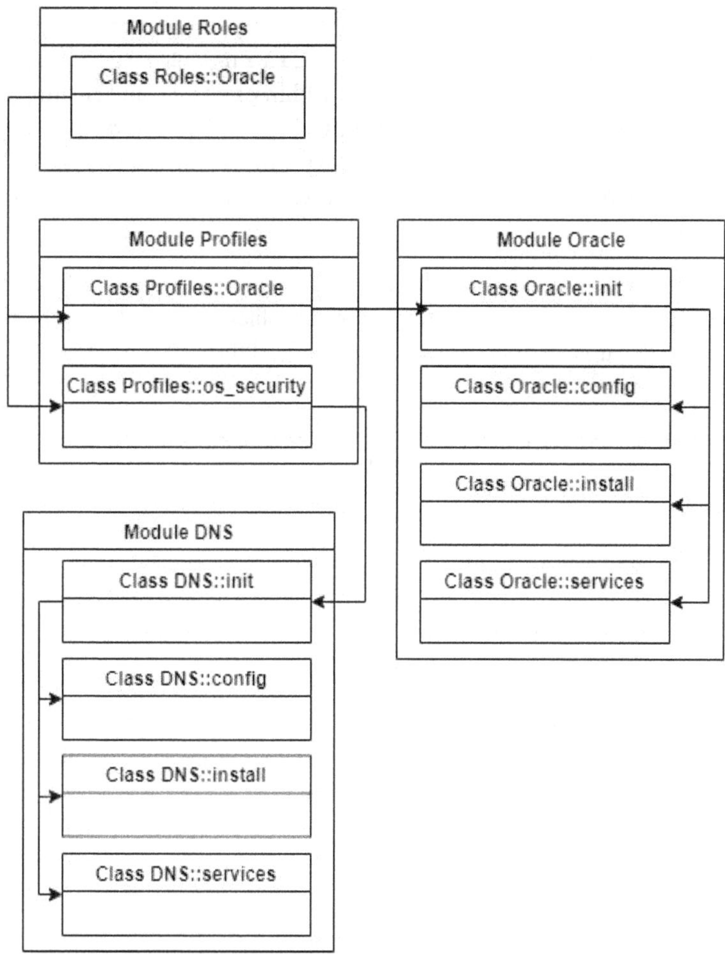

Figure 8.1 – An example of roles, profiles, and modules

The framework shown in the preceding figure is all about abstraction, so we decouple the business logic, implementation, and resource management and reduce the node-level complexity.

This pattern is not a requirement but provides suggestions on how you can structure code to avoid duplication and provide a model. In this scenario, several adaptations could be considered.

Using complexity escalation allows us to not create too much structure when we have little code initially. If there are only a couple of resources in a profile, then it may be easier to keep it that way and expand to a module when it becomes more complicated.

Depending on your organization's change management and delivery requirements, it may make sense to have multiple profile or role modules to allow for more granular control and access – for example, `teama_profiles` and `teamb_profiles`.

As discussed in *Chapter 3*, it is generally advised against using inheritance in Puppet code, but it may be worthwhile expanding the namespaces of a `profile` module by grouping manifests in a directory, such as creating an `exampleapp` directory within `profile` and creating `client.pp` and `server.pp` to represent the server and client versions (`profile::exampleapp::server` and `profile::exampleapp::client`, respectively). This could also be done for specific OSs. Before considering this approach, note that this structure is an edge case and carries a lot of risks when using inheritance.

If profiles are found to be too rigid to plan for in a changing stack or if adopting heritage servers means that it becomes necessary to drop certain parts of profiles or roles, then using parameters to make the profile more dynamic can allow classes to be defined through the parameter of the profile class, either by Hiera or default values.

As a simple example, the following code uses `include_classes` with the default values listed in the `exampleapp` module:

```
class profile::exampleapp(
  Array[String] $include_classes = ['exampleapp'],
) {
  include $include_classes
}
```

This would allow us to override the `include_classes` array from Hiera or the module data in the profile. This could be made tighter for inclusions by us only allowing classes from a set module:

```
class profile::exampleapp(
  Array[String] $include_classes = ['server'],
 ) {
$modules = $include_classes.map | String $module | {
  "exampleteam_exampleapp::${profile}"
}
include $modules
}
```

To add more structure to the parameters and make it clear in approval and code review processes, the class parameters could be broken out further. Here, we can add default, mandatory, additional, and knockout arrays, thus providing full flexibility:

```
class profile::exampleapp(
  Array[String] $include_default          = ['my_default'],
  Array[String] $include_mandatory        = ['my_base_profile'],
  Array[String] $include_additional       = ['my_test_default_
profile'],
  Array[String] $include_removal          = ['my_default'],
){
```

```
$profiles = $include_default + $include_mandatory + $include_
additional + $include_removal
include $profiles
}
```

It is possible to mix this pattern even further by, say, limiting multiple namespaces and having lists of class arrays for each namespace. This will be dictated by what approach will give your organization enough flexibility while making it clear what will be affected by the code and who should review it.

With this method, it may also be useful to define a `noop` flag using a parameter and then noop on resources. You could also do this via the `noop` function from `https://forge.puppet.com/modules/trlinkin/noop` to allow modules to be added and put into noop mode until they are accepted.

These adjustments to the patterns are more complex and involve reading the Hiera data to understand what a role and profile represent, but it will be up to your organization to decide which approach will work best. While a reduction in variation with rigid roles and profiles is ideal, this can lead to adoption resistance or issues with heritage if there are no appropriate ways to manage it.

Having reviewed the structure of modules that the roles and profiles pattern can create and the contents of such modules, we can see that this is a lot of content to manage manually by creating files and managing various testing tools. The next section looks at how to automate the life cycle of module creation and testing using the PDK.

Writing and testing a module using the PDK

The PDK was introduced to ease the effort required to consistently create the directories and files for modules and to also group some commonly used testing and validation tooling. We will be reviewing PDK version 2.7.1, the latest available at the time of writing. PDK installs its own Ruby gems and environment to provide the following tools:

Ruby Gem Name	Ruby Gem Purpose	Project Page
`metadata-json-lint`	Validates the syntax and lints `metadata.json` to style guidelines	`https://github.com/voxpupuli/metadata-json-lint`
`pdk`	Generates modules and module content with automated testing commands	`https://github.com/puppetlabs/pdk`
`puppet-lint`	Lints Puppet manifest code against the Puppet language style guide	`https://github.com/puppetlabs/puppet-lint`

`puppet-syntax`	Checks Puppet manifests, templates, and Hiera YAML for correct syntax	`https://github.com/voxpupuli/puppet-syntax`
`puppetlabs_spec_helper`	Provides the tooling necessary to test against different versions of Puppet	`https://github.com/puppetlabs/puppetlabs_spec_helper`
`rspec-puppet`	Compiles Puppet code and tests the expected behavior using a Puppet-specific implementation of Ruby RSpec	`https://github.com/puppetlabs/rspec-puppet`
`Rspec-puppet-facts`	Gives a method that provides facts for supported OSs using the output of `facterdb`	`https://github.com/voxpupuli/rspec-puppet-facts`
`facterdb`	Provides example output of facts for various OSs on different Facter versions	`https://github.com/voxpupuli/facterdb`

Table 8.1 – PDK gem list

The common misconception about the PDK is that it is packaging and installing these tools. What it is actually doing is running a `bundle install` in each module that's created. After, the PDK cache is saved, making it appear like PDK is packaging the tools.

Using the gems discussed in *Table 8.1*, the PDK can generate the following:

- Modules with complete module skeletons, metadata, and README templates
- Classes, defined types, tasks, custom facts and functions, and Ruby providers
- Unit test templates for classes and defined types

The PDK performs linting to check styles and best practices and to run syntax validation against the following:

- The `metadata.json` file; see *Table 8.2* for details
- Puppet manifest files (`.pp`) against specific Puppet versions
- Ruby files (`.rb`) against specific Puppet versions of Ruby
- EPP and ERB template files
- `Puppetfile` and `environment.conf`, which provides the module list for an environment and its environment settings, as will be discussed in *Chapter 11*
- YAML files

The PDK runs RSpec unit tests on modules and classes. This will be discussed in detail in the *Testing with RSpec using the PDK* section.

The PDK has build and release commands to make a `.tar` file for uploading to Puppet Forge and the Puppet debugging console.

To create a module, the `pdk new module` command (optionally with the module's name at the end) is run. Answer the questions regarding the module's name (if it wasn't provided, specify your Puppet Forge username, if you have one), who wrote the module, the license the code should fall under, and the OSs that will be supported. This process can be seen in the following screenshot:

```
> pdk new module
pdk (INFO): Creating new module:

We need to create the metadata.json file for this module, so we're going to ask you 5 questions.
If the question is not applicable to this module, accept the default option shown after each question. You can modify

[Q 1/5] If you have a name for your module, add it here.
This is the name that will be associated with your module, it should be relevant to the modules content.
--> packtexample

[Q 2/5] If you have a Puppet Forge username, add it here.
We can use this to upload your module to the Forge when it's complete.
--> davidsandilands

[Q 3/5] Who wrote this module?
This is used to credit the module's author.
--> davisandilands

[Q 4/5] What license does this module code fall under?
This should be an identifier from https://spdx.org/licenses/. Common values are "Apache-2.0", "MIT", or "proprietary"
--> Apache-2.0

[Q 5/5] What operating systems does this module support?
Use the up and down keys to move between the choices, space to select and enter to continue.
--> RedHat based Linux, Fedora, Windows

Metadata will be generated based on this information, continue? Yes
pdk (INFO): Using the default template-url and template-ref.
pdk (INFO): Module 'packtexample' generated at path 'C:/Users/David Sandilands/code/packtexample'.
pdk (INFO): In your module directory, add classes with the 'pdk new class' command.
```

Figure 8.2 – pdk new module questions

The answers provided for user details and licenses will be offered as defaults in future runs and can be checked by running `pdk get config` and reviewing the `user.module_defaults` settings.

> **Note**
>
> The `puppet module generate` command, which was used before the introduction of the PDK, was deprecated in Puppet 5 and removed in Puppet 6.

Once the answers have been entered and confirmed, a directory containing the module names will be created. It will contain the following content directories, which were previously discussed in the *What is a module and what is in it?* section:

- `data`
- `examples`

- `files`
- `Manifests`
- `spec`
- `tasks`
- `templates`

Using the default built-in template, it then creates the following additional configuration files and directories:

File/Directory Name	File/Directory Use
`appveyor.yml`	Appveyor CI integration configuration file
`CHANGELOG.md`	A change log that can be maintained
`.devcontainer`	How a container should be configured to test this module
`.fixtures.yml`	Test module dependencies configuration
`Gemfile`	Ruby gem dependencies
`Gemfile.lock`	Ruby gem dependencies
`.gitattributes`	Associates attributes and behaviors with file types
`.gitignore`	Files Git should ignore
`.gitlab-ci.yml`	Example configuration for using with GitLab CI
`metadata.json`	• Metadata, including questions filled out during creation
`.pdkignore`	• Files to ignore when building a package for Puppet Forge
`.puppet-lint.rc`	Configuration for the `puppet-lint` gem
`Rakefile`	Ruby task configuration
`README.md`	A template for a README page for the module
`.rspec`	Configuration defaults for `rspec` for unit testing
`.rubocop.yml`	Settings for Ruby style checking
`/spec`	A directory containing files for `rspec` unit testing
`/spec/default_facts.yaml`	Default facts available for all tests

/spec/spec_helper.rb	Entry point script for rspec that sets various configurations
.sync.yml	A file to customize the PDK template in use
.vscode	Configuration for VSCode, such as recommended extensions
.yardopts	Configuration file for Puppet Strings

Table 8.2 – PDK default template files and directories

For a pre-existing module, it is also possible to run `pdk convert` to adapt the module to the template. It will confirm the changes it would make before applying them.

The size of the contents of PDK has grown over time and the default template can contain a lot of unused files. It is possible to create a custom template simply by forking from `https://github.com/puppetlabs/pdk-templates` and following the README file to adjust the template as required. It can then be used in a module by using `--template-url` on the new module or convert commands. Alternatively, the `.sync.yml` files can be set to be deleted, unmanaged, or have settings changed. The following `.sync.yml` file example would set the `.gitlab-ci.yml` file so that it's not in the module. It would ensure the `.vscode` directory is not managed by PDK templates, avoiding any future updates. It would also disable legacy facts (global variables of facts, which were covered in *Chapter 5*):

```
common
  disable_legacy_facts: true
.gitlab-ci.yml
  delete: true
.vscode
  Unmanaged: true
```

The full settings that can be adjusted are documented in the PDK template README file: `https://github.com/puppetlabs/pdk-templates/blob/main/README.md`.

> **Note**
>
> If a configuration change is required across several existing modules, the `modulesync` module can be useful for managing this. It is available on the following web page: `https://github.com/voxpupuli/modulesync`.

Now that the contents of a PDK module and its tooling have been described in detail, we will describe the workflow of developing a module, as shown in *Figure 8.3*.

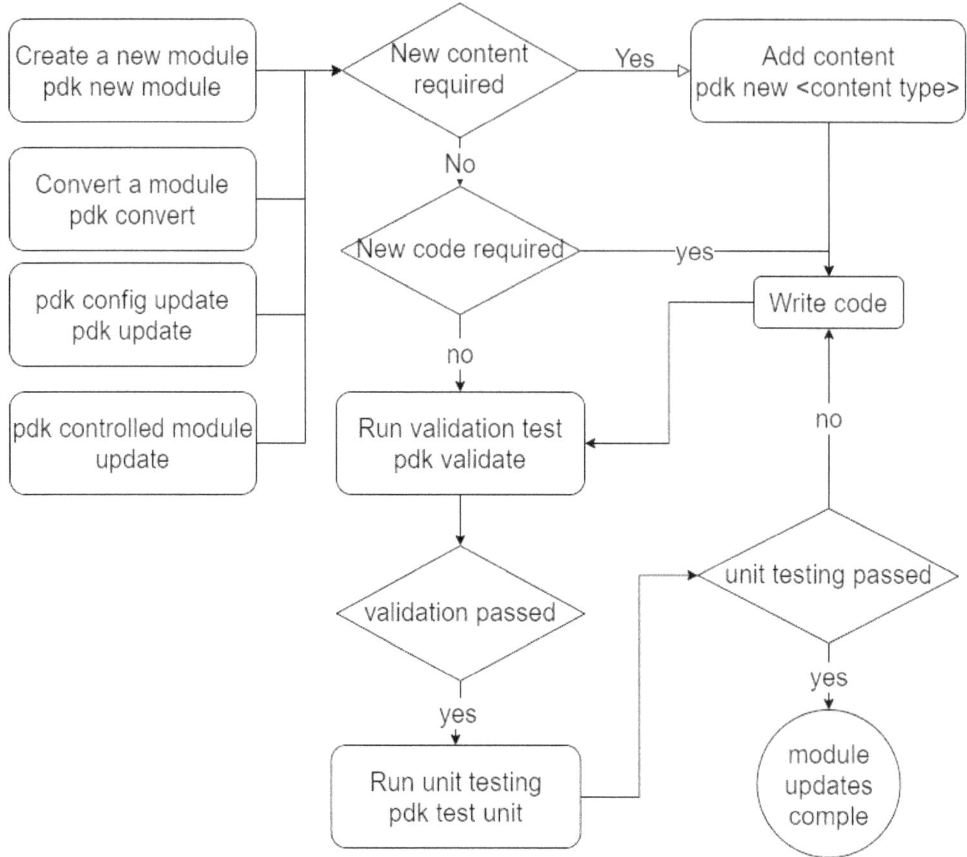

Figure 8.3 – The PDK workflow

As was discussed, by running `pdk new` to create a new module or `pdk convert` on an uncontrolled module, an initial module of PDK content and its settings is established. Using `pdk update`, the configuration of a module can be updated since we can change its settings, provide a new template, or change the PDK version.

The next step is to add any new content files that are required. This may include a `class`, `defined_type`, fact, function provider, task, or transport and can be done by using the `pdk new` command and the relevant content. This will create a file by using the template for the chosen content and a `rspec` test file. For content that's not available through PDK, such as external facts or plans, the files and tests must be created manually.

Once the file and test for the content are in place, your code should be added. At regular intervals, the code can be validated and tested via the `pdk validate` command, which checks the linting and syntax parsing. This command can also be used with the `-a` flag, which will attempt to auto-correct any errors. For linting errors, specific checks can be ignored in a section of a file by using inline comments or by surrounding areas of code with comments via lint:ignore:<rule name>. The following example shows how a line could be set to ignore the 140-character lint rule. In this case, the section of code would ignore the check for double quotes, which should only be used where both strings and variables are used in variable assignment:

```
$long_variable_text = "Pretend this is more than 140 characters" #
lint:ignore:140chars
  # lint:ignore:double_quoted_strings
  $variable1 = "don't do this"
  $variable2 = "this is just a simple example"
  # lint:endignore
```

If the check must be ignored in all code, the .puppet-lint.rc file could be updated by adding a flag such as --no-selector_inside_resource-check to ensure puppet-lint does not run the check to ensure selector code is inside a resource. The full list of puppet-lint checks can be viewed at https://github.com/puppetlabs/puppet-lint/tree/gh-pages/checks. Note that disabling checks should be avoided as much as possible as this may affect your module scoring or ability to get a module endorsed in Puppet Forge. This drives your code further away from recommended Puppet practices.

> **Note**
>
> http://puppet-lint.com/checks/ is not owned by Puppet and is out of date. Puppet forked the module to https://github.com/puppetlabs/puppet-lint, which is why it is recommended that https://puppetlabs.github.io/puppet-lint is used instead.

Once pdk validate is running cleanly, the pdk test unit command can be run to perform unit testing. The checks provided by the templates are basic and are meant to check if the code works at all; in the case of Puppet code, it ensures the code will compile. A powerful feature is that checks can be run against specific Puppet or Puppet Enterprise versions by using the --puppet-version flag or --pe-version – for example, pdk test unit -pe-version=2019.11 – so that code testing can take place before upgrades. In the next section, you will learn how to expand the rspec checks further.

Puppet Forge will be discussed in detail at the end of this chapter. Releasing to Puppet Forge is beyond the scope of this book but if the code is to be released for use in Puppet Forge, you can run the pdk build command to create a .tar file to be uploaded or the pdk release command to automate the process of uploading modules to Puppet Forge.

It is important to keep metadata.json up to date as it will restrict what testing takes place based on Puppet's supported version and is a key part of the documentation. The format and its options can be viewed at https://docs.puppet.com/puppet/latest/modules_metadata.html.

To view all the options available, you can review the full PDK command reference at `https://puppet.com/docs/pdk/2.x/pdk_reference.html`.

Having reviewed how to use the PDK to create and manage modules, as well as its validation and testing capabilities, let's learn how to perform full unit testing using RSpec.

Testing with RSpec using the PDK

To take this initial validation and compilation testing further at the unit test level, RSpec can be used to test the behavior and logic of modules, while ServerSpec can be used to test at the system integration level.

RSpec is a Ruby framework for testing Ruby code and is written in a **domain-specific language** (**DSL**) to make it easier to read. The `rspec-puppet` test is an implementation of RSpec, specifically designed for testing Puppet modules.

> **Note**
> It is important to note that the current project code is available at `https://github.com/puppetlabs/rspec-puppet`, forked from `https://github.com/rodjek/rspec-puppet`, with core guides and documentation available at `http://rspec-puppet.com/`.

When users start using RSpec, some may feel that it is just mimicking the Puppet code but in a different language. RSpec runs through the different logic and behavior of your Puppet code and ensures that the correct catalogs and output will be produced in various environments and cases. This protects against regressions when refactoring or upgrading to new Puppet releases. If the RSpec code is becoming a simple mimic of the code in the manifest, then the test scenarios are not being reviewed properly.

The advantage of this style of unit testing is that it allows you to test code without having to spin up any specific infrastructure or make any changes.

RSpec tests are contained in Ruby files in a module inside the `spec` directory, with directories containing the tests for different types of code, such as classes in the `spec/classes` directory, and defined types in the `spec/defines` directory.

We are ignoring the other possible test directories (the `types`, `type_alias`, and `functions` test directories) as creating them is beyond the scope of this book. However, most of what will be discussed here can apply to these types.

RSpec configuration is covered within the PDK and the files are created automatically with the `pdk new` commands. However, they can be added either when converting a module or using the PDK by adding the `--add-tests` flag to the `convert` command, `pdk convert --add-tests`, and with the `pdk new test --unit <name>` command, respectively.

Before we look at what a defined type and class will get by default from the PDK, we must run the `pdk new class exampleapp` and `pdk new define example_define` commands on the `exampleapp` module to create the main manifest and a defined type called `example_define`. This will result in a file called `spec/classes/exampleapp.rb` with the following contents:

```
require 'spec_helper'

describe 'exampleapp' do
  on_supported_os.each do |os, os_facts|
    context "on #{os}" do
      let(:facts) { os_facts }

      it { is_expected.to compile }
    end
  end
end
```

Further, `spec/defined/example_define.rb` can be created as follows:

```
require 'spec_helper'

describe 'exampleapp::example_define' do
  let(:title) { 'namevar' }
  let(:params) do
    {}
  end

  on_supported_os.each do |os, os_facts|
    context "on #{os}" do
      let(:facts) { os_facts }

      it { is_expected.to compile }
    end
  end
end
```

Breaking this down, the first step is to `require spec_helper`, which results in the `spec/spec_helper.rb` file being loaded. Because the `spec` directory is loaded into the path automatically, it only needs to state the title; this configures RSpec, which will be discussed in more detail later in this section. The next part, `describe`, is an RSpec keyword that's used to describe a group of tests. For both the `exampleapp` and `example_define` tests, the name of the class and defined class are described since there is only one basic group of tests for each.

> **Note**
>
> If you have used `puppet-rspec` previously, you may have set an additional type definition in the `describe` statement, such as `describe 'exampleapp', :type => :class do`. This is unnecessary due to the folders acting as auto identifiers of the type.

A defined type always needs a title and any parameters. Upon using the `let` keyword, a title is set, as well as parameters, which in this case are blank.

Both the `exampleapp` and `example_define` classes then perform a loop using the `on_supported_os` function, which is provided by the `rspec-puppet-facts` gem, taking the input from the `metadata.json` file, which contains details regarding the OSs that are supported and producing an array of facts in the `os_facts` variable. This is then passed to another `let`, which assigns these facts to the contents of the `os_facts` array.

The `it` keyword is known as an **example** in `rspec` terms and can be either a single line or encased within a `do` end block. This is a test case and contains an expect statement called `is_expected.to`, which is a verification step of a condition. This condition is expressed via a matcher. In this case, this will compile the Puppet code of the class and defined type and confirm that a catalog will be generated successfully.

> **Note**
>
> We recommend the styling guide available at `https://www.betterspecs.org/`, which is for the general Ruby RSpec style. We will be quoting recommendations from it throughout this chapter.

Having briefly examined the default compile test, let's look at each component and how to expand them further.

The describe and context keywords

One of the big confusions for many Puppet developers who have previously tried to use RSpec is understanding where to use the `describe` keyword and where to use `context`. They seem to be interchangeable, and this is for a very good reason. The `context` keyword is an alias of `describe`, so they are interchangeable, and your use only affects how your code reads.

Betterspecs recommends using `describe` to describe the method being tested. In terms of Puppet RSpec, this was why we saw `describe` with a class name of `exampleapp` and `exampleapp::example_define` as its defined class in the *Testing with RSpec using the PDK* section.

It is recommended that `context` be written in a style of *when*, *with*, or *without* situations, which should make it clear what scenario is being tested.

The style recommendation of this book is to write a single `describe` to match the Puppet type, such as `class`, and then `context` to match the scenarios to be tested.

The blocks of `describe` and `context` allow the situation being tested to be described and for us to set facts, variables, and parameters. Since they can be nested, it allows inheritance to take place, which will build up more detailed scenarios, or different logical routes to be tried, though care should be taken not to make the cases too hard to read.

The aim should be to test all cases. So, a plan should be made to test valid, edge, and invalid cases, allowing both positive and negative cases to be tested. As a simple example without any code tests or parameters set, the following code for the `exampleapp` class would look at the contexts for each supported OS, based on whether the `install` version is a middle value, a low edge version, or an invalid version:

```
describe 'exampleapp' do
  on_supported_os.each do |os, _os_facts|
    context "on #{os}" do
      context 'When install_version is 6' do
        it { is_expected.to compile }
      end
      context 'When install_version is 1' do
        it { is_expected.to compile.and_raise_error('unsupported
version') }
      end
      context 'When install_version is invalid string' do
        it { is_expected.to compile.and_raise_error('Invalid version
string') }
      end
    end
  end
end
```

Now that we have the basic structure of the scenarios to test, the next step is to use matchers to test what is produced in the catalog based on `context`.

Examples, expectations, and matchers

The sample `it` statements can be either a single line, as demonstrated in the *Testing with RSpec using the PDK* section, or can be broken up over multiple lines when the matcher that's used is too long to be on a single line. Using do and end, the same compile example could be expressed as follows:

```
it do
is_expected.to compile
end
```

In the general Ruby RSpec implementation, expectations have a broader choice, but in puppet-rspec, our expectations will be limited to just using the is_expected keyword. However, this can be negated by using not_to – for example, It { is_expected.not_to.

The matchers provide a variety of tests for testing various resource types. The matcher syntax is contain_<resource_type>('<title>').<options>.

For the compile matcher, we could be more explicit by adding the with_all_deps option to the compile – for example, it { is_expected.to compile.with_all_deps }. This would test that all the relationships in the catalog contain resources. Alternatively, we could look for a compile error with the and_raise_error('error_message') option, which will contain the message we expect to be thrown as a string – for example, it { is_expected.to compile. raise_error('lets cause failure' }.

The main set of matches is based on resource types using a pattern of contain_<resource_type>('<resource_title>') – for example,

it { is_expected.to contain_class('exampleclass::install') } and it { is_expected.to contain_service('httpd') }.

Rspec-puppet does not do class name parsing or lookup, so the matcher will only accept qualified classes without leading colons. So, install won't be found in exampleclass, but exampleclass::install will. If a resource type contains a :: symbol, this needs to be converted into a __ symbol, which will make it contain_exampleapp__exampletype.

Resource matchers can be further expanded by chaining them using the with, only_with, and without methods. This allows us to check the parameters of resources; with ensures the resource in the catalog has the parameters as specified, only_with ensures that only the parameters provided have been set and no others, and without accepts an array of parameters and ensures those parameters are not set. When using these methods, it is more readable to use an it do...end format, as shown in the following example:

```
it do
  is_expected.to contain_package('httpd').with(
  'ensure' => 'latest',
  'provider' => 'solaris',
  )
end
```

This can be shortened to only one parameter by following the method syntax for with and only_with:

```
<with_method>_<parameter name>
it {is_expected.to contain_server('exampleserver').only_with_
enable(true)  }
```

For without, the method accepts an array of parameters that should not be set on the resource:

```
it {is_expected.to contain_user('exampleuser').
.without(['managehome', 'home']) }
```

These methods can be chained together either as the same methods or as a mix:

```
it {is_expected.to contain_server('exampleserver').with_enable(true).
without_ensure }
```

A different kind of matcher for resources is using a count, which allows the have_<resource_type>_count syntax. For example, to verify if the total number of resources is 5 and the total number of classes is 4, the following code can be run:

```
it { is_expected.to have_resource_count(5) }
it { is_expected.to have_class_count(4) }
```

Having reviewed how to set examples, it is clear that for the describe and context keywords, parameters and pre-conditions will need to be set for there to be a testing scenario. For example, if the context is that the install version is 1, then the parameter install version will need to be set to 1.

Parameters and preconditions

In the default example for defined types, we explained how the let keyword can be used to set specify the title and empty parameters of a test instantiation of a defined type. However, these can also be used for other types, such as parameterized classes.

To populate parameters, an array of keys and values separated by a => symbol can be supplied in strings with an undefined value declared as :undef, which is translated to undef when it compiles the test. For example, to set param1 to the yup string and param2 to undef, the following let could be used:

```
let(:params) { {'param1' => 'yup', 'param2' => :undef } }
```

In addition to parameters, preconditions can also be set. So, if the manifest being tested is dependent on another class or variable being in the catalog, this could be added so it will be evaluated before the test class. For example, in the module pattern, we showed that the config class needed to be evaluated after the install class in the catalog but before the service class. This could be done using the following code:

```
let(:pre_condition) { 'include exampleapp::install' }
let(:post_condition){ 'include exampleapp::service' }
```

An array of strings can also be used if there are multiple conditions. If the test is for a specific node or environment, this can be set as follows:

```
let(:node) { puppet.packtpub.com' }
let(:environment) { 'production' }
```

The node should be a **fully qualified domain name** (**FQDN**).

Relationships

The relationships of resources can be tested with the that_requires, that_comes_before, that_notifies, and that_subscribes_to methods. It is not important if the Puppet code is using a require and RSpec is using that_comes_before or if the Puppet code is using directional arrows, so long as the variants are logically equivalent to each other since the test is on the catalog.

These methods are chained into the example with the requirement, but there are some differences between how relationships are declared in a Puppet manifest and how they are declared in a rspec test: the name should not be quoted, it cannot have multiple resource names under a single type, and if a class is referenced, there should be no leading :: to mark it as the top scope. As a simple example, a file called exampleconfig that requires the exampleapp package can be checked as follows:

```
it { is_expected.to contain_file('exampleconfig').that_
requires('Package[exampleapp]') }
```

To check that the exampleapp package was before the exampleapp::service and exampleapp::config classes, an array can be passed. However, note that they cannot be under one class:

```
it { is_expected.to contain_package('exampleapp').that_comes_
before('Class[exampleapp::service]','Class[exampleapp::config]') }
```

An example of a resource with parameters using it do...end that notified two files is as follows

```
it do
  is_expected.to contain_service('anotherapp').with(
    'ensure' => 'running',
    'enable' => 'true',
  ).that_notifies('File[config_a]', 'File[config_b]')
end
```

If the test is on something like a defined class that has require or before as part of its definition, this relationship can be set in parameters. However, the ref helper must be used to name the resource it is dependent on, using the ref('<type>','<title'>) syntax. For a defined type that requires the exampleapp package, the following code would add the relationship via parameters:

```
let(:params) { 'require' => ref('Package', 'exampleapp') }
```

Data from Hiera and facts

Data from Hiera and facts have a huge influence on the logic in our code, so it must be able to be supplied and customized to cover the different scenarios to be tested. As was shown in the default examples in the *Testing with RSpec using the PDK* section, the `rspec-puppet-facts` gem checks the `metadata.json` file to find the list of supported OSs. However, `metadata.json` does not have a way to provide architectures, and by default, `rspec-puppet-facts` chooses a default architecture depending on the OS, such as i86PC for Solaris or x86_64 for Fedora. If you want to be able to check additional architectures, you can pass hardware models in a comma-separated array. This will be combined with the following code:

```
  additional_archs = {
    :hardwaremodels => ['i386'],
  }
on_supported_os(additional_archs).each do |os, os_facts|
```

If it only makes sense to test a subsection such as a class that has been specifically made for an OS, then you can pass the relevant details using the `operatingsystem` and `operatingsystemreleases` parameters; this will override `metadata.json`:

```
    ubuntu = {
      supported_os: [
        {
          'operatingsystem'          => 'Ubuntu',
          'operatingsystemrelease' => ['18.04', '16.04'],
        },
      ],
    }

on_supported_os(ubuntu).each do |os, os_facts|
```

Using the `on_supported_os` method, this can only be set on all choices. If nothing is found, such as i386 on Windows 11, it fails to find it silently. View the `facterdb` module at `https://github.com/voxpupuli/facterdb` to see what is available.

It is not mandatory to use `on_supported_os` but without it, by default, there will be no facts. When you need to test data that doesn't exist in `facterdb`, it is possible to declare the facts using `let(:facts)` and the values you wish. For example, if you were testing what would happen with a theoretical RedHat 10 fact set, you would use the following code:

```
Context "when OS is redhat-10-x86_64" do
    let(:facts) do
      {
        :osfamily                  => 'RedHat',
        :operatingsystem           => 'RedHat',
```

```
        :operatingsystemmajrelease => '10',
        ...
    }
  end
```

Similarly, if additional facts were to be added to the `os_facts` variable in a nested `context`, the `merge` method could be used with the `super` method:

```
let(:facts) do
  super().merge({
    :student => 'david',
  })
end
```

> **Note**
> For structured facts, these merges can become more difficult. Voxpupli has an `override_facts` helper in `https://github.com/voxpupuli/voxpupuli-test` that can assist with this.

To add facts that can be consumed by the PDK for validation and testing the code, add a `spec/default_module_facts.yml`. This will contain YAML similar to the following:

```
---,
choco_install_path: C:\ProgramData\chocolatey
chocolateyversion: 0.9.9
```

The `default_facts.yml` file should not be edited as it is managed by the PDK and provides minimal facts for the PDK to run.

It is possible to add default facts via `.sync.yaml` either by adding a standard code block or by adding `default_facts.yml`, but this is needlessly complicated compared to `default_module_facts.yml`.

Any facts you provided with `let(:facts)` in a spec will merge on top of default facts.

In addition to these facts, three additional variables come from classification and external data sources: node parameters, which are global variables assigned from the classification to a node, trusted facts, which are variables assigned from within a Puppet client certificate, and trusted external facts, which are variables sourced from an external data source by a script. The full implementation of these will be described in detail in *Chapter 11* and *Chapter 14*.

All three types of variables can be added by using a `let` statement in the spec file or by setting them as defaults in `spec_helper`.

Trusted facts from Puppet 4.3 onwards will contain trusted fact keys (`certname`, `domain`, and `hostname`) that are populated based on the node name, as set with `:node`. However, trusted external facts and node parameters will be empty.

Trusted facts use `trusted_facts`, trusted external data uses `trusted_external_data`, and node parameters use `node_params`. For example, to declare trusted facts and trusted external data, the following `let` statements can be used:

```
let(:trusted_facts) { {'pp_role' => 'puppet/server', 'pp_cluster' =>
'A'} }
let(:trusted_external_data) do,
{
  pds: {
     puppet_classes: some_class,
     example: hiera_data,
   },
}
end
```

To set defaults, `.sync.yaml` can add additional lines by passing an array via `spec_overrides`; however, adding a `spec_helper_local.rb` file that contains the necessary lines will be easier than following the YAML syntax. Within a `Rspec.config` block, it is about following the `c.<fact_type> = {<fact/parameters_keys>}` format and using the fact/parameter name with `default_` at the beginning. So, to assign node parameters as defaults, `spec_helper_local.rb` can be updated as follows:

```
RSpec.configure do |c|
  c.default_node_params = {
    'owner'  => 'oracle',
    'site'   => 'Falkirk1',
    'state' => 'live',
  }
end
```

Similarly, trusted external data can be set like this:

```
Rspec.configure do |c|
  c.default_trusted_external_data = {
    pds: {
      puppet_classes: some_class,
      example: hiera_data,
    },
  }

end
```

Hiera will be covered in full in *Chapter 9*, but for now, it is adequate to know that Hiera provides a `hiera.yaml` file to help you learn how to look up the data and a configuration file. We have created a `hiera.yaml` definition at `spec/fixtures/hiera/hiera.yaml`, which would typically have a `datadir` defined at `spec/fixtures/hieradata`.

This configuration for Hiera can be set in two ways, as documented at `https://github.com/puppetlabs/rspec-puppet`. The first option is to use `let` and set the necessary variables, as follows:

```
let(:hiera_config) { 'spec/fixtures/hiera/hiera.yaml' }
hiera = Hiera.new(:config => 'spec/fixtures/hiera/hiera.yaml')
```

Lookups can then be performed as follows:

```
 primary_dns = hiera.lookup('primary_dns', nil, nil)
  let(:params) { 'primary_dns' => primary_dns}
```

Alternatively, the following could be added to `spec_helper_local.rb`. Here, automatic lookup of parameters would take place:

```
RSpec.configure do |c|
  c.hiera_config = 'spec/fixtures/hiera/hiera.yaml'
end
```

Having reviewed how to create tests for individual modules, one of the issues that you'll quickly find is that various resources, such as functions, are used within modules. These are dependent on the content of other modules. In this next section, you will learn how to use fixtures to make this content available for testing.

Managing dependencies with fixtures

`puppetlabs_spec_helper` can put dependent modules in `spec/fixtures/modules` for when an RSpec test unit is run. The `.fixtures.yml` file can specify `repositories:` for GitHub repository sources and `forge_modules:` for modules from Puppet Forge.

The main arguments that are taken are `repo`, which is either the Git repository link or the Puppet Forge module name, `ref` for a Git commit ID, or Forge module version number and `branch`, which is for a Git branch. The `ref` and `branch` arguments can be used together to revise a branch.

So, an example `.fixtures.yml` containing two Git repositories and two Forge modules would look like this:

```
fixtures:
  forge_modules:
    peadm: "puppetlabs/peadm"
    stdlib:
```

```
      repo: "puppetlabs/stdlib"
      ref: "2.6.0"
  repository:
     pecdm:  "git://github.com/puppetlabs/pecdm"
    Puppet-data-service:
      repo: "git://github.com/puppetlabs/puppetlabs-puppet_data_
service"
      Ref:  "feature_branch_1"
```

If there are no arguments other than `repo`, it can be shortened to one line, as shown here. If the fixtures file has changed, it is possible to run the `--clean-fixtures` flag with a `pdk test unit` command to ensure all contents are deleted.

More flags and options can be used with fixtures, as documented at `https://github.com/puppetlabs/puppetlabs_spec_helper#fixtures-examples`.

Coverage reports

It is possible to produce coverage reports by adding the following piece of code to `spec_helper_local.rb`:

```
RSpec.configure do |c|
  c.after(:suite) do
    RSpec::Puppet::Coverage.report!
  end
end
```

This checks whether Puppet resources have been covered and produces a percentage of resources covered and a list of untouched resources. The resource that's checked must be within the module being tested and not contain any dependencies brought in by fixtures. The resource coverage percentage can also be made into a pass or failure point by adding a percentage pass rate in brackets. For example, by updating the line to `RSpec::Puppet::Coverage.report! (100)`, this would ensure every resource (100%) is covered. This can sometimes be a motivator to push for RSpec use and coverage and only allow the resource percentage coverage to be reduced due to any particular issue or exception.

Further research and tools for RSpec

This section has tried to provide you with enough information that you can build meaningful `rspec-puppet` tests with facts data and dependencies. Also, note that normal Ruby code can be used, such as `case` or `if` statements and variables, and that there are many more options for advanced configurations in `spec_helper_local`, as documented at `https://rspec-puppet.com/documentation/configuration/`.

This book advises against Augeas use, but it is possible to test Augeas in RSpec. Details can be found here: `https://github.com/domcleal/rspec-puppet-augeas`.

Although it's beyond the scope of this book, when using custom functions and types, it is necessary to perform stubs and mocks, which can be done via `rspec-mocks`, as documented at `https://github.com/puppetlabs/puppetlabs_spec_helper#mock_with`.

It was mentioned at the start of the *Testing with RSpec using the PDK* section that for large manifests, having to type out all the RSpec for resources can be painful. However, several tools can do this for you. These include `https://github.com/logicminds/puppet-retrospec`, `https://github.com/enterprisemodules/puppet-catalog_rspec`, and `https://github.com/alexharv074/create_specs.git`; all of these can be used to generate RSpec from code or catalogs.

As with almost everything, it is possible to do all these tasks in YAML instead by using the `rspec-puppet-yaml` gem at `https://rubydoc.info/gems/rspec-puppet-yaml`. However, we would strongly advise against this.

For further research on RSpec, it can be useful to review the core RSpec documentation at `https://rspec.info/documentation/`.

Serverspec

Serverspec is an RSpec implementation that's designed to test at the server level once configuration management has been deployed. It is a tool that's independent of Puppet and doesn't integrate with PDK; it is typically added to a pipelining tool to run and requires you to remotely connect from a server to a test target. Many of the same principles and ideas that we saw in the RSpec apply. The documentation and a tutorial for this can be found at `https://serverspec.org/`.

Having learned all about how to create and test modules in this chapter, we can now look at how to use Puppet Forge to source pre-written modules.

Understanding Puppet Forge

Puppet Forge provides a rich resource of modules from Puppet, the Puppet community, and third-party vendors to reduce the amount of code your organization must write and maintain. It also allows you to contribute to projects or publish modules, allowing others to contribute to your projects.

It is important to understand the different types of authors, endorsements, and quality scores available in Puppet Forge to understand who is developing the modules, what you can expect from them, and how to make choices regarding the 7,000+ modules.

Anyone can register and publish modules. However, the Puppet company itself publishes under the `puppetlabs` username, while the **Vox Pupuli** community organization publishes under the `puppet` username. This confusion originates from Puppet originally being called Puppet Labs. This should not detract from the fact the Vox Pupli community develops to very high standards and works closely with Puppet, with both organizations contributing to one another. Full details about the Vox Pupuli community can be viewed at `https://voxpupuli.org/`, including how to contribute and be involved.

There are several other key consultancy contributors, such as `example42`, `enterprisemodules`, `camptocamp`, and `betadots`, who contribute modules and offer services. There are vendor organizations, such as `foreman`, `datadog`, `SIMP`, `cyberark`, and `Elastic`, that provide modules related to their products. Finally, individual contributors such as `saz` and `ghoneycut` have contributed several quality modules. Puppet has a Champions program, highlighting known contributors to Puppet, which can assist in understanding the reliability of module authors: `https://puppet-champions.github.io/profiles.html`.

> **Note**
>
> The process of releasing modules to Puppet Forge is beyond the scope of this book, but it can be reviewed at `https://puppet.com/docs/puppet/latest/modules_publishing.html` and used along with the `pdk build` and `pdk release` commands, as discussed in the *Writing and testing a module using the PDK* section.

In terms of understanding how to filter for modules to use while looking at the screen shown in *Figure 8.4*, which allows us to search for all the modules that are available in Puppet Forge, we have various options:

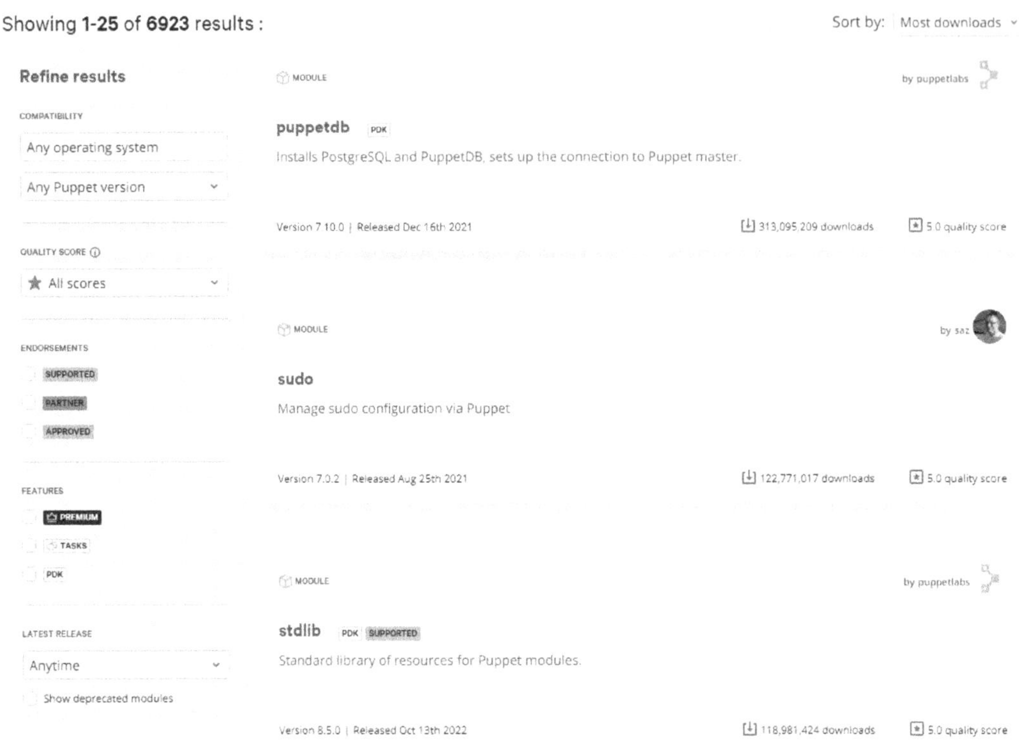

Figure 8.4 – Puppet Forge search screen

The most immediate valuable filter is **COMPATIBILITY**, which reflects the content of the `metadata.json` file for OS and Puppet version compatibility. The release date, latest release, and number of downloads can be key measures to show if this is a commonly used module and if it is being kept up to date.

Puppet implements an endorsement scheme that's managed by the **Content and Tooling Team** (**CAT**) with three different types: Partner, Approved, and Supported.

Approved modules pass specific criteria documented at `https://forge.puppet.com/about/approved/criteria`, which ensure the modules meet usability and quality standards. This can help you when you're trying to choose a reliable module or allow your team to aim for a standard and submit your modules via `https://github.com/puppetlabs/puppet-approved-modules`.

Supported modules follow the same standards as approved modules but are fully supported by Puppet or a Puppet-approved third-party vendor, allowing Puppet Enterprise customers to raise support cases if issues are experienced. Note that only the latest version of the module is supported, and Puppet Enterprise OS versions have limited windows of support beyond end-of-life dates. The full details can be viewed at `https://forge.puppet.com/about/supported`.

The third type of partner is when support and testing are provided but not by Puppet. For this support to be valid, a separate partner licensing scheme might be required.

As well as this endorsement approach, a score is put on each Puppet module. Since the mechanism for scoring was last updated, the details haven't been published in full and a breakdown of the scoring is not visible, but the module quality score is based on code style checks, compatibility tests, and metadata validation. This score gives you an idea of the module's overall adherence to Puppet code standards upon running `anubis-docker` to evaluate `https://github.com/puppetlabs/anubis-docker`.

Malware scanning was introduced in 2021 using VirtusTotal. A visible pass or failure of the module is visible on the **description** page; details are available via the **quality checks** tab. This is not intended as a replacement for internal malware-scanning processes but provides an extra level of protection. At the time of writing, this can only be on `puppetlabs` user modules, but this will be extended to Approved, Partner, and all future module releases at a later date.

Modules can be deprecated as new implementations come out or simply because the use case is no longer valid and won't be supported. These modules will be hidden by default but can be made visible by selecting the **show deprecated** option.

Premium modules were recently added with the release of the Puppet Comply product but they currently only apply to `cem_windows` and `cem_linux` modules, which can only be used upon purchasing Puppet Comply.

One area that had previously been neglected in Puppet due to its historic development focus on Linux is the Windows platform. Puppet Forge has a collection page (`https://forge.puppet.com/collections/windows`) that highlights modules designed for Windows, such as the Chocolatey package provider: `https://forge.puppet.com/modules/puppetlabs/chocolatey`. Another major development has been the auto-generation of PowerShell **Desired State Configuration** (**DSC**) Puppet modules. This automation looks at all the DSC content on PowerShell Gallery at `https://www.powershellgallery.com/` and wraps up the PowerShell code so that by including the Puppet module version of the DSC, you can download and install the code and then configure it using Puppet. These modules can be found under the DSC user at `https://forge.puppet.com/modules/dsc`; each module provides a reference to the available resource types that can be called. The modules include simple single actions such as `xInternetExplorerHomePage` for setting the home page for Internet Explorer and modules such as `xActiveDirectory`, which is used to deploy and configure Active Directory. `xInternetExplorerHomePage` is simple and has a single resource type called `dsc_xinternetexplorerhomepage` that can be used to set the default home page, like so:

```
dsc_xinternetexplorerhomepage { 'set home page':
  dsc_startpage => 'https://www.packtpub.com'
}
```

`xActiveDirectory` has various resource types to configure and deploy different aspects of Active Directory.

This has limitations since it is a fully automatic conversion and Puppet has no ownership of the DSC code. This makes testing limited and dependent on the quality of the code and documentation provided by the DSC code owner. You may also find some modules are deprecated in the PowerShell Gallery, so it is worth checking this. Also, note that due to a bug in `minitar`, only the Puppet Enterprise code manager can correctly unpack these modules from Puppet Forge directly. For open source users, refer to the module documentation instructions, which explain how to download the module from a web link to Puppet Forge and unpack the archive manually, ensuring that the module is installed and the DSC code is unpacked fully.

There are some further blogs and tools to be aware of, which are beyond the scope of this book but would be worth investigating for further information. To keep up to date with Puppet Forge and Puppet-managed modules, the CAT team runs a blog `https://puppetlabs.github.io/content-and-tooling-team/`. Puppet Forge also has an API, available at `https://forgeapi.puppet.com/`, that allows more programmatic queries to be run, and the `denmark` module, developed by Ben Ford, provides additional scans and checks to assist with reviewing modules: `https://github.com/binford2k/denmark`.

Lab – creating a module and testing it

In this lab, you will use the knowledge you've gained about module structure, the PDK, and testing to create and test a Grafana module. Then, using what you learned about Puppet Forge, you will explore the Forge site to choose modules:

- Using either the code you wrote for *Chapter 4* for the combined Grafana, Windows, and Linux class or the example answer at `https://github.com/PacktPublishing/Puppet-8-for-DevOps-Engineers/blob/main/ch04/all_grafana_data_types.pp`, create a new module called `packt_grafana`, with this Puppet code broken up into appropriate classes following the `init`, `service`, `config`, and `install` pattern (for this number of resources, a single class in the real world would be more appropriate, but this is just for practice). I recommend creating classes with `pdk new class`. Follow `https://puppet.com/docs/puppet/latest/puppet_strings_style.html` to ensure classes are fully documented and pass tests.
- Expand the default tests provided by the PDK and design the contexts to be covered while considering the parameters that could be passed and the OS choices available. Use the `https://github.com/PacktPublishing/Puppet-8-for-DevOps-Engineers/blob/main/ch08/.sync.yml` file in the module, which will include the gem file for `puppet-catalog_rspec`, and run `pdk update`. To generate some of the RSpec resources automatically, you can add `it { dump_catalog }` to each class spec file (you will need to define some parameters for this to work) and remove the line once you have got the output. Add a coverage test at 100% and ensure your tests achieve this.

- Using `pdk validate` and `pdk test unit`, correct the errors that can be found in the module, as shown here: `https://github.com/PacktPublishing/Puppet-8-for-DevOps-Engineers/tree/main/ch08/mistakemodule`.
- Go to Puppet Forge and decide which module you wish to use for the following tasks:
  - Configuring SSH on Ubuntu
  - Installing and configuring IIS
  - Configuring the time zone using DSC on a Windows machine (hint: It's not `xtimezone`; refer to `https://www.powershellgallery.com/`)
  - Install and configure Logstash

  See suggested answers at `https://github.com/PacktPublishing/Puppet-8-for-DevOps-Engineers/blob/main/ch08/module_choice.txt`

Summary

In this chapter, you learned how modules allow you to group code and data, making it easier to share and reuse code. We discussed that modules should focus on a clear single-use responsibility. We examined the directory structure of a module and highlighted where specific Puppet code and data were stored. A good starter manifest structure was shown, highlighting the main manifest (`init.pp`) that's used as an entry point, with parameters acting like public APIs to allow the module to be flexible and include the other classes required. We also saw that the `install.pp`, `config.pp`, and `service.pp` classes focused on installation, configuration, and services, respectively. In the case that the application becomes more complex than this, we discussed how a module can use classes and directories for different components.

Next, we looked at the PDK as a way to automate how modules are created and group common tooling to help us manage and test Puppet modules. We created a Ruby environment and installed the communities' most used development tools with configuration files in the module directory. The default template for producing modules was examined, as well as how to customize this by forking on `sync.yaml`.

After, we looked at the life cycle of development when using various PDK commands to create or convert a module, as well as adding different Puppet types such as classes or defined types, which create unit tests. We looked at the `pdk validate` command as a way to perform linting and syntax validation, as well as to autocorrect where possible with the `-a` flag. The templates created basic RSpec tests to check the compilation of catalogs. The PDK `build` and `release` commands were also mentioned as ways to bundle the PDK for Puppet Forge or to bundle and upload it as one command – `release`.

Next, you learned how to expand RSpec using `describe` and `context` to structure the test cases and expectations and matchers for defining individual tests. You learned that preconditions can be set via `let` statements, allowing dependencies for the class to be created in the test. You also learned how relationships can be defined by chaining the relevant function. You saw how `let` statements can be used to define facts, node data, trusted facts, and trusted external facts in data and that by using the `default_module_facts.yaml` and `spec_helper_local` files, defaults can be set for the module. After this, we covered Hiera, detailing how the configuration file can be set in a spec or via `spec_helper` and how lookups can be performed. For external dependencies, the `fixtures.yml` file was shown to be able to bring in module dependencies from Puppet Forge or local repositories to allow for catalog compilation. Coverage reporting was then added to the local spec helper, allowing unit tests to show what resources were not covered by tests and to put a pass percentage on the test. Then, we looked at some further RSpec tooling and sources, which allow you to generate RSpec code and some checks that are beyond the scope of this book. ServerSpec was then highlighted as a server-level testing framework that uses RSpec. It's independent of Puppet and beyond the scope of this book, but it's worth investing in and, ideally, adding to a pipeline.

Having shown you how to develop and structure your modules, you learned how to source modules from Puppet Forge, understand the different types of module support and endorsement available from Puppet, how scoring and scanning took place on modules, and ways to understand who contributors were and their place in the Puppet community. The Windows collection of modules was mentioned, as well as the PowerShell DSC collection, which provides automated wraparounds for modules in the PowerShell Gallery, allowing the content to be downloaded and used within Puppet code. The CAT team was mentioned as maintainers of Puppet Forge that support content with their blog publishing updates. The Denmark module was then highlighted as an additional way to score modules.

In the next chapter, you will learn how Puppet handles data and be introduced to Hiera and explore how it layers data into different scopes. We will discuss when best to use Puppet code, variables, and Hiera to store data and how to structure and feed this data to module parameters. We will also cover the correct ways to store data security at rest and in transport, as well as some common issues with using data in Puppet and how to approach them.

9
Handling Data with Puppet

This chapter will focus on how to handle data using Puppet. It will look at Hiera, Puppet's key-value data lookup tool, and how it ensures that Puppet's reusable code is made more configurable without burdening it with excessive logic and variables. The basic structure of Hiera will be reviewed, showing how it stores data in hierarchies that provide a rules-based key lookup without a lot of fuss and how it can look up keys in this data to return values using different backends, which are implementations such as YAML files of data or API calls to applications. The use of automatic parameter lookup will be discussed showing how this allows the parameterized profiles to receive data automatically and how the lookup function can be used in Puppet code directly to call data. We will briefly discuss the changes between Hiera 3 and Hiera 5 in terms of legacy Puppet. Then, the three Hiera layers will be reviewed in detail (the global, environment, and module layers), discussing how hierarchies and data should be managed in these different layers. The options available for lookup merging and priority behavior will be shown to highlight how data can either be found on the first match or by combining or merging different values found. We will then discuss when and where data should be used depending on the use case and best practice, and where the code should be kept in terms of directly in a control repo or in its own Hiera data repo. The security of data will then be discussed showing how data can be kept secure with different methods in storage, in transport, and while being used in Puppet code, highlighting the effects and limitations of using the `Sensitive` type, the `node_encrypt` module, and encryption of files via `eyaml`. Finally, some common issues and troubleshooting approaches/tools will be reviewed, showing how the `lookup` command can be optimally used to debug and explain values, and showing why we should never use global variables in hierarchies, how to avoid defaults, the dangers of using Hiera for classification, and how the **Hiera Data Manager** (**HDM**) can act as a tool to make your data more accessible.

In this chapter, we're going to cover the following main topics:

- What is Hiera?
- Hiera levels
- Deciding when to use static code or dynamic data
- Keeping data secure
- Pitfalls, gotchas, and issues

Technical requirements

Clone the control repo from `https://github.com/puppetlabs/control-repo` to your GitHub account as `controlrepo-chapter9` and update the following in this repo on the production branch:

- `Puppetfile` with `https://github.com/PacktPublishing/Puppet-8-for-DevOps-Engineers/blob/main/ch09/Puppetfile`
- `Manifests/site.pp` with `https://github.com/PacktPublishing/Puppet-8-for-DevOps-Engineers/blob/main/ch09/site.pp`
- Create a branch from production called `lab_error` and replace the following:
  - `data` with `https://github.com/PacktPublishing/Puppet-8-for-DevOps-Engineers/blob/main/ch09/data`
  - `hiera.yaml` with `https://github.com/PacktPublishing/Puppet-8-for-DevOps-Engineers/blob/main/ch09/hiera.yaml`

Build a standard cluster with two Linux clients and two Windows clients by downloading the `params.json` file from `https://github.com/PacktPublishing/Puppet-8-for-DevOps-Engineers/blob/main/ch09/params.json` and updating it with the location of your control repo and your SSH key from the control repo. Then, run the following command from your `pecdm` directory:

```
bolt --verbose plan run pecdm::provision --params @params.json
```

Let us first find out what Hiera is and why it is used.

What is Hiera?

So far, we have discussed how using Puppet creates stateful and reusable code and how, by using the roles and profiles method, parameters can be made available to make modules configurable. We also showed how to use those parameters in code, but to create a scalable, readable, and site-specific data source, Puppet uses a tool called **Hiera**. Without using Hiera data in Puppet code, it would require endless logic and variables to represent data variations required for node exceptions, location differences, OS version variations, organization differences, and many other circumstances.

Hiera is a data lookup tool that looks up values in files of JSON, HOCON, YAML, and EYAML, the built-in backends, or using custom backends that can call external sources such as websites or databases. It stores data in key-value pairs that can be looked up either explicitly via a function call in code or automatically using the automatic parameter lookup, which matches parameter names from classes to Hiera data values. As this name would suggest, Hiera is focused on using a hierarchy to find data, and the lookups follow a common default with a hierarchy of data sources that override as a more specific node match is found for the data. The hierarchies are configured in a `hiera.yaml` file; this YAML file lists out the levels in order of priority. This `hiera.yaml` file sets the version of Hiera to be used, which is required, although *5* is the only active version.

Using the built-in backends

For built-in backends in a hierarchy map, there will be a list of hierarchies, each of which will have the following:

- `name` - A readable label describing the level
- `datadir` - The base path relative to `hiera.yaml` where all data is stored
- `data_hash` - The Hiera backend/file type to use
- Either `path`, `paths`, `glob`, `globs`, or `mapped_paths` - The file path or paths to the data relative to `datadir`

A default map can also be created with these keys so that values don't need to be needlessly repeated in each layer of the hierarchy.

The `data_hash` lookup function key accepts `yaml_data`, `json_data`, and `hocon_data` as values but most Puppet implementations just use YAML data, so this book will default to the `yaml_data` backend.

The file path allows a hierarchy level to state a specific location for the data file of that hierarchy using variables interpolated in the code associated with the node, such as global variables associated with the **External Node Classifier** (**ENC**), discussed in *Chapter 11*, or facts and trusted facts. In YAML, the format is to use a percentage sign and a variable name within curly braces, `%{<variable_name}`, and to call a fact, the `facts` array is accessed using dots (`.`). So, `%{facts.application_owner}` would access the `application_owner` fact. Further dots can be used to access structured facts, such as `%{facts.os.family}` to access the `family` value within the `os` fact. Trusted facts similarly are accessed from the `trusted` array, such as `%{trusted.certname}`, and trusted external facts can be accessed using `%{trusted.external.pds.data}`.

So, a simple hierarchy could be created in the `hiera.yaml` file with the following piece of code:

```
---
version: 5
defaults:
  datadir: data
  data_hash: yaml_data
hierarchy:
  - name: "Node data"
    path: "nodes/%{trusted.certname}.yaml"
  - name: "Location"
    path: "location/%{fact.data_center}.yaml"
```

```
  - name: "Common data"
    path: "common.yaml"
```

This hierarchy would mean that a host with the `certname` trusted fact of `examplehost` and a `data_center` fact of `enterprisedc1` would first look in `data/nodes/examplehost.yaml`, then in `data/location/enterprisedc1.yaml`, and lastly, in the `/data/common.yaml` common file.

It is also possible to combine multiple variable interpolations together on a path, such as updating the location layer to differentiate on another fact – for example, assuming a `brand` fact existed and different brands within the organization would have variation for a data center, `path: "location/%{facts.brand}-%{fact.data_center}.yaml"`.

So, if `examplehost` had the `brand` fact set to `retail`, it would look in `data/location/retail-enterprisedc1.yaml`.

In these lookups, if it doesn't find a matching file for its level, it will return nothing and go to the next level. Using the `paths` path file variable instead would allow simplification. Since the only real difference between the hierarchy levels is the path, it could instead be declared with a single hierarchy and paths with an array of `paths`. For example, the hierarchy from the previous example could be reduced to one layer with `paths`:

```
hierarchy:
  - name: "YAML layers"
    paths:
    - "nodes/%{trusted.certname}.yaml"
    - "location/%{fact.data_center}.yaml"
    - "common.yaml"
```

If additional Hiera layers were required for a different backend, it would need to be understood that any hierarchy would have all its paths examined in order before moving on to the next hierarchy, which may prevent this simplification to maintain the correct order of hierarchy.

In this section, we will cover globs, only because they can be found in code bases, but they should *not* be used, as they make the data structure a lot more complicated than any environment truly needs.

The file path can use `glob` or `globs` to pass Ruby's style `Dir.glob` method. The full documentation of this can be viewed at `https://www.puppet.com/docs/puppet/latest/hiera_config_yaml_5.html#specifying_file_paths`. This allows the use of the following:

- An asterisk (`*`) as a wildcard
- Two asterisks (`**`) to match recursively through directories
- A question mark (`?`) to match any one character

- A comma-separated list within curly braces (`{this,that,or,not}`) for a literal match with any option in the list
- Sets of characters within square brackets (`[xyz]`) to match any one character in the given set
- A backslash (`\`) to escape special characters

For example, take the `facts.os.windows` fact and then match either from `display_id` (which was introduced in later versions of Windows 2019) or from `release_id` (which was introduced in Windows 2016 and deprecated in Windows 2019). This combination allows a consistent Hiera layer for a source that has changed repeatedly and needs a combination of facts to find different versions:

```
- name: "Windows Release"
  glob: "windows_release/{%{facts.windows.display_id},%{facts.windows.
release_id}}"
```

To create a layer containing network information for the network on the primary interface or the network domain, the following code could be created, which would search through any directory structure in the network folder to match:

```
- name: "Domain or Network"
    glob: "network/**/{%{facts.networking.domain},%{facts.networking.
interfaces.ethernet.bindings.0.network}}.yaml"
```

If multiple matches are found, the files will be searched in alphanumerical order. Also, multiple strings can be used in the search using `globs:` and passing an array of strings in a similar fashion to paths.

The final file path option is `mapped_paths`. This option works by providing a variable containing a collection of strings, a variable name (which maps each element of the collection of strings), and a template. For example, if a fact called `$oracle_sids` contained the `['ora1','ora2','ora3']` array, the following hierarchy would perform lookups in the `/oracle_dbs/ora1.yaml`, `/oracle_dbs/ora2.yaml`, and `/oracle_dbs/ora3.yaml` files:

```
- name: "Oracle sids"
    mapped_paths: [oracle_sid, sid, "oracle_dbs/%{sid}.yaml"]
```

Although we have taken some time to cover globs, it's important to reiterate that this should only be used to understand pre-existing complex data structures in code and for you to try and refactor and simplify. *This should not be used in new code bases.*

Having discussed the hierarchy in detail, it's now time to shift focus to the data used and how to call the lookups to the hierarchy. As was mentioned in the *Using the built-in backends* section, YAML is the most commonly used built-in data type and will be used in all examples, but the difference will only be in the presentation of the language rather than the actual structures used.

In the YAML data files, we create key-value pairs and keys with lists of values. The keys can just be single values but, more commonly, will be structured with the format <module_name>::<paramater_name>, where module_name can contain multiple segments reflecting a certain class namespace within the module.

To give an example of this, for the exampleapp profile module, a data file could contain the settings for the enable_service parameter to be true, it could contain an array of options of [opt1,opt2,opt3], and for a user's parameter, it could contain a hash of each user's settings to be created for exampleuser and anotheruser. This would look like this:

```
---
profile::exampleapp::enable_service: true
profile::exampleapp::options:
  - opt1
  - opt2
  - opt3
profile::exampleapp::users:
  exampleuser:
    uid: 101
    home: /app/exampleapp
    gig: 102
  anotheruser:
    uid: 201
```

Accessing data

The next point would be how this hierarchy and data is accessed in Puppet code and, as was mentioned at the beginning of this chapter, there are two ways Puppet looks up data in code: via the automatic class parameter lookup or via the Puppet lookup function. The recommended model involves driving virtually all data required via automatic parameter lookups to profile classes using the Role and Profile model (discussed in *Chapter 8*) and the Forge.

The automatic class parameter lookup works by taking any parameters of a class that has been included/declared as a resource and, first, checking whether the parameter has been set by the declaration, and if not, performing a Hiera lookup on each parameter of the <module_name>::< parameter_name> form. It is important to note that this is not a namespaced key itself in Hiera; it is just a string name and values can't be inserted into the data structure. In the case of using profiles and having a set profile module and an Oracle profile, this could look like profile::oracle::version. To set data for this, we might have a specific version for a server1.example.com node in a data/nodes/server1.example.com.yaml file, such as setting the version parameter for profile::oracle to Oracle 21c with the following line:

```
profile::oracle::version: 21c
```

If this lookup had failed, it would look to see whether any default value was set in the parameter in the class manifest, before then assigning it as `undef`.

The data found in Hiera by default will return as a string or an array of strings; we will show later how this can be converted.

> **Note**
>
> The automatic class parameter lookups do not work for defined resource types, only classes. To mimic the functionality, you can use an explicit `lookup()` call in your code.

The other mechanism in Puppet code is the `lookup` function. It is more direct and can be used within Puppet code; it is called with a key, which can be multiple segments, each separated by two colons (`::`), or it can be simple global values. The colons are used here simply for convention and do not drill down into a data structure. To look up the same Oracle parameter, the following example would assign it to an `oracle_version` variable:

```
$oracle_version=lookup(profile::oracle::version)
```

If the data is an array, it is possible to access a specific key using dot notation:

```
$exampleuser_id=lookup(profile::exampleapp::users.exampleuser.id)
```

It is possible to provide a default value if no value is returned by using the arguments in the function or an options hash (the full options can be viewed in the documentation at `https://www.puppet.com/docs/puppet/latest/hiera_automatic.html#puppet_lookup-arguments`) and providing a value to return – for example, to return the `no id found` string if the lookup in the previous example had returned no value, the following could be used:

```
exampleuser_id=lookup(profile::exampleapp::users.exampleuser.id,
{default_value => 'no id found'})
```

This will be discussed in more detail in the *Pitfalls, gotchas, and issues* section, but providing defaults is considered poor practice, as it hides failures and may make people assume a value has in fact been found and things are functioning correctly. It will also be noticed that the second and third arguments are marked as `undef`; these are for data type and the merge strategy, which will be discussed next.

> **Note**
>
> The `lookup` function replaces the legacy `hiera_<data_type>` and `hiera` functions that existed with Hiera 3. As these functions are deprecated, they should not be used as they can produce inconsistent results.

What has been discussed so far is the simplest case, where we expect to simply look up a value and find the first match. This is Hiera's default behavior and allows you to override values to varying degrees of specificity. Sometimes, though, you might want to return some combination of all the values present in all the layers of the hierarchy. Lookup options can be set in the data files to describe how this should happen.

The `lookup_options` reserved key allows for different merge behaviors to be set on lookups that are set against either a particular key or against a regular expression following this format:

```
lookup_options:
  <key name or regular expression>:
  merge: <MERGE OPTION>
```

The most common approach is to put this behavior in the `common.yaml` file, but if, say, a node override or some priority override may be more important, it can make sense to then put it into different levels of the hierarchy.

There are four merge behaviors that can be set with the data files:

- `first` - Return the first match in the hierarchy order
- `unique` - Return an array of all matching unique values in the hierarchy
- `hash` - Return a hash of shallow merged hash keys using the highest hierarchy key match
- `deep` - Return a hash of deep merged hash keys using the highest hierarchy key match

Hiera's default behavior, `first`, will look for the first value to match in hierarchy order. Assuming there is no other `lookup_option` value declared for the key, there is no need to implicitly declare it. But if, for example, `common.yaml` was set to `unique` and, for our node exception, we wanted to set only the values we had declared for `profile:oracle::limits`, we could set the following in our node's YAML data file:

```
lookup_options:
  profile::oracle::limits:
    merge: first
```

The `unique` keyword will find all matching keys and return a merged and flattened array. So, for example, if we wanted to install all requested Oracle versions in a profile, we could set the following:

```
lookup_options:
  profile::oracle::versions
    merge: unique
```

If the `11` value was found at the node level, `12` at the organizational level, and `11,13` found at the common hierarchy level, the returned value would be an array of `[11,12,13]`.

The `hash` keyword will merge hashes from all matching levels by merging the top-level keys of the hashes together. This essentially performs a shallow hash merge, which means that top-level keys are merged but the merge will not recursively descend and merge data structures nested underneath. This will keep the order in which the keys are written as matched from the lowest priority data source but will take the values from the highest priority source. It's easiest to think of this as it is adding the keys to a hash as it steps from highest to lowest levels. It will override and append values as it does so but it won't recursively merge the values in the keys. For example, imagine a lookup was performed on `profile::oracle::limits` and at the lowest level, `common.yaml` existed and contained the following:

```
lookup_options:
  profile::oracle::limits
    merge: hash
profile::oracle::limits:
  '*/nofile':
    soft: 2048
    hard: 8192
  'oracle/nofile':
    soft: 65536
    hard: 65536
  'oracle/nproc':
    soft: 2048
    hard: 16384
  'oracle/stack':
    soft: 10240
    hard: 32768
```

Then imagine that `/node/examplenode.server.com.yaml` had a higher priority due to the following `hiera.yaml` section:

```
hierarchy:
  - name: "Per-node data"
    path: "nodes/%{trusted.certname}.yaml"
  - name: "Common data"
    path: "common.yaml"
```

And `/node/examplenode.server.com.yaml` contained the following:

```
profile::oracle::limits:
  'oracle/nproc':
    soft: 4096
    hard: 16384
  'oracle/memlock':
    soft:  3145728
    hard:  4194304
  'oracle/stack':
    hard: 65536
```

The hash lookup on `profile::oracle::limits` would return the following:

```
profile::oracle::limits:
  '*/nofile':
    soft: 2048
    hard: 8192
  'oracle/nofile':
    soft: 65536
    hard: 65536
  'oracle/nproc':
    soft: 4096
    hard: 16384
  'oracle/stack':
    hard: 65536
  'oracle/memlock':
    soft:  3145728
    hard:  4194304
```

Note that, in this case, the `profile::oracle::limits.oracle/stack` key was taken from the highest priority so only the hard value was seen, and no recursive merge was performed. This shortcut syntax with a dot (`.`) can be used to access an element in hash or array, where, in the case of an array, the index number is used.

A `deep` merge combines any number of hashes or arrays but is able to merge values inside the hash or array recursively. This means that `hash` values are merged with another `deep` merge and arrays are not flattened and can contain nested arrays. If the previous example lookup options were configured as `deep_merge`, then the lookup would return both hard and soft limits for the `oracle/stack` key.

> **Note**
>
> Merging more than three nested levels of nesting in a hash will have a serious performance impact on Hiera and should be avoided.

There are also options that can be assigned to affect the merging of arrays. For instance, `sort_merged_arrays` will result in the merged array being sorted by key rather than the default behavior, where the array is ordered from lowest priority to highest, and `merge_hash_arrays`, where hashes within arrays will be deep merged if set to `true`. One final option allows `deep` merges to have a `knockout_prefix` key, whereby a key containing a value, normally as double dashes (`--`), is used as a prefix before the value and will cause a value to be removed instead of added.

For example, if the model given in *Chapter 8* for using flexible classes is implemented, using a `deep` merge and a `knockout` prefix would allow classes to be added or removed at each hierarchy level:

```
lookup_options:
 profile::base::extra_classes:
   merge:
     strategy: deep
     knockout_prefix: --
     sort_merged_arrays: true
```

Some example data could be for `node/example.server.com.yaml`, where the highest level of the hierarchy, `node`, contained the following code:

```
profile::base::extra_classes:
  - pci::dss
  - email
```

In contrast, `datacenter/europe.dc.1.yaml`, a lower hierarchy, contained the following:

```
profile::base::extra_classes:
  - email
  - gdpr
```

This would result in a lookup on `profile::base::extra_classes` containing `gdpr` and `pci::dss`, in that order, but not `email`.

So far, the examples have used the most common place to set `lookup_options` in `common.yaml`. But `lookup_options` performs a hash merge, which will take the highest order of each key found. So let's say, for example, that `/data/common.yaml`, the lowest level, contained the following code:

```
lookup_options:
  profile::base::extra_classes:
    merge:
      strategy: deep
      knockout_prefix: --
      sort_merged_arrays: true
```

And `/data/example.server.com.yaml`, at a higher level, contained the following:

```
lookup_options:
  profile::base::extra_classes:
    merge: first:
```

Then a lookup that matched the `profile::base::extra_classes` key in `/data/example.server.com.yaml` would use the first match lookup and not a `deep` merge.

Another lookup option is to use regex and the `convert_to` option, which converts values to something other than a string. One particularly useful example of this is when using values we wish to keep sensitive, we could simply add a regex string in the common level of the hierarchy, which would match all keys beginning with `profile` and a final key name that ended with `password` and ensure that the parameter was converted to `Sensitive`:

```
---
lookup_options:
  '^profile::.+::\w+_password$':
```

In the *Keeping data secure* section, there will be more discussion on securing data.

While it is possible to essentially override the lookup settings set in the data file in the `lookup` function itself, we would strongly recommend against this as it could be confusing to have the data saying one thing and the `lookup` function behaving otherwise. It could lead to changes in the data that have unexpected consequences for the `lookup` function. If it is required, the syntax can be found in the documentation at `https://www.puppet.com/docs/puppet/8/hiera_automatic.html#puppet_lookup`.

Interpolation is also available in Hiera data via both variables and functions. While this can be useful to avoid the repetition of data, it can also make the data vastly more complicated than we would want it to be, and in general, we would advise against it.

As with hierarchies using facts, `trusted` and `server_facts` can provide consistent variables, and the variables are interpolated in the same way, so a simple example would be to set a `config` file that uses the hostname as follows:

```
tivoli_config_file: '/opt/app/tivoli/client/%{trusted.hostname}.conf'
```

Hiera provides a limited number of special interpolation functions. They are not the same as Puppet functions. The following functions can be used to interpolate Hiera data:

- `lookup` (or `hiera`)
- `alias`
- `literal`
- `scope`

Using the same format as variables, a function can be declared as `${<function>(<arguments>)}`.

The `lookup` function allows a Hiera value to be looked up from within the data. This can be useful to prevent data having to be repeatedly entered and reduce maintenance since, if the data changed, it would only need to be changed in one place. For example, something like a repository server could vary depending on the client's location or be used repeatedly to provide the full location of packages. The following example shows how two binaries could provide their full paths using a lookup and reduce repetition:

```
profile::base::artifactoryserver: artifactory.example.com
profile::exampleapp1::binary:  %{lookup
(profile::base::artifactoryserver)}/exampleapp1.rpm
profile::anotherapp::binary:  %{lookup
(profile::base::artifactoryserver)}/anotherapp.rpm
```

This would also make maintenance much simpler; if the artifactory server was to change, only one line would need to be updated.

The `alias` function allows for data structures in Hiera data to be returned since `lookup` would only return a string. So, if the `base` profile had an `extensions` parameter that took an array of strings and we wanted to pass the same list of extensions to another profile, `exampleapp`, it would be coded like this:

```
profile::base::extensions:
   -  'option1'
   -  'option2'
   -  'option3'

profile::exampleapp::extensions: "%{alias(profile::base::extensions)}"
```

The `literal` function allows the escaping of the percentage sign (`%`) so that it does not assume it is for a variable or function to be interpolated. To do this, we use the `%{literal('%')}` function where a `%` sign is to be used. This can be useful in scenarios such as Apache configuration files or for Windows environmental variables; if, for example, we wanted to have the `%PACKAGEHOME%/External` string at `profile::nuget::`, then the following code could be used:

```
profile::nuget::
: %{literal('%')}{PACKAGEHOME} %{literal('%')}
```

The `scope` function is likely only to be used in legacy code. It really just interpolates variables and only had a use case when Puppet variables were dynamically scoped. The same Tivoli example in this section would be written as `tivoli_config_file: '/opt/app/tivoli/client/%{scope(facts.hostname)}.conf'`.

Using custom backends

In addition to the built-in backends described so far, custom backends can be written or downloaded from the Forge and configured into Hiera. It is beyond the scope of this book to write custom backends but Puppet's documentation covers how to write them at `https://www.puppet.com/docs/puppet/8/hiera_custom_backends.html#custom_backends_overview`.

Custom backends use one of three data types, selected based on their performance requirements for the type of data being accessed.

The `data_hash` backend type, as was seen for the built-in backend, is used for data sources that are cheap to read, such as files on a disk. This profile is used where the data is small, static, can be read all at once, and most of it gets used. It returns a hash of key-value pairs.

The `lookup_key` type is used for data sources that are expensive to read, such as secure HTTP API connections. This profile is used where the data is big, only part is used, and it can change during compilation. It returns a key pair. The most commonly used custom backend is `hiera-eyaml` for encrypting Hiera, which will be covered in detail in the *Keeping data secure* section.

The `data_dig` backend type is used for data sources that access arbitrary elements of a collection, such as a database. With a similar profile to `lookup_key` but accessing subkeys of elements to return a key pair, the function will dig into a dotted key.

A final data type to mention is `hiera3_backend`, which was only relevant as a stepover from legacy Puppet setups; this book will not cover this configuration, but details can be found in the Puppet documentation at `https://www.puppet.com/docs/puppet/8/hiera_config_yaml_5.html`. The Puppet documentation advises how to migrate from Hiera 3 backends if you encounter them in legacy code at `https://www.puppet.com/docs/puppet/8/hiera_migrate.html`.

> **Note**
>
> From a user perspective, Hiera version 5 is an evolution of Hiera 3, with Hiera 4 as an experimental version, but Hiera 5 was fully implemented in Puppet itself while Hiera 3 was its own standing implementation. Puppet 7 and below has a dependency on a Ruby gem for Hiera version 3 to support any legacy Hiera 3 backends where *Hiera:Backend* was extended. This dependency was removed in Puppet 8.

These data types can then be combined with the file paths, as already discussed with the built-in backends, but with the additional paths of `uri` and `uris` to allow the direction to URIs such as web sources.

The `options` parameter then allows a hash of anything required by the custom backend, such as credentials or key information, and the content will depend on the implementation.

Most modules will explain in their README file how to use the `options` parameter. For example, `https://forge.puppet.com/modules/petems/hiera_vault/` is a Hiera backend for HashiCorp's Vault; building on their example, the following code shows an example assuming that the keys would all start with `secret_`, come from a `vault.example.com` server, and have mounts for two teams (`digital` and `trade`), which used the node name, location, and `common` for their secret hierarchy:

```
hierarchy:
  - name: "Vault secrets"
    lookup_key: hiera_vault
    options:
      confine_to_keys:
        - "^secret_.*"
      ssl_verify: false
      address: https://vault.example.com:8200
      token: notreallyatoken>
      default_field: value
      mounts:
        digital:
          - %{::trusted.certname}
          - %{::trusted.extensions.pp_region}
          - common
        trade:
          - %{::trusted.certname}
          - %{::trusted.extensions.pp_region}
          - common
```

Another example is `https://forge.puppet.com/modules/tragiccode/azure_key_vault/`, allowing access to secrets in Azure, which, if we were to create a lookup based on the department assigned to the server looking for keys starting with `secret`, would look like the following:

```
- name: 'Department Azure secrets'
    lookup_key: azure_key_vault::lookup
    options:
      vault_name: "%{trusted.extensions.pp_department}"
      vault_api_version: '2023-02-04'
      metadata_api_version: '2023-02-11'
      key_replacement_token: '-'
      confine_to_keys:
        - '^secret_.*'
```

In *Chapter 13*, the **Puppet Data Service** (**PDS**) will be examined, along with a series of backends useful for extending Puppet's data access.

Now that we have reviewed how Hiera works, let us look at how it works in the different layers of Puppet.

Hiera layers

Hiera has been discussed just in the context of the levels in a single hierarchy but there are three layers of hierarchy, each of which contains its own configuration of levels. When a lookup is performed by Puppet as part of a Puppet run, it will look through each of these layers, examining the levels of hierarchy within each.

The global layer is the first layer and is configured by default in `$confdir/hiera.yaml`, usually `/etc/puppetlabs/puppet/hiera.yaml`. Hiera version 3 only works at this layer, and its existence is more just a leftover for compatibility purposes. Puppet's documentation suggests its only purpose should be for Hiera 3 compatibility and acting as a global override, but we would advise you to *not use it at all* since it exists outside of the code deployment and control processes, which will be reviewed in *Chapter 11*. This would leave control of the file localized to the Puppet server, which would only be desirable if you wanted to step around the code deployment process.

The environment layer is the next and main layer of data, and it is configured inside each environment usually at a path such as `/etc/puppetlabs/code/production/hiera.yaml`. Environments and control repos will be discussed in complete detail in *Chapter 11*, but to understand the context here, an environment is a set of Puppet modules and manifests at fixed versions for a specific group of Puppet nodes, and a control repo is a module structure used to manage the environments, containing a file called a Puppetfile detailing the sources of the modules, at which version they should be deployed, and where they should be deployed.

A choice needs to be made as to whether the `hiera.yaml` file and data will be contained in the control repo together, or whether to have a separate control repo and modules containing Hiera data. This is configured by the control repo deploying the modules typically into a data directory in the environment and ensuring that Hiera uses that data path in its `hiera.yaml` file. This separation can make sense when the control of a set of data needs to be managed by a certain team or group and containing it within the control repo would allow too much access/visibility. For example, if our `hiera.yaml` file was configured to use data as a source path, we could add Hiera data from a module into that path with an entry into a Puppetfile:

```
mod 'exampleorg_hieradata',
  :git      => 'https://<your_git_server>/exampleorg/hieradata.git',
  :install_path => 'data'
```

The final layer is the module layer, and this is configured by a `hiera.yaml` file inside each module, typically with a data folder in the module too. So, when deployed in an environment on a server, the `hiera.yaml` file would be in a location such as `/etc/puppetlabs/code/environments/production/modules/example_module/hiera.yaml`. The best use for the module layer is to set defaults for the parameters of all classes in the modules, being careful to keep them relevant to the focus of the module, and not external organizational data, which would be better placed into the environment layer. An example of setting defaults can be seen in the `puppetlabs/ntp` module, available at `https://github.com/puppetlabs/puppetlabs-ntp`, which sets defaults based on the OS version. The `hiera.yaml` file could also be configured to allow for increasing granularity of specific OS versions, going from defaults and a general OS family such as Windows to a specific full OS version, such as AlmaLinux-8.5, as shown in the following code:

```
hierarchy:
  - name: 'Full Version'
    path: '%{facts.os.name}-%{facts.os.release.full}.yaml'
  - name: 'Major Version'
    path: '%{facts.os.name}-%{facts.os.release.major}.yaml'
  - name: 'Distribution Name'
    path: '%{facts.os.name}.yaml'
  - name: 'Operating System Family'
    path: '%{facts.os.family}-family.yaml'
  - name: 'common'
    path: 'common.yaml
```

> **Note**
>
> The module layer is often seen as an alternative to the `params.pp` class, which used to be part of the module pattern and contained default values and Hiera lookup calls. It was used before the modern Hiera layers existed with automatic parameter lookups.

You can only bind data keys in the module's namespace, so in the `exampleapp` module, only `exampleapp::key` values could be set, not a global key such as `key1` or another module such as `anotherapp::key`. This can lead to another pattern option particularly useful for in-house written modules, whereby this limitation is used to allow application teams to have full control of their environmental data for modules without being able to affect other modules. This might be a consideration for the profiles modules owned by a particular team who wishes to manage expectations.

`default_hierarchy` is sometimes known as the fourth layer and is only available in the module layer; it essentially involves declaring a `default_hierarchy` key within the module hierarchy. The key difference with this layer is that it will only be called if there is no match within the other three layers, so there is no merging behavior:

```
default_hierarchy:
  - name: 'defaults'
    path: 'defaults.yaml'
    data_hash: yaml_data
```

> **Note**
>
> `default_hierarchy` produces the same behavior as the `params.pp` approach did since any match in the three Hiera layers will ignore and not merge any matching values.

Having reviewed these layers, this leads to a question of how should the hierarchies be constructed. Hierarchies can be made complicated very quickly but we should remember that the underlying approach is that they should be made to run from the most specific data for a node down to general data. They should be as short as possible since data files are easier to work with, and the more evaluations of hierarchies you create, the greater the impact on Puppet's infrastructure performance. Too many backends (particularly customized backends) will create complications and external dependencies, which can break Puppet compilation. The Roles and Profiles method should allow less data to be managed in Hiera, and if built-in facts are not enough, custom facts can be created and multiple facts can be used in a path together.

The global level lends itself to just being structured on the name of the node and data common to all nodes since it would only be used for overrides outside of Puppet code environment control.

For the environment layer, the common structure of node data to look at is as follows:

- The name of the node
- The node owner

- The node's purpose
- The location of the node
- Data common to all nodes

This could lead to a simple hierarchy such as the following:

```
- name: "Node data"
  path: "node/%{trusted.certname}.yaml"
- name: "Org data"
  path: "node/%{facts.org}.yaml"
- name: "Application-Tier"
  path: "app_tier/%{facts.app_tier}.yaml"
- name: "Datacenter"
  path: "datacenter/%{facts.datacenter}.yaml"
- name: "Common data"
  path: "common.yaml"
```

The module layer, as discussed, then just becomes a focus for defaults for values often based on facts such as OS version and platform.

> **Note**
>
> Do not use the `environment` fact itself in any hierarchy. Use the environment layer for environment-based data.

Lab – add data to a module

In this lab, download and update the Grafana module from *Chapter 8* to contain defaults in Hiera data instead of on the parameters.

To do this, let us assume the `common.yaml` file will contain all the present defaults in `init.pp`.

For Red Hat, we will have the following: `download_source = 'https://dl.grafana.com/enterprise/release/grafana-enterprise-8.4.3-1.x86_64.rpm'package_provider ='yum'`.

While for Windows, we will have the following:

```
download_source = 'https://dl.grafana.com/enterprise/release/grafana-enterprise-9.4.1.windows-amd64.msi'
package_provider = 'windows'
```

You can refer to `https://github.com/PacktPublishing/Puppet-8-for-DevOps-Engineers/tree/main/ch08/grafana`.

An example answer is at `https://github.com/PacktPublishing/Puppet-8-for-DevOps-Engineers/tree/main/ch09/grafana`.

Later, in the *Keeping data secure* section, it will be shown how the password can be properly secured and not just in plain text in the YAML file.

In this section, we have seen how to use the three layers of Hiera and how to structure the hierarchy in these layers. Now, we will look at when data should be used in Hiera and when it should just be used in code directly.

Deciding when to use static code or dynamic data

Having viewed all the possibilities of managing data structure and looking over the code examples covered in this book, it probably raises the question about when to write code and when to use data. *Figure 9.1* highlights a decision tree to follow:

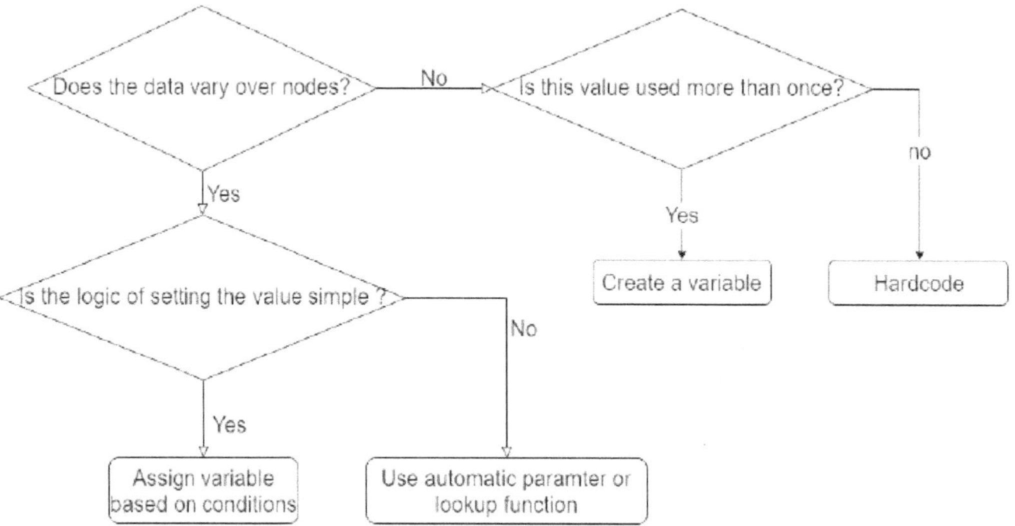

Figure 9.1 – Data or code decision tree

The first key thing is if the data doesn't vary over nodes and it's only used once, the simplest thing is to hardcode the data in Puppet code – for example, just directly setting the owner of a file as `exampleuser` in a file resource attribute.

If a value is used multiple times, then there is clearly a value in assigning a variable and using this variable where it is required. This simplifies maintenance if the value needs to be changed but does mean you have to keep track of variables when reading the code.

If, on the other hand, there is variation across nodes and overriding the value on certain conditions, the first question should be about how complex the logic is. If it is as simple as a single check, then the gain from abstracting into Hiera is not big; the issue with abstracting values into Hiera is that they are no longer clearly visible looking at the code and require translation and thought. So, if simple conditional logic can be used, it's generally better to keep the values in code.

Once logic becomes more complicated and can vary based on combinations of conditions, then we can use Hiera data and an auto parameter lookup, or if it was found to be required, the `lookup` function.

It is also best throughout this to use the simplest method available at the time and escalate through the levels of complexity as code changes and grows. Creating complex data structures and performing abstraction for the future simply creates complexity and requires more work without really gaining benefits.

Keeping data secure

One of the key elements of managing data is ensuring that secret data is kept secure, and this can be challenging with Puppet when this data must be stored, transferred across the infrastructure to the client, and used within Puppet code to set the state. In this section, we will discuss the methods available to secure data, what levels data can be secured at, and the limitations of the methods used at each level.

The most common first step is to secure data in storage. This can be achieved using `hiera-eyaml`, a custom Hiera backend available at `https://github.com/voxpupuli/hiera-eyaml`. This module creates `pkcs7` keys, which are then used to encrypt and decrypt data. Having followed the instructions in the module to create and distribute keys, a hierarchy can be created, such as the following:

```
hierarchy:
  - name: "Hiera data in yaml and eyaml files committed to the
control-repo"
    lookup_key: eyaml_lookup_key
    options:
      pkcs7_private_key: /etc/puppetlabs/puppet/eyaml/private_key.
pkcs7.pem
      pkcs7_public_key:  /etc/puppetlabs/puppet/eyaml/public_key.
pkcs7.pem
    paths:
      - "nodes/%{trusted.certname}.yaml"
      - "location/%{facts.whereami}/%{facts.group}.yaml"
      - "groups/%{facts.group}.yaml"
      - "secrets/nodes/%{trusted.certname}.eyaml"
      - "os/%{facts.os.family}.yaml"
      - "common.yaml"
```

It can simplify the hierarchy to note that the `eyaml` backend can read YAML files too, and there's no reason to separate `yaml` and `eyaml` files into different hierarchies assuming their path and options are the same, as shown in the previous example.

`hiera-eyaml` is fine for simple encryption and limited numbers of users involved with encrypting secrets, but for larger setups, using `gpg` keys with `https://github.com/voxpupuli/hiera-eyaml-gpg` becomes more practical rather than sharing signing keys amongst multiple teams. Once the configuration and key management are done, this simply varies by using `gpg_gnugpghome` options rather than `pkcs7` key options, such as the following:

```
options:
  gpg_gnupghome: /opt/puppetlabs/server/data/puppetserver/.gnupg
```

An alternative to these encrypted data file approaches is if an appropriate secure key store exists, such as HashiCorp Vault, or a cloud-native key store, such as Azure Key Vault, then using a backend that can access these services will ensure data is securely stored.

Regardless of the backend choice, this will only ensure the data is secured in storage. As was discussed in the *Accessing data* section, by default, Hiera will return a string when accessed by Puppet code. `lookup_options` can be used to convert the parameter type to `Sensitive` in Puppet 5.5 and above, and care should be taken to ensure all secure parameters are covered either via wildcards or explicit naming.

Care must be taken to use the `Sensitive` data type well; it can be easy to either mistakenly keep it secured so the value can't be used where it is needed or accidentally expose it when using the `unwrap` function.

When using `file` and `content`, for example, the following attempt to put `secret_value` into a `/etc/secure` file would be exposed on a file diff, which, as discussed in *Chapter 3*, is when a comparison of changes to files is recorded into the report logs:

```
file {'/etc/secure':
  ensure => present,
  content => ${secret_value},
}
```

This could be prevented by setting the `file_diff` parameter to `false` or setting the server not to use file diffs.

Similarly, for templates, care must be taken. If using Puppet 6.2 or greater, then templates will work directly with Sensitive values and you can simply use the Sensitive value in a template:

```
file {'/tmp/test1':
  ensure => present,
  content => (epp('example.epp', { 'password' => $secure_password })),
}
```

For versions below Puppet 6.2, you would need to unwrap the variable in the template and then mark the contents as Sensitive, as in this example:

```
content => Sensitive(epp('example.epp', { 'password' => unwrap($secure_
password)})),
```

Using `Sensitive` well keeps the data out of the logs, but unfortunately, not the catalog file itself, and if you are using PuppetDB, catalogs will be stored there too. In this case, using the `node_encrypt` module available at `https://forge.puppet.com/modules/binford2k/node_encrypt` allows for any secret to be encrypted in the catalog using the clients' keys, and using a `Deferred` function decrypts them at the time of catalog application. This keeps secrets out of the catalog and the report produced after a catalog is applied.

Assuming the instructions to configure `node_encrypt` have been followed on the infrastructure, this means the line assigning values to the `content` parameter in the previous piece of code could be updated to invoke the `node_encrypt::secret` function as follows:

```
content => (epp('example.epp', { 'password' => $secure_password })).
node_encrypt::secret,
```

> **Note**
>
> The current version of `node_encrypt` relies on `Deferred` functions, which became available in Puppet 6, so version 0.4.1 needs to be used to work on older versions, and you would use the `node_encrypt::file` type instead of the `file` type to encrypt file resources.

This section has shown how to keep data secure in storage, transport to catalog, and report processing, and some of the issues that can be experienced. In the next section, we will discuss general issues and problems when handling data in Hiera.

Lab – use eyaml to store a secret

In this lab, the `puppet-hiera_eyaml` module has been used to set up `eyaml` with default `pkcs` keys, with a global Hiera setup to look at the node name, OS, and common values. In `site.pp`, a Hiera lookup is performed to look up the value of `secret::examplefiles`, which is used as content to create a `/var/tmp/secret_example` file on the Puppet primary server. The lookup has a default of not set. In this lab, you will encrypt a secret and add it within the OS level so the content of the file changes.

SSH to the primary server and elevate to root:

```
ssh centos@<primary_host>
sudo su -
```

Run the `eyaml encrypt -p` command from within the `/etc/puppetlabs/puppet` directory and enter a secret of your choice at the prompt:

```
cd /etc/puppetlabs/puppet
eyaml encrypt -p
```

Copy the output after the string starting with `ENC[` and paste it into the data section at `/etc/puppetlabs/puppet/data/os/RedHat.eyaml` so it contains something like this:

```
---
secret::example: ENC[PKCS7,<long string of chars>]
```

Run `puppet agent -t` and observe the change in the `/var/tmp/secret_example` content to the content you set.

This was a very simple example and it should be noted, as was highlighted in the *Hiera layers* section, that you would more likely be using an environment hierarchy and keeping your data secure, as was shown in the *Keeping data secure* section, by using the Sensitive option in the lookup `options` parameter. Additionally, the public key used for `eyaml` could be copied to a desktop to encrypt secrets, if that was secure enough for your organization's policies.

Now that we have fully reviewed the Hiera configuration, we will show how we can understand issues with lookup and data.

Pitfalls, gotchas, and issues

When working in large datasets with multiple levels and layers, it can become complicated to understand why certain answers have been generated or where errors have been inserted. This section will focus on approaches to understanding and debugging data lookups and tools that can make the data more visible.

Hiera problems tend to fall into a few categories: syntax, formatting, backend communication and performance issues, hierarchy ordering mistakes, and many others.

The `puppet lookup` command is the best way to test Hiera data and is, in effect, like the `lookup` function used in Puppet code. Using this on the primary server, the basic syntax of this command is `puppet lookup <key> --node <node_name> --environment <environment_name>`.

This command will return the value, if found, or nothing. It is important to understand the effect of the various flags available to the command to return more detailed information. A common mistake is to use the `--debug` and `--explain` flags together; they shouldn't be used together as the former is focused on high levels of logging to allow you to understand why errors such as syntax, formatting, or the backend are being generated, while the latter is focused on showing how a value was reached, where Hiera looked, and what it found.

For example, an `explain` lookup on `motd::content` might look like the following:

```
puppet lookup --explain motd::content --node node-name --environment
production

 Searching for "lookup_options"
  Global Data Provider (hiera configuration version 5)
    Using configuration "/etc/puppetlabs/puppet/hiera.yaml"
    Hierarchy entry "Classifier Configuration Data"
      No such key: "lookup_options"
  Environment Data Provider (hiera configuration version 5)
    Using configuration "/etc/puppetlabs/code/environments/production/
hiera.yaml"
    Merge strategy hash
      Hierarchy entry "Yaml backend"
        Merge strategy hash
          Path "/etc/puppetlabs/code/environments/production/data/
nodes/pe-server-0-540983.05eqwrwxv1ourfszstaygpgbth.zx.internal.
cloudapp.net.yaml"
            Original path: "nodes/%{trusted.certname}.yaml"
            Path not found
          Path "/etc/puppetlabs/code/environments/production/data/
common.yaml"
            Original path: "common.yaml"
           Found key: "motd::content" value: "test"
```

Looking at the output from debug, we see far more information with regards Facter and other system work going on, as can be seen from the command and sample output as follows:

```
puppet lookup motd::content --node node-name --environment production
--debug

Debug: Facter: Managed to read hostname: pe-server-0-d6a9f5 and
domain: vhcpsckl41fedgadugqovud0sa.cwx.internal.cloudapp.net
Debug: Facter: Loading external facts
Debug: Facter: fact "domain" has resolved to:
vhcpsckl41fedgadugqovud0sa.cwx.internal.cloudapp.net
Debug: Lookup of 'motd::content'
  Searching for "lookup_options"
```

```
    Global Data Provider (hiera configuration version 5)
      Using configuration "/etc/puppetlabs/puppet/hiera.yaml"
      Hierarchy entry "Example yaml"
        Merge strategy hash
          Path "/etc/puppetlabs/puppet/data/nodes/pe-server-0-d6a9f5.
vhcpsckl41fedgadugqovud0sa.cwx.internal.cloudapp.net.eyaml"
```

Without a node being provided, the lookup will assume the lookup is for the server you are running the command from, and the environment will default to `production`.

In terms of syntax and formatting problems, one of the most common errors is when the opening `---` of the YAML file is malformed. This can happen in a couple of ways:

- A space is inadvertently added to the start of the line or a Unicode character conversion takes place, changing it to –. In this case, an error in `debug` will look like this:

  ```
  Error: Could not run: (<unknown>): mapping values are not allowed
  in this context at line 2 column 8
  ```

- If a space is inserted within the dashes, such as `-- -`, then an error in `debug` will be seen like this:

  ```
  Error: Could not run: (<unknown>): did not find expected '--'
  indicator while parsing a block collection at line 1 column 1
  ```

Another common syntax mistake is using key-value pairs without a space between the colon symbol (`:`) and the value; so `key: value` and `key : value` are valid but `key:value` is not and it will error in debugging like so:

```
Error: Could not run: (<unknown>): mapping values are not allowed
in this context at line 3 column 10
```

If tabs are used instead of spaces for indentation, then in debugging, an error will be caused such as the following:

```
Error: Could not run: (<unknown>): found character that cannot start
any token while scanning for the next token at line 4 column 1
```

For formatting, using single quotes in data with variables will result in a literal string of the variable name being returned instead of interpolation.

File permissions can also be an issue and, therefore, it is worth ensuring you are running the lookup commands as the same user, as Puppet will be typically running under the `pe-puppet` or `puppet` user.

Using `--debug`, it can be useful to see whether custom backends are the areas that experience issues, errors, or slowdowns. In general, we would recommend examining patterns such as the PDS and external data providers.

Be careful to note this will not debug the actual data but only the `hiera.yaml` file, data files that are not valid YAML will just be ignored, which can be seen using `--explain`.

In terms of hierarchy problems, this is where the `--explain` flag will prove most useful since it will step through explaining the configuration files used, the hierarchies found, the merge strategy, and the paths examined in detail so that it becomes clear how it stepped through the hierarchy and how it may not be working as expected.

Depending on what variables are being used in your hierarchy, it may be required to use the `--compile` flag since, by default, when using Puppet `lookup`, it will not perform a catalog compilation, so only the `$facts`, `$trusted`, and `$server_facts` variables will be available. We strongly advise against using arbitrary values from manifests as these can vastly overcomplicate the lookup and produce unpredictable results.

From this, it can be seen that you always want to use the `Facter` array, to avoid the risk of module variables and top-scope `Facter` variables clashing.

Some other options can be useful to test what would happen if you changed the configuration, such as changing the merge strategy with the `--merge` flag or by providing updating facts using `--facts`, for example.

The full command reference of options for lookups can be seen at `https://www.puppet.com/docs/puppet/latest/man/lookup.html`.

If updating the global Hiera file, be careful to restart the **Puppet Server Services** to ensure it is re-read.

Having touched on the point previously in the *Accessing data* section, we do not recommend using defaults on `lookup` functions. Data defaults in modules or profiles should be meaningful as well. So, providing the default config file location makes sense for a module if you expect most users just to use it, but if it is being added just to avoid failure in lookup, that can be a serious mistake and will mask issues in Hiera data or code that won't be noticed, as the code is successfully applied with defaults. The key thing to avoid is passing a default value that then requires lots of logic in Puppet code to work out how to translate it.

Classification in Hiera is possible since some users choose to look up Hiera data and include classes in the `site.pp` file. Modules such as `https://github.com/ripienaar/puppet-classifier` focus on this sort of approach. There is a balance of coding structure to consider, as can also be seen in our flexible roles and profiles approach. By putting too much data into Hiera, it can abstract away from clear coding since the data is then not directly visible in the code. So, it is best to consider whether the complexity elevation is worth it.

One of the issues of Hiera can be its structure, which makes it inaccessible to less involved users. To make Hiera data more visible, **Betadots Hiera Data Manager** (`https://forge.puppet.com/modules/betadots/hdm`) is an excellent option as it allows graphical search, updates, and deletion of Hiera data. However, in production environments, this should be limited to just viewing data.

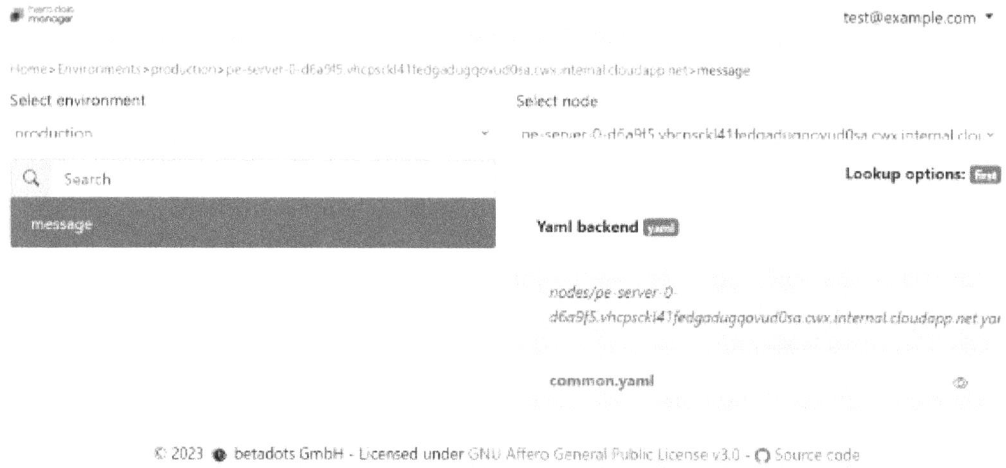

Figure 9.2 – Hiera Data Manager example lookup

Another option to make Hiera data more accessible for self-service is the PDS, which will be discussed in detail in *Chapter 13*.

Lab – troubleshoot Hiera

Troubleshoot the Hiera data in the production environment:

1. SSH to the primary server, elevate to root, and deploy the `lab_error` environment:

```
ssh centos@<primary_host>
sudo su -
puppet code deploy environment lab_error --wait
```

2. Perform a lookup with the `debug` flag of the `profile::error::example` key on the primary server in the `lab_error` environment and work through the errors found, correcting them in your control repo and running the `code deploy` command from the previous step:

```
puppet lookup profile::error::example --debug --environment lab_error
```

3. Resolve the errors in the data in `controlrepo-chapter9/data` and `hiera.yaml`.

4. Run the same command with `explain` to understand how it gets to the current solution and why it is not finding a value based on its `os.family` fact:

```
puppet lookup profile::error::example --debug --environment lab_
error
```

5. Update the Hiera data in the `control` repo branch, `lab_error`, and redeploy so that the lookup now finds the value for the `os.family` fact of the primary node:

```
puppet code deploy environment lab_error --wait
```

```
puppet lookup profile::error::example --debug --environment lab_error
```

Check the commented solutions at `https://github.com/PacktPublishing/Puppet-8-for-DevOps-Engineers/tree/main/ch09/data_solutions`.

As part of creating the lab environment in the *Technical requirements* section, HDM has been installed using the `puppet-HDM` module. Try using HDM to view the data following these steps:

1. Open a web browser at `http://<public IP of puppetserver>:3000`.
2. Complete the signup details to create an admin user (the details are not important).
3. Click **Create a new user** on the next page and enter `non-admin user` (the details again are not important).
4. Click the admin user name on the top right, log out, and then log back in as the non-admin user you created.
5. Select `environment production` and `lab_error` in turn.
6. Explore the Hiera keys and values visible to HDM in each environment.

Summary

In this chapter, we examined how Puppet can handle data using the Hiera tool, reducing how much complexity would need to be put into code to represent a node, data center, organizational, OS, and other configuration differences. Hiera was shown to be a tool based on hierarchies of data that allowed us to access different files based on facts. It had built-in backends for data to be stored in YAML, JSON, HOCON, and EYAML files. The data structure was shown; we examined how values could be put into data files and how lookups can be performed; the types of merge were examined here as well as how special setups such as `knockout` prefixes can be used in arrays.

We then showed how some custom backends can be used that have data types on different profiles; typically, these are specific integrations such as Vault or EYAML from the Forge, or in-house developed integrations to access data.

We then covered how Hiera worked over three layers – global, environment, and module – showing how global layers had little purpose in a modern Puppet setup but can be used as an override system, environment as the main source for data, and module allowing for defaults to be set on modules. Some common approaches to structuring hierarchies were then discussed, including an approach that stepped through the name of the node, the node owner, the node's purpose, the location of the node, and common to all nodes' data.

A review of how to make decisions on whether to use data in code or in Hiera showed that it depended on how flexible data needed to be, and this can vary from static data that is hardcoded in Puppet code to more advanced and flexible data requiring the full hierarchy to be described accurately. It was advised not to build ahead but to refactor as required so as not to make data more complicated than it needs to be.

We then discussed how to keep data secure in storage and transport, and when being used in Puppet catalogs, reports, and PuppetDB. We saw how to use `eyaml` to secure data in storage by encrypting values with the more flexible PGP approach, allowing multiple keys and teams. Then, the `Sensitive` value was shown to ensure values were not exposed in logs or code. This did not prevent values in catalogs and reports, and the `node_encrypt` module was shown to allow resources and values to be encrypted and be applied at configuration time using `Deferred` functions.

Approaches to debugging and troubleshooting were then reviewed, highlighting the difference between `--explain` and `--debug`. The former allows an understanding of how the hierarchy was reviewed and the latter returns errors such as syntax and failures with backends. The advice was given to be careful with using Hiera as a classifier, as this would abstract classification information away from code, but highlighted that the PDS did use this approach in later chapters.

In the next chapter, having reviewed the Puppet language in detail, the focus will change to the Puppet infrastructure. We will examine the open source components that make up the Puppet platform, how they make themselves available to the system via APIs, and how they communicate and log. The full Puppet agent life cycle will be examined, looking at the process of agent registration and communication with the platform. PuppetDB and PostgreSQL will be seen to allow the storage of data such as facts, reports, and catalogs, allowing discovery and examination with the **Puppet Query Language** (**PQL**). Compile servers will then be discussed as Puppet's method of scaling horizontally.

Part 3 – The Puppet Platform and Bolt Orchestration

In this part, you will understand how Puppet is structured as a platform, how the various components work together and communicate, and the common architecture approaches used to deliver scale. We will then show the various methods that can be used to classify which code is applied to servers and how code is versioned and deployed to infrastructure. Bolt will be introduced as Puppet's way of running procedural scripts and code, which can be traditional scripts in various languages or plans based on the Puppet language. We will then review how you can monitor, tune, and integrate Puppet infrastructure with various tools and third-party products.

This part has the following chapters:

- *Chapter 10, Puppet Platform Parts and Functions*
- *Chapter 11, Classification and Release Management*
- *Chapter 12, Bolt for Orchestration*
- *Chapter 13, Taking Puppet Server Further*

10
Puppet Platform Parts and Functions

So far, we have discussed Puppet as a language, but in this chapter and the following chapters, we will start to focus on Puppet as a platform and the infrastructure and components of the platform.

In *Figure 10.1*, the full architecture of services involved in Puppet Server and the Puppet client, to be discussed in this chapter, is shown. These services focus on how Puppet code is enforced on servers:

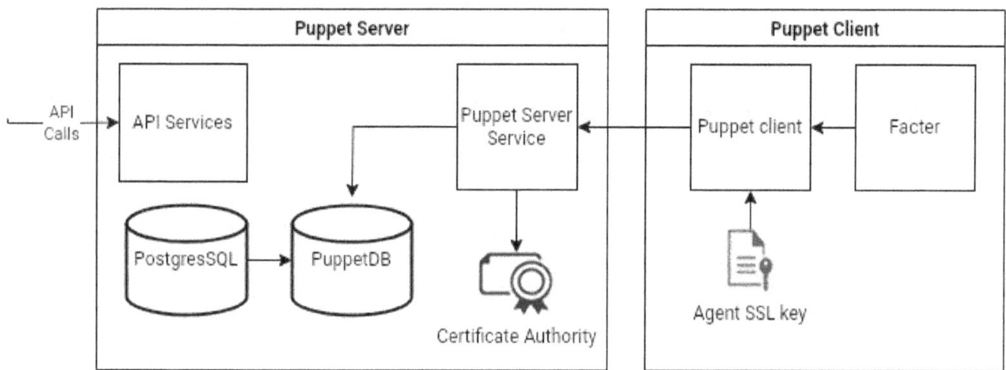

Figure 10.1 – Puppet server and client components

We will start by highlighting that we do not run through installation methods in this book. There are several open source projects to base automation on for open source Puppet and Puppet Enterprise; throughout this book, we have used the peadznd pecdm modules as the most automated mechanism for installing **Puppet Editor** (**PE**). As components are discussed, it will be noted how the versioning of Puppet packages can differ, and we'll look at some related install versions, as well as the key users, directories, configuration files, and services installed.

First of all, we will examine the core services provided by the Puppet Server. These services include catalog compilation to receive requests from clients, process their current state, and determine how they should be configured based on Puppet code. A **certificate authority** (**CA**) allows agents to register and communicate with the Puppet server securely. It also includes some associated API services to access, request, and control those services.

Having established how the server functions, we will then show how the Puppet agent communicates with the server, requesting to have a key signed by the CA, what the communication for a catalog compilation involves, and how the agent processes and stores the returned catalog.

We'll then view how PuppetDB is used to store, facts, catalogs, and events and how we can access this information via both APIs and **Puppet Query Language** (**PQL**). The relationship between PuppetDB and PostgreSQL will be examined as a frontend application to a backend database architecture, and we will also discuss how other data is stored in PostgreSQL by the Puppet services directly.

It will then be shown how the compilation can horizontally scale to compile catalogs of hundreds of thousands of servers using compile servers.

Throughout these topics, subtle differences between how PE and Puppet open source are set up will be highlighted.

This chapter will not cover the PE-specific features of the orchestrator, the PE console, or the supported architectures (which can allow for these services to be split out into more scalable infrastructure); these will be covered in *Chapter 14*.

In this chapter, we're going to cover the following main topics:

- Puppet platform installation and versioning
- Puppet Server
- Puppet agent-to-server life cycle
- PuppetDB and PostgreSQL
- Scaling with compilers

> **Note**
>
> As part of an effort to remove harmful terminology from its products, Puppet dropped the use of the terms *master server* and *compile master* and now uses *primary server* and *compile server*. As these names were quite embedded, there will be some places where classes or configuration settings do still refer to *master*.

Technical requirements

Clone the control repo from https://github.com/puppetlabs/control-repo to your GitHub account in a repo called controlrepo-chapter10.

Build a large cluster with three compilers and three clients by downloading the params.json file from https://github.com/PacktPublishing/Puppet-8-for-DevOps-Engineers/blob/main/ch10/params.json and update it with the location of your control repo and your SSH key for the control repo. Then, run the following command from your pecdm directory:

```
bolt --verbose plan run pecdm::provision --params @params.json
```

Puppet platform installation and versioning

This book makes the choice not to go into the methods of installing Puppet; there is little to add to the installation instructions for open source, documented at https://puppet.com/docs/puppet/latest/server/install_from_packages.html, and any further choice of automation will depend heavily on the use case of your organization and available tooling and product sets you want to integrate with.

For open source Puppet, there are a number of projects automating the deployment, configuration, and integration of Puppet, such as example42's psick (https://github.com/example42/psick) or the Foreman project (https://github.com/theforeman/foreman-installer), which has a specific module for installing Puppet Server (https://forge.puppet.com/modules/theforeman/puppet) that can be used even outside of Foreman to install Puppet. Dashboards similar to what has been provided by the PE setup can also be found in projects such as Puppetboard (https://forge.puppet.com/modules/puppet/puppetboard) or Puppet Summary (https://github.com/skx/puppet-summary).

For PE, although manual instructions are available at https://puppet.com/docs/pe/2021.7/installing_pe.html, the automation choice is clear with the Puppet-supported peadm module; in *Chapter 12*, we will review how the module is used for the lab along with pecdm as a Bolt project.

Key points to recognize with the packages installed are that Puppet repositories provide set versions of Ruby, OpenSSL, Hiera, and Facter to use for different versions of Puppet and that packages such as puppetserver may not match the Puppet version being installed—for example, Puppet 7.17 will have Puppet server version 7.8 installed; these associated versions are available in the release notes. For PE, you can see all the underlying open source package versions in the documentation at https://puppet.com/docs/pe/2021.7/component_versions_in_recent_pe_releases.html#component_versions_in_recent_pe_releases.

Puppet Server

In historic versions of Puppet, Ruby-based solutions such as WEBrick or Passenger were used for running the Puppet service, but in all modern versions of Puppet, to improve scaling and performance, Puppet Server is run as a Clojure and Ruby application on a **Java Virtual Machine** (**JVM**). Puppet Server has a number of related services that share state and route requests between them. These services run inside a single JVM process, using the Trapperkeeper service framework, which is a Clojure framework for hosting long-running applications.

Puppet Server is installed via the `puppetserver` package in open source Puppet and the `pe-puppetserver` package in PE. This will create a system service of the same name and configuration files that, by default, will be placed in `/etc/puppetlabs/puppetserver/conf.d` in **Human-Optimized Config Object Notation** (**HOCON**) format.

> **Note**
>
> The Puppet `hocon` module is the best way to automate the management of HOCON files (`https://forge.puppet.com/modules/puppetlabs/hocon`).

Next, we'll look at the services that make up Puppet Server.

The embedded web server

Puppet contains a Jetty-based web server in the JVM that sets up the mount points and communications necessary for web requests to take place between components and to access the APIs.

The `webserver.conf` file sets the main configuration for the web server, such as the file location of **Secure Sockets Layer** (**SSL**) keys, the port, and the host IP, which should only ever need adjusting if using an external CA. The `web-routes.conf` file sets the mount points for web API access by mounting the handlers, as shown in the following example file:

```
# Configure the mount points for the web apps.
web-router-service: {
    # These two should not be modified because the Puppet 4 agent
expects them to
    # be mounted at these specific paths.
    "puppetlabs.services.ca.certificate-authority-service/certificate-
authority-service": "/puppet-ca"
    "puppetlabs.services.master.master-service/master-service": "/
puppet"

    # This controls the mount point for the Puppet administration API.
    "puppetlabs.services.puppet-admin.puppet-admin-service/puppet-
admin-service": "/puppet-admin-api"
}
```

The core mount points that can be seen in this file required for client-to-server communication are listed here:

The `puppet-ca` mount point is used by clients to communicate with the CA service and check or make a **certificate signing request** (**CSR**).

- `master-service` provides a mount point used by clients to request catalogs that are compiled via JRuby interpreters.
- The request logging configuration set by default in `webserver.conf` is at `/etc/puppetlabs/puppetserver/request-logging.xml` and determines how HTTP access requests are logged. By default, messages will be logged to `/var/log/puppetlabs/puppetserver/puppetserver-access.log`.

This section should have given you an understanding of how the embedded web service sets up a web server in the JVM with the mount points necessary for requests to be made to the different components of Puppet Server and how it will log these requests. Now, we will see the two core APIs accessed via the endpoints made available by the mount points, Puppet API via `/puppet` and `/puppet_ca`, and then the Admin API via `/puppet_admin_api`.

The Puppet API service

The Puppet API service is made up of two endpoints created by the embedded web server—`/puppet` for configuration-related services and `/puppet-ca` for the CA.

Both are versioned with a string such as `/v3` and authorization is controlled via the `auth.conf` file, a HOCON formatted file. It is unlikely you will need to edit this file unless requiring more advanced access to integrate services, but to show an example of content, the following code allows Puppet nodes to request their own catalog from the API:

```
{
    # Allow nodes to retrieve their own catalog
    match-request: {
        path: "^/puppet/v3/catalog/([^/]+)$"
        type: regex
        method: [get, post]
    }
    allow: "$1"
    sort-order: 500
    name: "puppetlabs v3 catalog from agents"
},
```

> **Note**
> More detailed instructions for customization authorization are available at `https://github.com/puppetlabs/trapperkeeper-authorization/blob/main/doc/authorization-config.md`.

The Puppet agent on all modern versions of Puppet 5 to 8 uses `/puppet/v3` endpoint services to manage clients. The `v3` API has two types of endpoints—**indirectors** and **environment** endpoints.

Indirectors take the form `/puppet/v3/<indirection>/<key>?environment=<environment>`.

Here, the indirection value is the indirector requested, the key is the key relevant to the call to the indirector, and the environment is the environment that should be used for this request. For example, to request a catalog be compiled, a client would construct the following:

```
/puppet/v3/catalog/pe.example.com?environment=production
```

The following indirectors exist under `/puppet/v3/` for clients:

- `Facts`: The `facts` endpoint allows setting facts for the specified node name
- `Catalog:` Returns a catalog for the specified node
- `Node`: Returns node information such as classification
- `File bucket file`: Manages the contents of a file bucket
- `File content`: Returns file content such as files in modules
- `File metadata`: Returns the metadata of a file such as the permissions of a file in modules
- `Report`: Allows the storing of Puppet reports for nodes

The following indirectors exist under `/puppet/v3/` for the server:

- **Environment classes**: Returns all the classes that can be parsed in the requested environment
- **Environment modules**: Returns information about all the modules found in an environment, such as their names and versions
- **Static file content**: Returns the file content of a specific version of a file resource in an environment

The separate environment endpoint that was not an indirector allows a simple call to `/puppet/v3/environments` that returns all environments known to the server. In the next chapter, we will talk about environments in more detail.

Tools and services can also access these same endpoints to examine data, and a `v4` API exists with a catalog endpoint that allows for more extensive use of PuppetDB to manipulate facts and the catalog. It is used by tools such as `octocatalog-diff (https://github.com/github/octocatalog-diff)`, which can generate, compare, and manipulate catalogs.

The `/puppet-ca` endpoint follows a similar format using `v1` and indirectors, as follows:

- **Certificate**: Returns the certificate of a specified name
- **Certificate Clean**: Revokes and deletes a certificate
- **Certificate Status**: Requests the status of a certificate or a CSR
- **Certificate Revocation List**: Requests the **Certificate Revocation List** (**CRL**) file

As an example, to request a certificate for `server.example.com`, the following endpoint would be hit: `/puppet-ca/v1/certificate/server.example.com`.

These actions will be discussed in more detail in the *CA* section of this chapter.

In this section, we did not go into full detail about each endpoint and making API calls to them, but later in the chapter, where we look at the client-to-server lifecycle, we will follow the logging of calls and highlight their use to show how these APIs are used by Puppet. Full details of the endpoints can be viewed at `https://puppet.com/docs/puppet/latest/http_api/http_report.html`.

The Admin API

The Admin API has just two endpoints at `/puppet_admin/v1/`, as follows:

- **Environment cache**: Used to clear the cache of environment data
- **JRuby pool**: Used to clear the JRuby pool or retrieve a Ruby thread dump of running JRuby instances

Both endpoints are for more in-depth development work, so are beyond the scope of this book but help complete the picture of the Puppet server components. Details of these endpoints can be viewed at `https://puppet.com/docs/puppet/latest/server/admin-api/v1/jruby-pool.html` and `https://puppet.com/docs/puppet/latest/server/admin-api/v1/environment-cache.html`.

CA

By default, Puppet uses its own in-built CA and **public key infrastructure** (**PKI**) to secure all SSL communications.

Two commands are used to interact with the Puppet CA setup—`puppetserver ca` for server-side actions such as signing or revoking certificates and `puppet ssl` for agent-side tasks such as requesting and downloading certificates. These commands make calls to the `puppet-ca` endpoint via the CLI.

> **Note**
>
> Despite the introduction of the `puppet-ca` endpoint, the five commands of the previous `ruby ca` implementation were still available until their removal in Puppet 6: `puppet certificate`, `puppet cert`, `puppet certificate_request`, `puppet ca`, and `puppet certificate_revocation_list`. They have all been replaced by the `puppetserver ca` and `puppet ssl` commands. Even if you are using Puppet 5, it is strongly advised not to use these Ruby commands as using both API and Ruby implementations simultaneously can corrupt the CA.

While the automation of installation discussed in the introduction should cover the initial setup, it is worth knowing whether the CA setup has been performed by running `puppetserver ca setup`. Before the `puppetserver/pe-puppetserver` service has started, it will create a separate root CA and an intermediate signing CA. If the `puppetserver/pe-puppetserver` service is started before this step, it will create a single combined root and signing CA, which was the prior way Puppet operated. Unless you have a specific need for a single certificate, this should be avoided. From PE 2019.x and Puppet 6.x, these certificates last 15 years; previously, this was 5 years, and it's important to understand that upgrading Puppet versions does not extend the CA.

> **Note**
>
> Extending an expired CA is possible via the `ca_extend` module (`https://forge.puppet.com/modules/puppetlabs/ca_extend`).

The keys and certificates created in this step will be created in a directory called `/etc/puppetlabs/puppetserver/ca` for Puppet 7 and above or one called `/etc/puppetlabs/puppet/ca` for Puppet 6 and below. There is a **symbolic link** (**symlink**) for `/etc/puppetlabs/puppet/ca` to the new location to avoid confusion. The directory will contain the following:

- `ca_crl.pem`: The CRL
- `ca_crt.pem`: The CA-signed certificate public certificate
- `ca_key.pem`: The CA private key
- `ca_pub.pem`: The CA public key
- `inventory.txt`: A list of certificates the CA signed with their serial numbers and expiry dates
- `requests`: Unsigned CSR files

- `root_key.pem`: This is the root key used to sign the CA certificate if using a separate root CA and an intermediate CA
- `serial`: This file contains an incrementing counter of the new serial number for certificates
- `signed`: This folder contains all signed CSR files

In addition to these files, an infrastructure CRL can be maintained, which by default is not used in open source Puppet but is used in PE. To have a smaller CRL, the `infra_inventory.txt` file is managed to contain the Puppet infrastructure servers; when revoked, these systems are added to `infra_crl.pem`. This is enabled by setting `infra certificate-authority.enable-infra-crl` to `true` in the `puppet.conf` file. We will talk in more detail about the `puppet.conf` file later in this chapter. This approach means the Puppet clients only need to receive the small infrastructure CRL, which is important for estates with a high churn of servers. The following files will be maintained:

- `Infra_inventory.txt`: A list of certificates the CA signed for infrastructure servers
- `Infra_serials`: This file contains an incrementing counter of the new serial number for infrastructure servers
- `Infra_crl.pem`: The CRL of infrastructure servers

If your organization requires the use of an external CA, it is possible to use your organization's own root CA and import it using the `puppetserver ca import` command (the full process is outlined at `https://puppet.com/docs/puppet/latest/server/intermediate_ca.html`), leaving Puppet to act just as an intermediate CA. Alternatively, the CA service can be disabled by deploying a single externally generated root and signing CA, as outlined at `https://puppet.com/docs/puppet/latest/config_ssl_external_ca.html`. This book recommends against using this approach as it would require automating the distribution of certificates, which Puppet services no longer perform.

When an agent makes a request to the CA, a CSR is sent and the signing policy by default has to wait for a manual signing with the CSR stored in the `requests` folder. Waiting requests can be reviewed by running `puppetserver ca list` and then signed by running `puppetserver ca sign --certname < certname to sign >`. All certificates that have been signed can be viewed by running `puppetserver ca list --all`.

If you are using PE, certificate signing can be performed and viewed on the PE web console, as pictured in *Figure 10.2*:

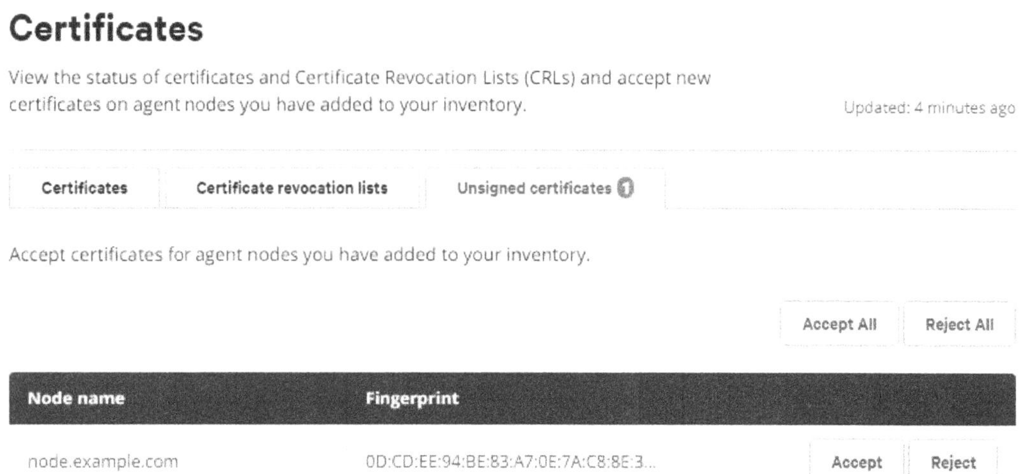

Figure 10.2 – PE console certificate signing

Certificates can be revoked using the `puppetserver ca revoke --certname < certname to revoke >` command, and to clean up and remove a revoked certificate from the CA, you can then run `puppetserver ca clean --certname < revoked certname >`.

It is common when manual auto signing is used for a workflow that tools such as VMware's **vRealize Orchestrator** (**VRO**) will make calls to the CA API as part of the deployment and decommissioning of servers.

To automate this process, auto signing can be configured in three ways. In what is known as **naïve signing** whereby `autosign = true` is added to the `master` section of `puppet.conf`, this change causes the CA to sign any request and should never be used in a production environment.

The second way is to create an `autosign.conf` file at `/etc/puppetlabs/puppet/autosign.conf`. In this file, there can be server names or domain name globs where each line represents a node name or a domain that can be auto-signed. For example, let's say a file had the following content:

```
server1.puppet.com
*.example.com
```

This would mean that `server1.puppet.com` and any server in the `example.com` domain would be auto-signed.

The third method is to set the `autosign` value in the `puppet.conf` file to be equal to a script. This script can be in any language and will receive as its first argument the certificate name and then the CSR contents as standard input. The script should then end with a zero-return code to sign the script or a non-zero code to not sign it. This leads to a common method of CSRs containing a secret to check in the script, or in the public cloud, tags can be used. It is beyond the scope of this book to discuss writing these scripts, and while Puppet only provides a description of how to construct these scripts at `https://puppet.com/docs/puppet/latest/ssl_autosign.html#ssl_policy_based_autosigning`, Amazon gives an excellent example at `https://aws.amazon.com/blogs/mt/aws-opsworks-puppet-enterprise-and-an-alternate-implementation-for-policy-based-auto-signing/`.

This section has laid out how a CA is configured and run as a Puppet server. Later in this chapter, the full lifecycle of agents will be reviewed, showing how the client creates a CSR and uses the CA to finish the services offered by Puppet Server and looking at JRuby interpreters.

JRuby interpreters

JRuby is a Java implementation of Ruby allowing for the use of Ruby on JVMs; this allows for greater scalability than with traditional Ruby deployments such as Ruby on Rails as most Ruby interpreters aren't capable of thread safety and use locks to run one thread at a time. Puppet Server has a pool of JRuby interpreters/instances that are available to perform various application work such as compiling catalogs and handling reports. The number of interpreters in the pool reflects how many Ruby application actions can be run simultaneously and can be configured with the `max-active-instances` parameter in the `puppetserver.conf` file, in PE via Hiera in the console, or in code via `puppet_enterprise::master::puppetserver:: jruby_max_active_instances`. We will be looking at this in more detail in *Chapter 13*, where we discuss the metrics and tooling to review and set this sizing.

Having discussed the components of Puppet Server, we will now look at the configuration such as users, logging, and filesystems to understand where these services can be customized and what is required by them.

Configuration and logs for Puppet Server

We briefly touched on certain configuration files and the settings available as we discussed each component, but we will give a summary here. For most of the configuration files, it is unnecessary to customize them, and most defaults will meet your requirements.

For PE, the `pe-puppetserver` Puppet Server service will run under a `pe-puppet` account, while on open source Puppet, the `puppetserver` service will be run under the `puppet` account. In both accounts, they will have a `nologin` shell set so that the user just provides an account to run the service and own relevant files for the service.

The following configuration files and application directories will be created and used:

- `/etc/puppetlabs/puppetserver/bootstrap.cfg`: A file containing a list of services that Trapperkeeper should start up; these are the handlers mounted by the embedded web server.
- `/etc/puppetlabs/puppetserver/request-logging.xml`: A file defining how HTTP access requests are logged.

- `/etc/puppetlabs/puppetserver/conf.d`: This directory contains the following main configuration files for components in HOCON format:
  - `global.conf`: This file sets global configuration settings for Puppet and by default just contains the logging config file location.
  - `webserver.conf`: This file configures the embedded web server with details such as port and logging.
  - `web-routes.conf`: This file sets mount points for Puppet's web services.
  - `puppetserver.conf`: This file sets the configuration for the core Puppet server application such as the number of `jruby` instances running.
  - `auth.conf`: This file sets the access permissions for endpoints mounted by `web-routes.conf`.
  - `ca.conf`: This file configures settings for the CA.
  - `products.conf`: An optional file that can set product settings such as analytics data and update checks.
- `/etc/puppetlabs/puppetserver/ssl/ca`: Certificates and keys related to the Puppet CA (`/etc/puppetlabs/puppet/ssl/ca` in Puppet 6 and below).
- `/opt/puppetlabs/puppet/lib/ruby/vendor_gems`: Puppet Server puts Ruby gems related to the operation of the CA in this directory.
- `/opt/puppetlabs/server`: This directory contains the `JRuby-gems` and binaries for running Puppet Server.
- `/var/run/puppetlabs/puppetserver/puppetserver.pid`: This file contains the PID of the running Puppet process.
- `/etc/puppetlabs/puppet.conf`: This file holds the configuration for both the Puppet client and Puppet Server on the primary. These settings can be viewed by running `puppet config print`.

The vast majority of settings in the files will be used at default values unless external integrations such as the external root CA are required and are only worth mentioning as a reference to understand the setup of the Puppet. A full reference and options for settings can be reviewed at `https://puppet.com/docs/puppet/latest/server/configuration.html` for `/etc/puppetlab/puppetserver`-based settings.

> **Note**
>
> If you have chosen one of the open source Puppet automation tools/modules discussed in the introduction, it may allow the setting of configuration values on installation.

PE users should be aware due to extra automation of configuration that a lot of those settings such as those in `puppetserver.conf` are configured via Hiera, and the documentation at `https://puppet.com/docs/pe/2021.7/config_puppetserver.html` should be followed for the configuration.

The configuration for tuning these settings will be looked at in more detail in *Chapter 13*.

The full options for the settings for `/etc/puppetlabs/puppet.conf` can be reviewed at `https://www.puppet.com/docs/puppet/latest/config_file_main.html`; the file itself provides sections that can configure the Puppet server, the Puppet agent, and how `puppet apply` runs. The sections are `main`, which provides default values, `agent`, which provides settings to the Puppet client, `user`, which provides settings for when using Puppet `apply`, and `master/server` for applying settings to the Puppet server.

Since Puppet 6, it has been possible to use a `server` section instead of a `master` section, but many automation tools have not caught up with this change, and as they are not interchangeable terms and could create confusion, be careful to only use the term relevant to your implementation.

Puppet applies settings from the `master/server`, `apply`, or `agent` section first, then falls back to the `main` section and, if it finds no setting, will use a default.

Let's look at some example content of a file on a `peadm` built Puppet lab server:

```
[master]
node_terminus = classifier
storeconfigs = true
storeconfigs_backend = puppetdb
reports = puppetdb
certname = pe-server-davidsand-0-cffe02.tq2kpafq5bsehkpub4ur5a35ya.
xx.internal.cloudapp.net
```

The square brackets indicate the name of a section and then a set of key-value pairs. The settings here show the certificate name (`certname`) of our Puppet server and also show that it sends reports to PuppetDB via the `reports` setting, that it is set up to store catalog, node, and fact information with `storeconfigs` set to `true`, that these will be stored in PuppetDB, and that `storeconfigs_backend` is set to PuppetDB. Finally, `node_terminus` is set to `classifier`, which reflects how the primary server should classify clients. This will be discussed in greater detail in the next chapter.

The best way to view and manipulate the settings including defaults not set by `puppet.conf` is by using the `puppet config` command, which can show all settings. By running `puppet config print all` known, the settings will be printed, or an individual setting can be printed by detailing the section and value to print via `puppet config print --section master certname`. The `puppet config` command can also add or remove values using the `set` or `delete` options and selecting a section key and value to perform an action on. For example, the following commands will delete `storeconfigs` from the `master` section and change the certificate name to `newname.example.com`:

```
puppet config delete --section master storeconfigs
puppet config add --section master certname newname.example.com
```

These commands will automatically add a section if it's not already in the file, but the Puppet service would need to be restarted for any changes to take place.

We will work with more examples of manipulating the `puppet.conf` file as we look at the agent lifecycle in the next section, but the full options and syntax for the `puppet.conf` file can be viewed at `https://puppet.com/docs/puppet/latest/config_file_main.html`.

Puppet Server by default keeps logs at `/var/log/puppetlabs/puppetserver` in the following files:

- `Puppetserver.log`: This is where the primary server logs activity such as compilation errors and warnings
- `Puppetserver-access.log`: This is where requests to HTTP endpoints are logged
- `Puppetserver_gc.log`: This is where logs of garbage collection are gathered

Now that we have reviewed the Puppet server components fully, we will look at the Puppet agent configuration and lifecycle, which will show how these services are used by a client, and how to monitor and review the logging of a cycle.

The Puppet agent-to-server lifecycle

This section will look at how the Puppet agent makes requests to the Puppet server components we have run through and how it secures its communications before requesting configuration to enforce from the Puppet servers. It should be noted the Puppet servers themselves also contain Puppet agents.

The installation of Puppet agents is detailed at `https://puppet.com/docs/puppet/latest/install_agents.html#install_agents` for open source and `https://puppet.com/docs/pe/2021.7/installing_agents.html#installing_agents` for PE. Integrating this install with your server deployment workflow and ensuring the necessary configuration is placed at `/etc/puppetlab/puppet.conf` is critical for automation.

> **Note**
>
> The `puppet_conf` module provides tasks to manage Puppet configuration files (`https://forge.puppet.com/modules/puppetlabs/puppet_conf`).

Most of the settings will depend on your environment setup, but for most environments, the defaults will be taken with the critical setting of ensuring that the server setting in the `agent` section is set so that the agent knows which Puppet server to contact – open source Puppet or PE-Puppet. The PE service can then be started under the root user. This will contact the Puppet server every 30 minutes by default or can be triggered by running the `puppet agent -t` command.

Figure 10.3 shows the workflow of this Puppet certificate process as the client ensures it has the signed SSL certificate to ensure secure communication with the Puppet server:

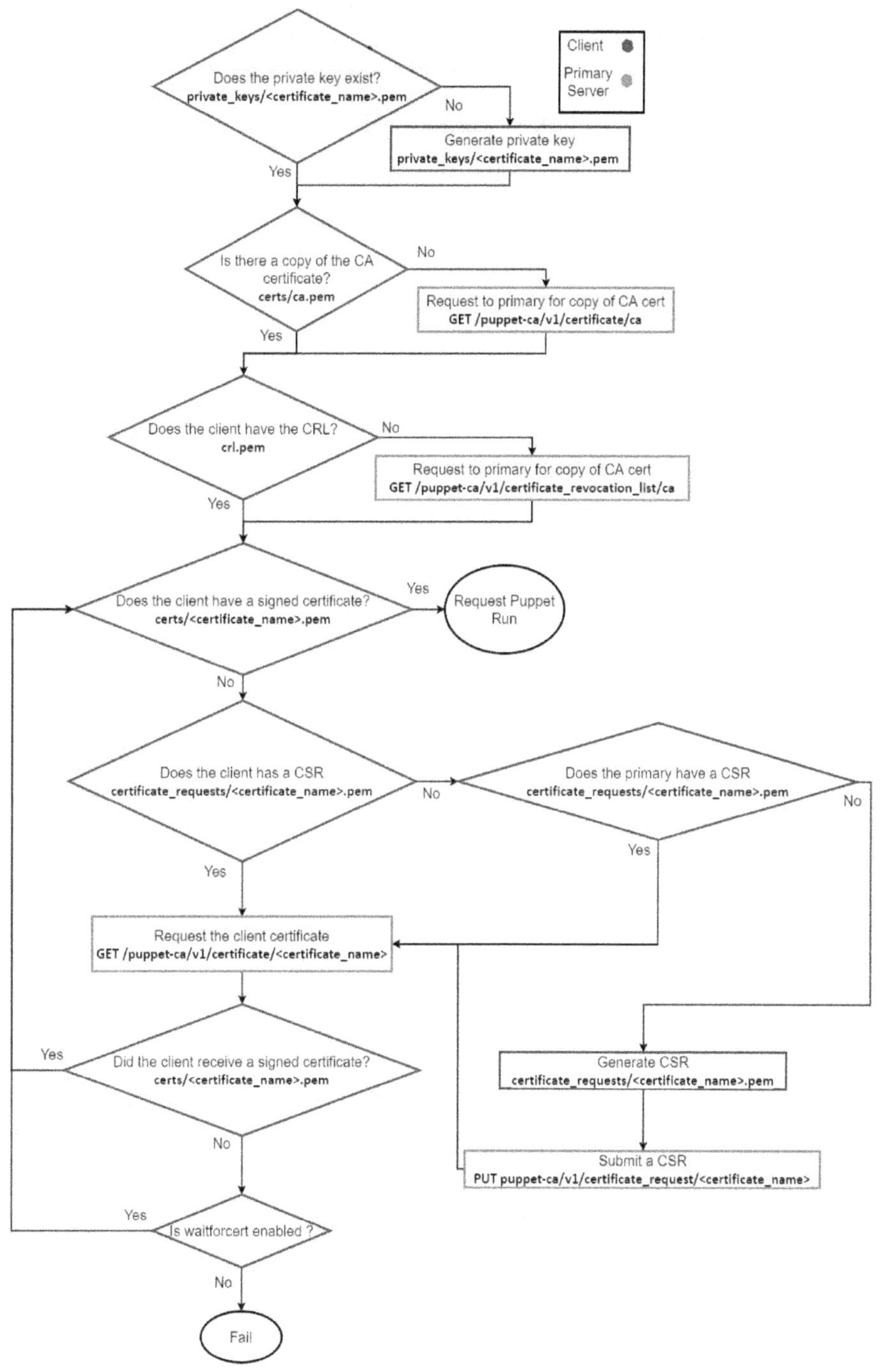

Figure 10.3 – Puppet client certificate workflow

The first step is to validate the certificates. In the `ssl` directory titled `/etc/puppetlabs/puppet/ssl`, the following files will already exist or be created in this process:

- `private_keys/<certificate_name>.pem`: The private key used to create a CSR
- `certs/<certificate_name>.pem`: The signed certificate returned for this client
- `certs/ca.pem`: A copy of the CA certificate sent from the Puppet server
- `crl.pem`: The CRL from the Puppet server
- `certificate_requests/<certificate_name>.pem`: The CSR to be sent to the Puppet server, which is deleted once a signed certificate is received

In addition to this directory, it is possible to create a `/etc/puppetlabs/puppet/csr_attributes.yaml` file and include the trusted facts to be created in the CSR. This will cause the trusted facts to be included in the certificate for the client when the CSR is signed by the Puppet server.

Using trusted facts can be useful to ensure hard classification information is not changed, such as a production server being reclassified for development, or the role being changed, since both could result in lower levels of security. The **organization ID** (**OID**) numbers translate into names, which can be reviewed at `https://puppet.com/docs/puppet/latest/ssl_attributes_extensions.html`. This file must exist before the CSR is created; otherwise, the only way to change the CSR or certificate is to start again.

As *Figure 10.3* shows, if a private key doesn't exist, the client makes a new key before checking for local copies of `ca.pem` and `CRL.pem`, making a request to the server, and downloading if either is absent. The next step is to then check whether a signed certificate exists and request it from the client if it does not. If a signed client certificate exists, it can continue to request node data; otherwise, it will create a CSR file and send it to the primary server. If the `waitcert` setting is enabled in `puppet.conf`, the client will then wait for the CSR to be signed by the server and check the status with the primary server every 2 minutes On future runs, the client will present its signed certificate to the server as proof of its identity.

Having secured communication, the first step is to perform plugin sync from the server to the client, which ensures all facts, functions, resource types, resource providers, and Augeas lenses are downloaded to the client using the `file_metadata` endpoint.

Once this has been completed, the client runs `facter`, sends the output to the Puppet client, and requests a catalog from the Puppet server using the `\catalog` endpoint. A copy of this catalog is stored in the `cache` directory (configured using the `vardir` parameter in `puppet.conf`) on the client, which by default is `%PROGRAMDATA%\PuppetLabs\puppet\cache\client_data\catalog\<certname>.json`. (`PROGRAMDATA` is generally `C:\Program Data\` on Windows and `/opt/puppetlabs/puppet/cache/client_data/catalog/<certname>.json` for Linux- and Unix-based systems.) The client then receives this catalog and implements the steps, enforcing the state described in the Puppet code, or if the client is set to be run in no-op mode, it will simulate the catalog. The client generates a report, which is then sent back by default to the Puppet server using the `report` endpoint. This can be configured to send the report to other report processors such as Splunk, which will be discussed in *Chapter 13*.

In addition to the catalog in the `client_data` folder, several other useful files for investigation are generated in the `cache` directory:

- `lib`: This is a cache for various plugins synced by plugin sync from the primary server.
- `facter`: This will contain custom facts.
- `facts.d`: Here, external facts are cached by plugin sync from the primary.
- `reports`: This contains the last generated report file.
- `state`: This directory contains the files and directories associated with the state of previous Puppet runs:
  - `classes.txt`: A list of classes that were included in the last catalog applied
  - `graphs`: If the `graph` option is used during a Puppet run, the generated .dot graph files of resources and dependencies will be saved here
  - `last_run_report.yaml`: This is a full report of all resources and how they were checked or changed during the catalog enforcement
  - `resources.txt`: A list of resources that were included in the last catalog applied
  - `state.yaml`: A list of all resources and when they were last checked or synced, used with features such as `audit`

Several directories and files have been ignored as they are either for legacy purposes or for practices that this book does not recommend, such as `filebucket`. A full listing can be seen at `https://puppet.com/docs/puppet/latest/dirs_vardir.html`.

> **Note**
>
> The cached catalog will be used in the event of the client losing communication with the Puppet infrastructure to ensure it continues to enforce its last known state.

The last step of the agent-to-server cycle is to send the event report to the Puppet Server. These reports will reflect events of what happened for each resource in the catalog. These events can have one of the following states:

- Failure – This will be an event with errors in applying the catalog or issues such as dependencies or an issue with that particular resource
- Corrective – The resource was in the correct state in the previous run but has had to be corrected

- Intentional – The resource had to be created or corrected but was not in the correct state in the previous run
- Unchanged – The resource is in the correct state and requires no change

Unchanged events are not reported in Puppet 8 by default. This change was made to reduce the store space required for storing reports. This can be changed by setting `exclude_unchanged_resources=false` in each agent's `puppet.conf` file.

The report events will also reflect what mode the client agent is running or whether a resource is set to be applied differently from the client. While the same event states still apply, each event will report if the event took place in execution mode or in no-op mode. As was previously discussed in *Chapter 3*, no-op mode means the resource is only effectively tested to see whether the resource will need to be changed to meet its declared state. In *Chapter 15*, we will discuss how this can be useful in heritage envionments where we want to see how big the configuration drift is and choose a progressive approach to get there to avoid causing issues on production systems.

In terms of accessing these reports, we will see in *Chapter 13* how report processors can be used to send them to third-party tools and in *Chapter 14* how Puppet Enterprise provides an event viewer interface as part of its graphical console.

Lab – monitoring certificate signing logging

To better understand the process, we will now describe how we can monitor the process of a Puppet run by removing the certificates of our node and re-registering. During the registration, we will monitor the logs to see the API requests made through this process and note the steps of the process. Here are the steps:

1. Open SSH terminal sessions to the Linux client and two separate SSH terminal sessions to the primary Puppet server.
2. On the Linux client, run the following command:

   ```
   puppet ssl clean
   ```

3. On one of the server sessions, run `puppetserver ca clean --certname <instance name>` (note that this should be the certificate name, which can be checked via `puppet config print certname` on the node).
4. On the Linux client, move the `ssl` directory to a backed-up location using the following command:

   ```
   mv /etc/puppetlabs/puppet/ssl /etc/puppetlabs/puppet/ssl.old
   ```

5. On one of the Puppet server's sessions, run `tail -f /var/log/puppetlabs/puppetserver/puppetserver-access.log`, and on the other, run `tail -f /var/log/puppetlabs/puppetserver/puppetserver.log`.

6. On the node, run `puppet agent -t` and see the calls on the Puppet server sessions.
7. On the web console, under the **Certificates** section, select the **Unsigned** tab, sign the certificate request, and then run `puppet agent -t` on the client. Note the new calls on the server in the `access.log` and `puppetserver.log` files and how this relates to the steps discussed in this section.
8. View the catalog received for the client and investigate the other files in the cache.

> **Hint**
>
> Using a tool such as `jq` can make viewing JSON much easier (`https://stedolan.github.io/jq/download/`).

To view an example output of logging for this lab, see the following files:

`https://github.com/PacktPublishing/Puppet-8-for-DevOps-Engineers/blob/main/ch10/puppet_access_log_extract` shows the access logs with comments explaining the output

`https://github.com/PacktPublishing/Puppet-8-for-DevOps-Engineers/blob/main/ch10/puppet_server_log_extract` shows the Puppet server log with comments explaining the output

`https://github.com/PacktPublishing/Puppet-8-for-DevOps-Engineers/blob/main/ch10/puppet_client_terminal.txt` shows the client terminal and commands entered

`https://github.com/PacktPublishing/Puppet-8-for-DevOps-Engineers/blob/main/ch10/puppet_server_terminal.txt` shows the server terminal and commands entered

PuppetDB and PostgreSQL

PuppetDB allows for the collection of Puppet data and advanced features such as exported resources. In open source Puppet, it is entirely optional, while PE installs PuppetDB by default. The following is kept by PuppetDB:

- The last facts from the nodes
- The last catalog compiled for each node
- 14 days (default) of event reports for each node
- Exported resources

PuppetDB is a Clojure frontend application running on a JVM, using PostgreSQL as a backend database. This common architecture is where the backend database just provides the tables, and the frontend database contains the application objects, giving some key advantages compared to a single database. It eases the updating process of PuppetDB since the actual data can be left in the backend table, and it also allows great scalability—as we will see in the last section of this chapter, *Scaling with compilers*—where PuppetDB can be scaled horizontally by running PuppetDB on many compiler servers, as a result reducing the load on the primary server PuppetDB service.

Information on the installation and configuration of PuppetDB is provided at `https://forge.puppet.com/modules/puppetlabs/puppetdb`. PuppetDB is likely to be included in any automation you choose and is part of PE.

PostgreSQL creates a `pe-postgres` user for PE or a `postgres` user for open source Puppet, which is created as a user to run the PostgreSQL database. This user will use a `nologin` shell and own relevant files for running Postgres. The following directories are used by PostgreSQL:

- `/opt/puppetlabs/server/apps/postgresql/{version}`: To install the database application
- `/opt/puppetlabs/server/data/postgresql/{version}`: To contain the data files of the database
- `/var/log/puppetlabs/postgresql/{version}`: To contain the logs of the database

PuppetDB creates a `pe-puppetdb` user for PE or a `puppetdb` user for open source Puppet, which is created as a user to run the PuppetDB database under with a `nologin` shell and to own the relevant file for running PuppetDB. As PuppetDB is a Clojure application running on the JVM; it is very similar to puppet web server in its structure with a handler mounted at a `/pdb` endpoint and an `auth.conf` file defining who can access this endpoint. The following directories are used by PuppetDB, and some key files are highlighted:

- `/etc/puppetlabs/puppetdb`: This directory contains configuration files for PuppetDB, including the following:
    - `bootstrap.conf`: The `bootstrap.conf` file the lists services that should be started in the Trapperkeeper framework
- `/etc/puppetlabs/puppetdb/conf.d`: This directory contains configuration files in an `ini` format:
    - `auth.conf`: Configures authorization for who can access the endpoints made available
    - `routing.ini`: Configures which handlers should be made available at endpoints
- `/opt/puppetlabs/server/apps/puppetdb`: This directory contains the application binaries for PuppetDB

- `/opt/puppetlabs/server/data/puppetdb`: This directory contains the data of PuppetDB

It is beyond the scope of this book to look at the in-depth configuration of `PuppetDB`, but you can refer to `https://puppet.com/docs/puppetdb/latest/configure.html` for more information. However, in *Chapter 13*, we will look in more depth at how to monitor, review, and tune PuppetDB and PostgreSQL performance and how modules such as `https://forge.puppet.com/modules/puppetlabs/pe_databases` can assist with maintenance.

For now, we will review how the data can be accessed by using PQL with HTTP calls to the `/pdb` endpoint or via the `puppet query` command line to make calls to the endpoint.

> **Note**
>
> The **Abstract Syntax Tree** (**AST**) query language is also available as a format to use for queries. However, with PQL available, it has little use now but can be reviewed at `https://www.puppet.com/docs/puppetdb/8/api/query/v4/ast.html`.

PuppetDB is structured into entities to allow for accessing different types of data. Here is a list of each entity and a brief description of what the endpoint contains:

- `aggregate_event_counts`: Aggregate counts of the `event_counts` entity
- `catalogs`: The catalogs stored for each node
- `edges`: Edges are relationship information in catalogs such as *contains* or *requires*
- `environments`: The environments known to PuppetDB
- `event_counts`: Event counts about various resources in reports
- `events`: Events reflect the actions performed for a resource returned by a report
- `facts`: The facts returned for each node
- `fact_contents`: This entity is structured to access fact content more easily
- `fact_names`: All known fact names
- `fact_paths`: Similar to the `fact_names` entity but provides further granularity for structured facts
- `nodes`: Node information
- `producers`: Producers are the servers that compiled the catalog and sent the report
- `reports`: Reports contain the outcome of applying a catalog
- `resources`: Resource information in catalogs

To begin looking at PQL queries, the simplest way is to return all the data in an entity. This can be done by simply listing an entity name and empty curly braces. For example, to return all node data, it would be `nodes {}`; to search for nodes with particular parameters within the curly braces, we use attribute names and the value they should equal (=), contain (~), be less than (<), or greater than (>). For example, to return nodes whose last report status was unchanged, the query would be `nodes { latest_report_status = "unchanged"}`.

We will not list the output for any of these queries as they can be verbose, but you will try to make examples in your lab at the end of this section.

These attribute statements can be further negated with `!`, chained with `and/or`, and parenthesized with brackets `()` to contain different statements. For example, to make a more complicated query to find whether a particular file was declared with the wrong permissions, we could run this PQL query:

```
resources { (type = "File" and title = "/etc/motd") and ! (
parameters.mode = "0644" and parameters.owner ="root") }
```

On the command line, this can also be run via the `puppet query resource {'latest_report_status = "unchanged"}'`.

PuppetDB queries can also be used in Puppet code with the PuppetDB function. Here's an example:

```
$changed_nodes = puppetdb_query(node[certname]{ resource {'latest_
report_status = "unchanged"}}) .map |$value| { $value["certname"] }
notify {"Nodes changed":
    message => "The following nodes changed on their last run
${join($changed_nodes, ', ')}",
}
```

In all these examples, it has been assumed certificates are set up for secure SSL communication either directly on Puppet infrastructure or with clients running the query. If using default locations, the `puppet query` command picks up the certificates automatically but can also be set like so:

```
puppet query '<PQL query>' \
  --urls https://puppetdb.example.com:8081 \
  --cacert /etc/puppetlabs/puppet/ssl/certs/ca.pem \
  --cert /etc/puppetlabs/puppet/ssl/certs/<certname_of_local_host>..
pem \
  --key /etc/puppetlabs/puppet/ssl/private_keys/<certname_of_local_
host>..pem
```

Web points can also be accessed via `curl` or equivalent commands, like so:

```
curl -X GET <fqdn_of_puppetDB_host>https://<fqdn_of_puppetDB_
host>:8081/pdb/query/v4\
  --tlsv1 \
  --cacert /etc/puppetlabs/puppet/ssl/certs/ca.pem \
```

```
    --cert /etc/puppetlabs/puppet/ssl/certs/<certname_of_local_host>.pem
  \
    --key /etc/puppetlabs/puppet/ssl/private_keys/<cert_name_of_local_
  host.pem \
    --data-urlencode 'query=<PQL query>'
```

To allow queries to be made directly from a desktop or other nodes, the Puppet client tools can be used. The setup instructions for installing on Open Source Puppet are detailed at `https://puppet.com/docs/puppetdb/latest/pdb_client_tools.html` and Puppet Enterprise has instructions at `https://www.puppet.com/docs/pe/2021.7/installing_pe_client_tools.html`.

Alternatively, the SSL authentication can be deactivated to allow unauthenticated queries following the instructions at `https://puppet.com/docs/puppetdb/latest/configure.html#jetty-http-settings`. This book would strongly advise against this as it would open the data for anyone to access on your network.

> In this section, we showed some of entities and queries it was possible to use with PQL. It would be impractical to go through all the possible options available to these entities and the range of options available to PQL, but the full details can be seen in the documentation at `https://puppet.com/docs/puppetdb/latest/api/query/v4/entities.html`. Additionally further examples of PQL queries can be seen in the documentation at `https://puppet.com/docs/puppetdb/latest/api/query/examples-pql.html`, and the Vox Pupuli community is building useful examples on its web pages at `https://voxpupuli.org/docs/pql_queries/`.

Lab – querying PuppetDB

SSH to the primary server and query PuppetDB for the following information:

- List the memory size of all the compiler servers (*hint*: compiler servers all have a trusted fact and Facter has a memory fact)
- List all the services being enforced on the Puppet server
- List the start and end times of the latest report of each server

Example answers can be found at `https://github.com/PacktPublishing/Puppet-8-for-DevOps-Engineers/blob/main/ch10/PQL_samples_answers.txt`.

> **Note**
>
> Be careful with these queries in production systems working at scale; some endpoints such as reports could contain a lot of data, and a query may put a lot of stress and load on a system.

Scaling with compilers

1. The review of Puppet platform components so far assumes that all components are present on a single primary server. However, as the number of managed nodes increases, it becomes impractical for a single server to handle them. According to Puppet's documentation, a primary server can manage up to 2,500 clients on default settings. To handle the growing number of nodes, Puppet uses horizontal scaling, which involves using Puppet compile servers. In *Figure 10.4*, a subset of primary services is shown to be moved onto compile servers. These servers can be configured in a round-robin selection in the client's configuration file or placed behind a load balancer. This enables multiple nodes to work together to compile catalogs while still allowing certain services to run on the primary server. According to Puppet's documentation, with the default compiler settings, up to 3,000 clients can be served per compiler:

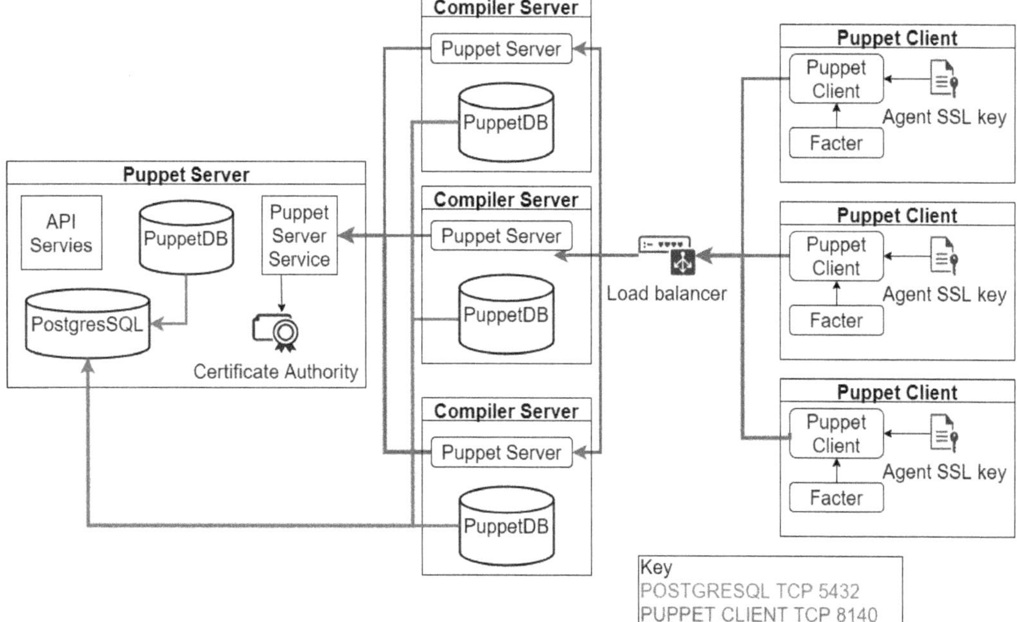

Figure 10.4 – Puppet compiler services

A compile server hosts a subset of services that are present on the primary server, such as Puppet Server and PuppetDB. This enables remote completion and synchronization of catalog compilation requests, thereby increasing the number of JRuby instances required for compiling catalogs.

1. The most widely used approach for directing client requests to compile servers is to utilize a hardware - or cloud-based load balancer. As there are several load balancer options available, Puppet does not provide explicit instructions on configuration. However, it recommends using the `/status/v1/simple` endpoint to check the health of compile servers. If the load balancer does not support HTTP health checks, checking whether the host is listening for TCP connections on port `8140` can provide a limited check.

2. There are alternatives to a load balancer, such as using DNS SRV records, which is detailed at `https://puppet.com/docs/puppet/latest/server/scaling_puppet_server.html#using-dns-srv-records`, or using a DNS entry with round-robin settings, as detailed at `https://puppet.com/docs/puppet/latest/server/scaling_puppet_server.html#using-round-robin-dns`, but as these tend to be much less frequently used, we will not go into detail in this book.

> **Note**
>
> In the `puppet.conf` file, it is possible to add a list of servers in the client server value to contact but this list would work only in the event of failures and would not try to balance out connections.

3. With compile servers, the CA remains on a single Puppet primary server and is referred back to when clients send their CSR or certificates for checking.
4. As stated at the beginning of the chapter, we will refrain from delving into the installation process in detail, as it would not add much value to Puppet's own instructions, available at `https://puppet.com/docs/puppet/latest/server/scaling_puppet_server.html` for open source and `https://puppet.com/docs/pe/2021.7/installing_compilers.html` for PE. However, it is essential to note that compile servers may require `dns_alt_names` to be added to their `puppet.conf` file if load balancers are being used in TCP proxying mode or a DNS round-robin method. This is necessary to enable all server names that may be used in requests through the load balancer.
5. Even with a load balancer enabled, it is possible to just target compile servers directly by running `puppet agent -t server=<server to send request>`.
6. In *Chapter 13*, we will provide more detailed information on how to monitor and manage server settings for scalability, and in *Chapter 14*, we will discuss Puppet's reference architectures for achieving scalability. However, it is important to note that there may be latency issues if compile servers are located too far away from the primary server. Therefore, it is recommended to keep them within the same region in cloud terms, as per best practices.

Lab – viewing compiler and load balancer configuration

The deployed lab environment consists of three compile servers. You can view the reports they are compiling and how pecdm configured the load balancer, as follows:

1. Log in to the web console and review the report runs of the Puppet instance server. In the **Metrics** section of a report, look for the **Report submitted by** section and note that this may vary in different reports. If there are few reports available, enter the **Jobs** section and run Puppet several times on your instance node to generate more reports.

2. View how PECDM created the Azure load balancer in the Terraform module at `https://github.com/puppetlabs/terraform-azure-pe_arch/blob/main/modules/loadbalancer/main.tf`.

Summary

In this chapter, we learned about the services provided by the Puppet server and how the embedded web server attaches handlers to mount points, which can then be requested via HTTP requests to endpoints.

It was shown that the `/puppet` endpoint provides services for configuration requests and how indirectors or environments can request specific components such as requesting a catalog from a server. The `/puppet-ca` endpoint similarly used indirectors to allow for requests to the CA. The `/puppet-admin-api` endpoint was then shown to allow for clearing the environment cache and JRuby instances as more advanced administrative actions.

It was then shown how Puppet creates a CA server with a root CA and an intermediate CA to sign or can run in legacy mode with a single combined CA. The options for using externally provided certificates were then discussed. The process of signing certificate requests was shown, with the `puppetserver certificate` command for managing certificates and requests and the `puppet ssl` command for managing agent certificate management. It was then shown how this process could be automated with auto signing, which could auto-sign everything, based on naming or based on a script running and viewing the certificate request.

JRuby interpreters were discussed, showing how JRuby is an implementation of Ruby on Java and capable of running Puppet's Ruby components, such as compiling Puppet code, in a scalable and concurrent way.

An overview of the user, service, and configuration files and logging was shown, examining the server side of `puppet.conf` and how to configure and view settings in the file and defaults using the `puppet config` commands.

Having reviewed the components of Puppet Server, the Puppet client lifecycle was then viewed, seeing how the agent makes CSRs to the CA and sends facts and a request for a catalog. The logs were viewed to show where requests re made and how this can be tracked through requests. It was shown how the client could be configured via `puppet.conf` and how additional information could be added to the CSR.

PuppetDB and PostgreSQL were then explored as a frontend/backend database architecture that can store reports generated from applying Puppet catalogs along with the latest facts and events from nodes. We reviewed the file directories and logging locations and then saw how PuppetDB could be queried on the API, command line, and Puppet code using PQL.

Compilers were then shown to be able to allow Puppet Server to scale horizontally by allowing Puppet Server and the PuppetDB services to be put onto multiple servers, which could be load-balanced for clients.

In the next chapter, we will show how Puppet classifies clients requesting catalog compilations so that it knows which version of code to apply and which classes. We will show how environments allow multiple versions of code to exist on the primary server and how to use a control repo to manage the modules and versions that should be included.

11
Classification and Release Management

The focus of this chapter will be on how Puppet deploys code and classifies this code to servers. Environments will be examined first, showing how this creates isolated groups of servers with particular versions of modules. We will discuss how this can provide both static and temporary environments. We will show how modern Puppet uses directory-based environments to have environment code in a specific location that **Puppet Server** can automatically discover. The methods a primary server can use to classify nodes will be discussed, at the most basic level using node definitions in the `site.pp` main manifest file or a collection of manifests, using Hiera lookups within these node definitions or with an **External Node Classifier** (**ENC**) script run by the primary server. The implementation of the **classification service** for Puppet Enterprise will be discussed, showing how it builds on top of these solutions using its own ENC script and the additional feature of node groups in the web console.

The Puppet agent run will be looked at in detail to show the steps involved and how data is loaded, cached, and refreshed when a catalog is compiled.

It will then be shown how to use the control repo structure with Puppetfiles to manage modules to deploy code into environments using `r10k` or `g10k`, with a discussion of various methods to synchronize code depending on the configuration of the local infrastructure. The PE-specific implementation, **Code Manager**, built on `r10k`, will then be discussed.

Having reviewed the technical structures for classification and release management, focus will then be put on the challenges and limitations of using this with regulated processes and multiple teams.

In this chapter, we're going to cover the following main topics:

- Puppet environments
- Understanding node classification
- Puppet runs
- Managing and deploying Puppet code
- Lab—classifying and deploying code

Technical requirements

Clone the control repo from `https://github.com/puppetlabs/control-repo` to your `controlrepo-chapter11` GitHub account and update the following files in this repo:

- `Puppetfile` with `https://github.com/PacktPublishing/Puppet-8-for-DevOps-Engineers/blob/main/ch11/Puppetfile`.
- Build a standard cluster with three clients by downloading the `params.json` file from `https://github.com/PacktPublishing/Puppet-8-for-DevOps-Engineers/blob/main/ch11/params.json` and updating it with the location of your control repo and your SSH key for the control repo. Then, run the following command from your `pecdm` directory:

```
bolt --verbose plan run pecdm::provision --params @params.json
```

Puppet environments

Puppet environments are a way to define specific versions of modules, manifests, and data to be used for groups of servers. Unfortunately, *environment* is a general technology term used for other purposes in organizations, and can easily be confused. The best advice would be to always say **Puppet code environment** if used in discussions outside of a purely Puppet context. This prevents a Puppet environment being associated directly with anything else.

Modern Puppet environments are dynamic directory-based, which means the Puppet server—or, in the case of `puppet apply`, the client—will look for the assigned environment to exist within a directory. Several variables set the location of related directories, including the `environments` directory itself, and we strongly recommend leaving all these settings at default to avoid confusion and issues. We will now look at the levels of code directories and paths within an environment.

Environment directories and paths

The first level is the code and data directory set by the `codedir` variable in `puppet.conf`, defaulting to `/etc/puppetlabs/code` for Unix and `%PROGRAMDATA%\PuppetLabs\code` for Windows (this is normally `C:\ProgramData\PuppetLabs\code`). Puppet Server does not use the `codedir` setting in `puppet.conf` and uses `jruby-puppet.master-code-dir` in `puppetserver.conf`, so both would need to be set if changed.

> **Note**
>
> Prior to `Puppet 3.3`, environments were declared in the `puppet.conf` file and each environment had to be declared in a section with `modulepath` and `manifests` variables set. This is still technically possible in Puppet today if `codedir` was not set but there is no reason to implement this approach.

The code and data directory contains two directories. First, there is a module directory to provide global user modules included in the default `basemodulepath` variable in `puppet.conf`. This `basemodulepath` variable by default contains `$codedir/modules:/opt/puppetlabs/puppet/modules` on Unix and `$codedir\modules` on Windows. The extra directory for Unix is used by the PE Server installation to place modules used to configure PE. These modules are prefixed with `pe` to avoid confusion with any modules that are already in use in environments.

The second directory is an environment directory; by the default setting of `environmentpath` in `puppet.conf`, this is `$codedir/environments` and is where environments will be viewed.

> **NOTE**
>
> The `codedir` directory is used to contain global Hiera data and configuration and, by default, `hiera_config` settings. If it finds a `$codedir/hiera.yaml` file, it will override the default `$confdir/hiera.yaml` file, which is now standard, as discussed in *Chapter 9*.

Within the `environments` directory, each environment to be created will have a directory that can have a name containing lowercase letters, numbers, and underscores. Each environment directory can contain the following:

- Puppet modules in directories specified by `$modulepath`
- Hiera data configured in a `hiera.yaml` file in the directory
- Classification data in a manifest or set of manifests in a directory specified by `$manifest`
- Environment configuration data in an `environment.conf` file in the directory

Having reviewed the directories and paths of environments, we will now look at the environment configuration files in more detail.

Environment configuration files

Environment configuration data can be set in an `environment.conf` file within the `environment` directory; this file has an INI-style format like `puppet.conf` but with no sections.

By default, if the `modulepath` environment variable is not set in `environment.conf`, it will be set to `$environmentpath/$environment/modules:$basemodulepath`.

So, in Unix-based systems, by default this will be the following:

```
/etc/puppetlabs/code/environments/$environment/modules: /opt/
puppetlabs/puppet/modules
```

In Windows systems, it will be this:

```
C:/ProgramData/PuppetLabs/code/environments/production/modules;C:/
ProgramData/PuppetLabs/code/modules
```

Remember to use a semicolon (`;`) to separate Windows directories in a list and a colon (`:`) for Unix systems.

In the *Managing and deploying Puppet code* section, we will discuss how the modules are deployed into this directory and how to list the contents of each directory in `modulepath`.

> **Note**
>
> Never set a `modulepath` variable to read from another environment directory. In the *Puppet runs* section, we will discuss the potentially inconsistent effects of environment data being cached and refreshed.

The `manifest` variable can be a single manifest file or a directory containing multiple manifests that will be read in alphabetical order. Puppet will see this variable as containing a directory if the path ends with a forward slash (`/`) or a full stop (`.`) and will recognize if it is a directory. If there is no setting in `environment.conf`, the default will be a directory at `$environmentpath/$environment/manifests`, which is `/etc/puppetlabs/code/environments/$environment/manifests` for Unix-based systems and `C:/ProgramData/PuppetLabs/code/environments/$environment/manifests` for Windows-based systems. The directory environment will never use the global `manifest` setting from `puppet.conf`. In the next section, we will go into further detail about how node definitions and Hiera lookups can be used to classify servers in this environment with these manifests.

The `environment_timeout` variable states how long Puppet Server will cache a particular environment, overriding what is set. Puppet advises not to set this in `environment.conf`, only using the global version in `puppet.conf`, and to only use `0` or `unlimited`. The role of caching will be discussed further in the *Puppet runs* section of this chapter.

The `config_version` variable can set a script to run once the catalog has compiled and return the output as part of the logging. If not set by default, a script will return the time the catalog was compiled in the Unix epoch format (the number of seconds that have elapsed since January 1, 1970 midnight UTC/GMT). For the default epoch script, the output will appear as follows:

```
Info: Applying configuration version '1663239677'
```

A more useful example will be shown when using Git-based deployment solutions in the *Managing and deploying Puppet code* section.

> **Note**
>
> `environment.conf` and the `config_version` script can use the `basemodulepath`, `environment`, and `codedir` global variables.

Now that we have reviewed the configuration of environments, it is useful to understand how we can validate the configuration and the types of environments deployed.

Environment validation and deployment

The settings discussed in `puppet.conf` and `environment.conf` can be checked by using the `puppet config print` command, deploying the `--environment` flag to look at a particular environment and `--section` for a particular section in `puppet.conf`. For example, to check the `codedir` variable in `puppet.conf` and the `modulepath` variable in the production environment, the following commands could be run:

```
puppet config print codedir
puppet config print --environment production modulepath
```

By default, Puppet Server will create a production environment but a Puppet client running `apply` will not. For both scenarios, production is the default environment Puppet will run from. In the next section of this chapter, we will show how servers get classified into other environments.

There are three environmental strategies: permanent, temporary, and organizational silos. Permanent environments are long-lived and the environment naming typically matches the server's use, such as if the server is a product or development server. Temporary environments are those that can be used in situations such as testing changes before they are promoted, while organizational silo environments reflect divided infrastructure where different teams such as Windows and Linux teams own different servers and have different environments. These strategies can be mixed together as required to meet your organization's approach.

Now that we've learned about Puppet code environments, we will see how to classify clients based on their use in an environment and set of modules from that environment.

Understanding node classification

Classification of a node involves finding which environment a node should use, which classes should be applied to a node, and which parameters should be applied to a node. The ideal scenario is to have a single role class applied to a host, but the business logic can be more complicated. This applies to both agent runs to the Puppet Server and `puppet apply` runs.

Having defined what node classification is, we will now look at the methods that can be used for classification, taking node definitions first as the simplest approach.

Node definitions

The most basic method of node classification is using a **node definition**, which is a section of Puppet code allowing matching against node names to assign classification information and top-level variables to servers but not the environment. If only using node definitions, the client's requested environment based on `puppet.conf` will be used. The node name will be the same as the `certname` setting from `puppet.conf`, which by default is the node's **fully qualified domain name** (**FQDN**).

The syntax of a node definition is set out here:

- The `node` keyword
- A node name as a quoted string, **regular expression** (**regex**), or `default`
- A mixture of the following Puppet code items within curly braces ({ }):
    - Class declarations
    - Variables
    - Resource declarations
    - Collectors
    - Conditional statements
    - Chaining relationships
    - Functions

It is recommended to keep node definitions down to a minimum and only use class declarations and variables. If any manifest contains a node definition, then the node definitions must match all nodes, or compilation for nodes that do not match will fail. This is normally made safe by ensuring there is a default definition even if the default definition contains no code.

A node will only match one node definition, and this is prioritized by the following:

- An exact name match
- A regex match (multiple regex matches are unpredictable, and only one will be used)
- `default` (the keyword that nodes will match if they have failed to match any other definition)

> **Note**
> A prioritization step before `default` would look for any partial matches of the hostname if `strict_hostname_checking` were set to `false` in the `puppet.conf` primary server. To avoid this insecure matching, it is set to `true` by default in Puppet 5.5.19 + and 6.13.0+, and in Puppet 7 onward was removed as an option.

For example, the following code will classify `server1.exampleapp.com` to the `role::oracle` class and `server2.exampleapp.com` and `server3.exampleapp.com` to the `role::apache` class. Any other servers that end with `exampleapp.com` will be classified to `role::example_common_windows` or `role::example_common_linux` depending on the OS family, such as `server5.exampleapp.com`, and any other node will be classified to `role::common`, such as `server1.anotherapp.com`:

```
node /.exampleapp.com$ {
  if $facts['os']['family'] {
    include role::example_common_windows
  else
    include role::example_common_linux
  }
}
node 'server1.exampleapp.com' {
  include role::oracle
}
node 'server2.exampleapp.com','server3.exampleapp.com' {
  include role::apache
}
node default {
  include role::common
}
```

It is a default to have a `site.pp` file in a `manifest` directory to keep things simple, but multiple manifests in this directory can contain node definitions that could be used to organize the files based on organization, use case, or ownership. It is clear having many node definitions simply will not scale; a recommended way of keeping node definitions simple is to use a default definition that looks at the certificate of the node to have a `pp_role` extension that contains the name of the role, as shown in this code example:

```
node default {
  $role = getvar('trusted.extensions.pp_role')
  if ($role == undef) {
    fail("${trusted['certname']} does not have a pp_role trusted
fact")
  }
  elsif (!defined($role)) {
    fail("${role} is not a valid role class")
  }
  else {
    include($role)
  }
}
```

Using the `getvar` function to avoid issues with hosts without certificates and the `defined` function to confirm the declared role is visible in the environment, it will include the role declared in the certificate.

Any code applied outside of the node definitions will apply to all nodes, but setting uncontrolled global defaults like this is not a recommended approach. In the previous code block, role classes were used, but any class could be included for exceptions.

A local `puppet apply` call will not look for manifests in the `manifest` variable setting from `puppet.conf` but is expected to do what is passed on the command line either via the `-e` flag or by passing a specific manifest file.

Having looked at the code-driven approach of node classification, we will now look at how Hiera data can be used to classify nodes.

Classifying nodes with Hiera

A more data-driven approach can be made in the default node definitions with Hiera arrays using the `lookup` function. While the `lookup` function could be used outside of the node definition, we recommend avoiding this to ensure, if any other node definition were specifically added for a node, it would have the expected result of only applying the node definition and not a less predictable mix.

The first step would be, as we saw in *Chapter 9*, to ensure an appropriate Hiera hierarchy is in place for each environment, assuming a simple hierarchy of node, OS, and defaults in the `hiera.yaml` environment, as shown here:

```
datadir: data
data_hash: yaml_data
  - name: "Node data"
    path: "nodes/%{trusted.certname}.yaml"
  - name: "OS defaults"
    path: "os/%{facts.os.family}.yaml"
  - name: "Common data"
    path: "common.yaml
```

We could then add a lookup within a `default` node definition:

```
node default {
lookup( {
  'name'            => 'classes',
  'value_type'      => Array,
  'default_value'   => [],
  'merge'           => {
    'strategy' => 'unique',
  },
} ).each | $classification | {
  include $classification
}
```

While it would seem more appropriate to call the variable `class`, this is not possible due to `class` being a reserved word.

Data in the environment-level Hiera could then be added to a `common.yaml` file to ensure that by default, servers get the `core` role:

```
---
classes:
  - role::core
```

Then, we create an `os/RedHat.yaml` file in the data file containing the following code:

```
---
classes:
  - role::core::redhat
```

This would ensure all servers from the Red Hat family such as CentOS would get assigned the `role::core::redhat` class. To assign a particular role to a server, we create a `node/exampleapp.example.com.yaml` file, containing the following code:

```
---
classes:
  - role::docker
```

This would assign the `role::docker` class to the `exampleapp.example.com` node.

To allow exceptions and a more complex combination setup, hashes instead of arrays could be used, changing the lookup in `site.pp` from a unique to a deep merge strategy and the data from an array to a hash:

```
node default {
lookup( {
  'name'           => 'classes',
  'value_type'     => Hash,
  'default_value' => []
  'merge' =>
    'strategy' =>  'deep',
}).each | $classification | {
  include $classification
}
```

In this case, we could use keys within Hiera that are only visible in Hiera to take over the role construct and use profiles directly, setting a `common.yaml` file to ensure the default classification gets a core profile and a security profile:

```
---
classes:
base profile: profile::core
security profile: profile::security
```

Then, for a specific server `exampleapp.example.com`, the `security_profile` variable could be set in `node/exampleapp.example.com.yaml`:

```
---
classes:
security_profile: profile::security::legacy
```

This would override the security profile key and result in `exampleapp.example.com` being classified as `profile::security::legacy` and `profile::core`.

More complex Hiera-based key lookups could be constructed to look up based on Facter values, but since this is not a recommended approach in this book, enough detail has been shown to understand how Hiera can be used. It is worth seeing the `psick` module `https://forge.puppet.com/modules/example42/psick` by example42 that uses the Hiera approach and can be used to have a preset and staged way of including modules in the Linux case. Including the `psick` class and simply setting the Hiera keys with hashes would be enough to classify a host:

```
psick::firstrun::linux_classes
psick::pre::linux_classes
psick::base::linux_classes
psick::profiles::linux_classes
```

Having reviewed the code and data approaches for classification in detail, we will now cover the more advanced approach of using ENC scripts.

ENC scripts

An ENC is a script that the Puppet Server or a `puppet apply` call can run. The requirements of the script are to take an argument of the `certname` of the client and return either a nonzero return code for unknown nodes or a YAML output containing the classes, parameters, and environment for catalog compilation. Inside this ENC, it is possible to access various external data source references such as PuppetDB or internal data sources of your organization. It is not important which language the ENC is written in.

An example output would look like this:

```
---
classes:
  role::core::windows
  sqlserver_instance:
    features:
      - SQL
    source: E:/
    sql_sysadmin_accounts:
      - myuser
parameters:
  dns_servers:
    - 2001:4860:4860::8888
      - 2001:4860:4860::8844
  mail_server: mail.example.com
  vault_enabled: true
environment: uat
```

In this example, it can be seen the server will have the `role::core::windows` class applied and the `sqlserver_instance` class with associated parameters, a list of parameters that will be global variables in the catalog, and an environment of **user acceptance testing** (**UAT**).

It would normally be better to pass class parameters via Hiera data but this is just to demonstrate what is possible in the ENC output.

To configure the use of ENC scripts, two variables must be set in `puppet.conf`: first, `node_terminus`, which defaults to `plain`, to only use the manifests to define classification. Setting `node_terminus` to `exec` causes the second variable, `external_nodes`, to be checked, which should be set to the location of a script. For example, the Foreman project uses an ENC that is defined in `puppet.conf` by its configuration module as follows:

```
node_terminus = exec
external_nodes = /etc/puppetlabs/puppet/node.rb
```

The contents of the script can be seen here: `https://github.com/theforeman/puppet-puppetserver_foreman/blob/master/files/enc.rb`.

The configuration module for placing this script can be found at `https://forge.puppet.com/modules/theforeman/puppetserver_foreman`.

Developing ENCs is beyond the scope of this book, and it would be advisable to avoid the complexities involved in accessing external data this way as it can be expensive to access.

We have covered how ENC scripts work, but PE uses its own type of ENC script with additional features.

PE classifier

PE provides its own ENC classifier that accesses the classification service API, which is a Clojure app, and stores node group information in the **PostgreSQL** classification database.

This is configured by setting `node_terminus = classifier` in `puppet.conf` by the installer and should not be changed as it will not be supported.

> **Note**
>
> `node_terminus` for PE used to be called `console` on PE 4 and previous versions.

The node groups come in two types: environment and classification. The environment groups are used to assign environments to nodes, while the classification nodes are intended for assigning classes and adding parameters and variables. The node groups can be viewed and configured from the PE web console in the **Node groups** section.

All node groups can contain rules to define based on facts or by directly naming servers to be contained by a node group. They can contain any classes with any defined class parameters that will be classified to these matching nodes, parameters known as configuration data that act like Hiera data, acting as overrides and taking precedence over Hiera, and variables that are declared as global variables for the group.

> **Note**
>
> Older versions of PE did not enable configuration data by default and a section in `/etc/puppetlabs/puppet/hiera.yaml` had to be added:
>
> ```
> hierarchy: - name: "Classifier Configuration Data" data_hash:
> classifier_data
> ```

By default, as pictured in *Figure 11.1*, PE will have an **All Nodes** node group as a containing parent node group for all configurations, and beneath that split out to **All Environments**, an environment group acting as a parent group for all declared environment groups, and **PE Infrastructure**, a classification group used for configuring the PE architecture:

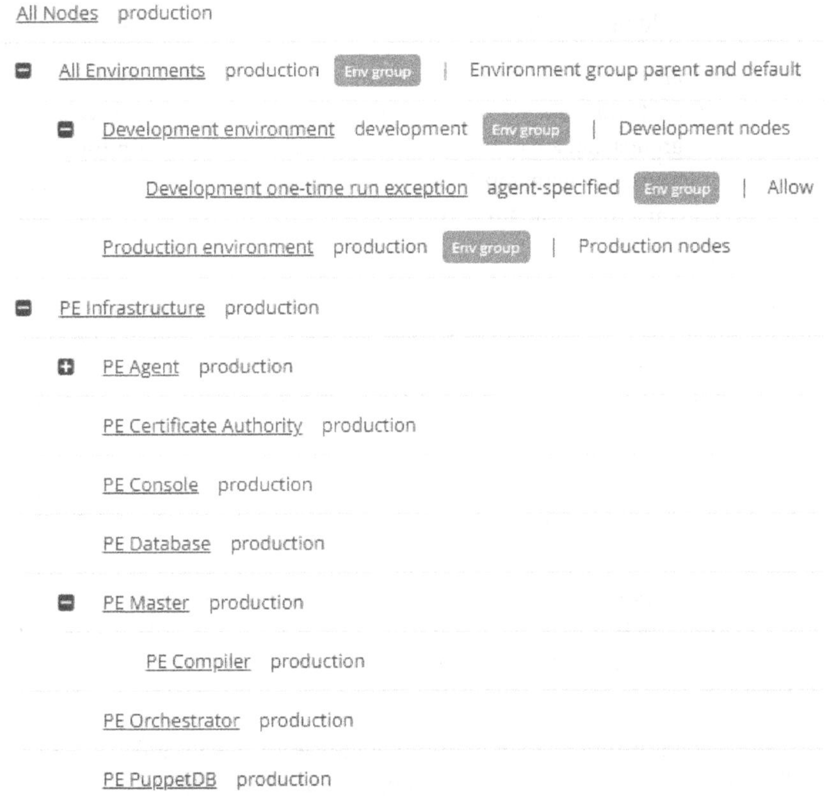

Figure 11.1 – PE default node groups

The environment groups are marked in *Figure 11.1* as **Env group**. The environment node groups by default use the `trusted.extensions.pp_environment` fact in a rule to match production or development into groups of the same name and ensure the named environment is assigned. If `trusted.extensions.pp_environment` is not set, the **All environments** node group will act as a catch-all to assign production as the default environment. Using the `pp_enviroment` trusted fact prevents the server from being moved to another environment without regenerating the server certificates, which will require access to both the client and primary server. The **Development one-time run exception** node group (called **Agent-specified** in previous PE versions) sets rules to allow development servers to run environments specified by the client. This can be useful when developing modules in feature branches, allowing testing to take place simply by running `puppet agent -t --environment=myfeaturebranch`.

Approaches to the development and deployment of environments will be discussed further in the *Managing and deploying Puppet code* section, but it may prove necessary to have more environment levels between production and development, in which case the recommended approach would be to create a node group of that environment name under all environments and create a rule matching `trusted.extensions.pp_environment` with your set environment name.

Environment groups should be kept simple, so avoid assigning any class parameters or variables.

When classification groups are nested, they inherit the definition from their parent group. When creating a group structure, it will make sense to start with a general layer of configuration and narrow it down to classification groups that are more specific. This can be seen in the **PE Infrastructure** node groups, which start by ensuring general top-level parameters such as `puppet_master_host` are set, which apply to all Puppet infrastructure hosts, and then have specific services and functions such as a compiler or PuppetDB, which will be configured only on a subset of nodes.

It can be confusing because this inheritance applies to rules as well, so if the parent rule has already set a rule limiting nodes, the child node groups' rules will be combined with the parent node group. This also applies to the pinning of nodes; you cannot just ignore rules and pin any server visible to the primary server. It is also important to note if the child node group contains no rules, it will not apply classification, even that inherited from a parent group.

Further confusion can arise from the purpose of the environment variable in classification node groups; this does not define where the assigned classes will run from but, in fact, tells the node group the environment in which to look for available class names. This can create issues if node groups are shared between development and production nodes and new classes are initially introduced to a development environment before being promoted to production, so it can often be the case that it makes the most sense for application node groups to use the lowest level of environment to have full visibility of classes.

To keep things simple, it is recommended to use straightforward classification roles that are kept as children of all nodes and simply have rules matching `trusted.extensions.pp_role` to a specific class role name, and then assign that role class to the classification role group.

To automate the creation of node groups, the `node_manager` module (`https://forge.puppet.com/modules/WhatsARanjit/node_manager`) can be used to manage them through Puppet code, which is how the `peadm` module itself configures Puppet node group information. For example, `peadm` ensures that nodes with the `puppet/puppetdb-database` trusted extension are assigned to the **PE Database** node group with the following code:

```
node_group { 'PE Database':,
  rule => ['or',
    ['and', ['=', ['trusted', 'extensions', peadm::oid('peadm_role')],
'puppet/puppetdb-database']],
    ['=', 'name', $primary_host],
  ]
}
```

> **Note**
>
> The `node manager` module has the `purge_behavior` setting, which, if set to `none` for resources, ensures only the specific changes you wish to make are applied to node groups. By default this is set to `all`, removing any settings you have not declared.

Alternatively, the APIs can be used to perform backups and restores of node group data, saving to a file with `/classifier-api/v1/groups` and restoring with `/classifier-api/v1/import-hierarchy`. `Peadm` implements backup and restore classification tasks using these APIs: `https://github.com/puppetlabs/puppetlabs-peadm/tree/main/tasks`.

> **Note**
>
> Since PE version 2019.2, a `$pe_node_groups` top-scope variable that returns all node groups is available.

A further method to use external data to add classes with the **Puppet Data Service** (**PDS**) will be shown in *Chapter 13*. But having reviewed the various methods of classification, we will now discuss best-practice approaches to classifying nodes.

Recommended approach

A mixture of ENCs and node definition approaches can be used as it will merge the information, but this can make it harder to understand where classification has taken place. It would be best practice to choose one option if possible or at least to be clear on the purpose of each mechanism, such as node definition to match roles based on certificate and Hiera to match node exceptions.

Presuming classification has not already been chosen by your organization or is specific within your configuration model, such as using Foreman or `psick`, we recommend the simple pattern of assigning a default node definition based on the `pp_role` extension in the certificate for open source Puppet: use node groups matching the `pp_role` extension against node group role, and `pp_environment` against the environment to be used for PE. This is what Puppet Support expects and is the built model, but it limits the use of any variables or configuration data within the Hiera data setup.

The other mechanisms in sections *Node definitions* and *Classifying nodes with Hiera* were discussed since in many organizations, classification will already be in place and will not be easily changed and therefore must be understood. It is important to know if complex classifications must be produced; this can mean data is not being put in the right place or—worse—Puppet is not being used well and too many variations of servers are being produced. When we maintain tight standards with minimal exceptions, servers can be disposed of and rebuilt easily, reducing operational complexity and the cognitive load of support teams.

Now that you have understood how servers are classified to an environment and to classes, we will show how different data is loaded and cached during Puppet runs.

Puppet runs

In this section, the steps of a Puppet run and classification will be detailed. For the case of Puppet runs, a `puppet apply` command should be considered as the equivalent of a Puppet server and client on the same node.

When a catalog request is made by a client, four things are sent to the server:

- The node name
- The node's certificate (not sent for `apply`)
- Facts
- The requested environment

The node name is the `certname`, and along with the requested environment is embedded in the API request made—for example, `/puppet/v3/catalog/exampleserver.example.com?environment=uat`.

The certificate can contain extensions, which will be turned into trusted facts.

After the server receives the agent data, it asks the configured node terminus for a node object. In the case of `plain`, this will be blank, or for `exec` or `classifier`, YAML output will be returned containing classes, parameters, and environment.

By default, `puppet.conf` sets `strict-environment-mode` to `false`, and this returned environment will override the agent request; if it is set to `true`, the catalog compilation will fail. The `agent_specified_environment` fact will appear if the agent specified an environment on the Puppet run.

The variables will then be set from the facts as both top-scope variables and the `$facts` hash, extensions in the certificate as trusted facts in the `$trusted` hash, and parameters returned from the node terminus as top-scope variables.

The main manifest will then be evaluated, looking for it to be defined by the environment configuration first and then the client's `puppet.conf` file if it is unset. If any node definitions exist, Puppet will attempt to match the `certname` and fail compilation if it does not.

Any resources outside of the node definition are evaluated and added to the catalog and any classes. As was noted in the *Node definitions* section, it is not recommended to declare anything outside of node definitions. The matching node definition will then evaluate the code, overriding any top-scope variables with variables declared in the node definition, adding resources to the catalog, and loading and declaring classes in the node definition.

Puppet will then load the manifest containing classes declared in the main manifest using the `modulepath` variable configured for the environment. As a class is loaded, the code is evaluated and resources are added to the catalog, and any classes declared within them will be loaded and evaluated.

Puppet then loads and evaluates the classes that were returned from the node object.

Having seen how Puppet classifies nodes and how agent runs process these classification methods, it is now time to see how the environments are managed and deployed to the primary server to make the right versions of code available to the nodes.

Managing and deploying Puppet code

By default, just creating the folders and dropping module contents into place combined with the `puppet module install` command to automate pulling from the Forge API is enough to make modules visible in environments and to allow them to be wrapped up in package management to create versions. But this is not an approach that we recommend as it centralizes the deployment of modules and environments, most likely making a single team a gatekeeper. We will see that control repos provide more flexible control.

The most common approach is to use a Git repository known as a control repo. Puppet provides a template for this repository at `https://github.com/puppetlabs/control-repo`.

> **Note**
>
> The Puppet Forge author example42 provides its own templated control repo for use with its integrations and pre-designed implementation approaches: `https://github.com/example42/psick`.

Puppet's control repo template contains many of the directories and files discussed in the first section of the chapter, along with Hiera data and some additional files specific to module deployment. *Figure 11.2* shows the contents of the Puppet control repo:

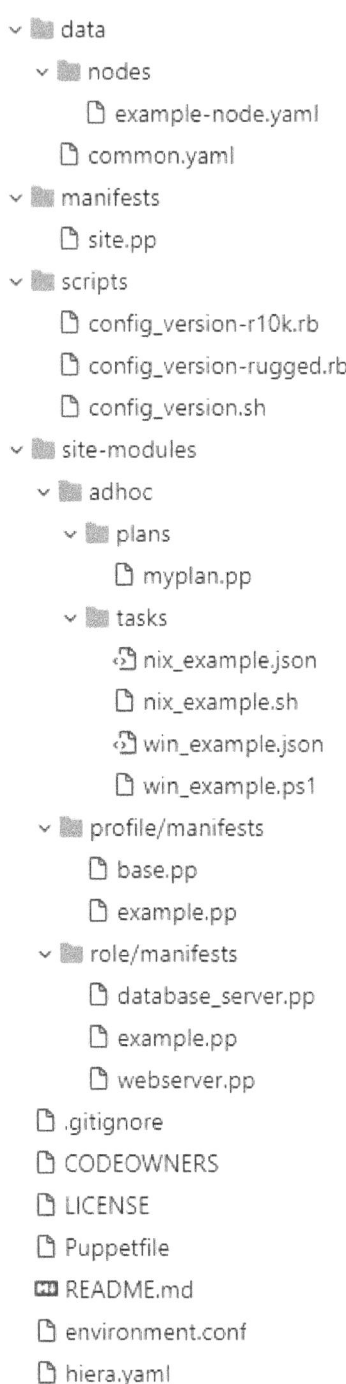

Figure 11.2 – File structure of the Puppet control repo template

In the first section of this chapter, *Puppet environments*, we discussed many files and directories, with `environment.conf`, config version scripts, and the `manifests` directory for classification. Also visible is the Hiera configuration in `hiera.yaml` and a data directory showing a simple initial two layers of nodes, to match specific node names and common data, to act as a default for nodes that do not match. The `site-modules` directory intends to show that ad hoc plans and tasks can be deployed as part of this control repo as well as potentially give a home to roles and profiles. The `scripts` directory is also worth reviewing to see in the config version script at `https://github.com/puppetlabs/control-repo/blob/production/scripts/config_version.sh` how it will add Git revision control information about the environment to the run. The part that we have not reviewed is the Puppetfile file.

The Puppetfile file is a Ruby-based **Domain-Specific Language** (**DSL**) that provides a way to declare which modules should be downloaded to an environment, where to source them from, and which version to use. It is also possible to override module location settings by declaring `moduledir` as a variable or the `installpath` parameter on a particular module. We do not recommend this as good practice as it can be confusing to users unfamiliar with your environment and, if set to be outside the environment directory, can affect caching and make the environment inconsistent. This will be discussed later in this section.

Puppetfile module declarations at their simplest level contain the following:

- `mod` keyword
- A name in single quotes
- Optionally a comma, then a version number or the `:latest` keyword

For example, the following code block assumes Puppet Forge as the source and installs the latest version of `dsc-octopusdsc` if the module is not present, but will not result in the module being updated:

```
mod 'dsc-octopusdsc'
mod 'puppetlabs-chocolatey', '6.2.0'
mod 'puppetlabs-stdlib' , :latest
```

This piece of code will install `puppetlabs-chocolatey` to the fixed version 6.2.0 and will install `puppetlabs-stdlib` and keep updating it to the latest version. It is important to note this will not result in Puppet Forge dependencies being installed—this must be managed within the Puppetfile manually. Looking at module documentation on the Puppet Forge you will see example code on how to add the modules to Puppetfiles.

To access modules within other Git repositories, the `git` option and the HTTP address to the repository should be given. These can then be paired with one of the following options to clone a specific version of the Git repository:

- `ref`, with a reference to a tag, a commit, or a branch
- `tag`, with a specific tag
- `commit`, with a specific commit reference
- `branch`, with the name of a branch or the `:control_branch` keyword (which will automatically look up the control repo's branch name)
- `default_branch`, a branch to use if all the preceding options fail

The following code demonstrates how the `git` options in the preceding list can be mixed and matched:

```
mod 'exampleorg-examplemodule1',
  :git => 'https://internalgitservice.com/exampleorg/examplemodule1',
  :tag =>  'v.0.1'
mod 'exampleorg-examplemodule2',
  :git => 'https://internalgitservice.com/exampleorg/examplemodule2',
  :commit => '68a140bd096a55019b3d5c8c347436b318779161'
mod 'anotherorg-anothermodule',
  :git => 'https://internalgitservice.com/anotherorg/anothermodule',
  :branch => :control_branch,
  :default_branch => 'main'
```

This code block takes `examplemodule1` at `tag` version `v.0.1` and `examplemodule2` at `commit` version `68a140bd096a55019b3d5c8c347436b318779161` from the same Git organization. For `anothermodule`, if a branch with the same name as the environment that we are trying to deploy exists, it will use that; otherwise, it will clone at the `main` branch.

In air-gapped environments where access to the Puppet Forge API is limited or in regulated environments where it is an audit requirement to have a company-stored copy of all code, it may prove necessary to download copies of code from the Forge and use it from your organization's own Git system. In this case, it is strongly advised that you follow the project URL on the module page, perform a Git clone of the source of the Puppet Forge module, and then change the remote directory to your own Git repository copy. This ensures the commit history is maintained and on a regular basis, you can simply clone the code again and add new commits to your own local repository.

Regardless of how Forge modules are downloaded, if they are not coming directly from the Forge at their latest version it is important to frequently check versions and make this part of a regular cycle to test and update. This ensures you are getting the latest features and fixes and means you avoid having to perform large version upgrades that are harder to test. Following the **Content and Tooling** (**CAT**) team at `https://puppetlabs.github.io/content-and-tooling-team/blog/` can help keep track of module releases.

> **Note**
>
> JFrog Artifactory users can use a Puppet Forge plugin to synchronize and host modules internally, as documented at `https://www.jfrog.com/confluence/display/JFROG/Puppet+Repositories`.

With this structure to manage several environments, it is simply a case of creating branches on the Git repository, with each branch representing an environment that can have its own independent content to be deployed.

To manage deployment, the standard system used for **Open Source Puppet** is known as `r10k`, and the system used for **Puppet Code Manager** for PE is based on `r10k` but has further integrations for PE.

The installation instructions for `r10k` are straightforward and available direct from the repository at `https://forge.puppet.com/modules/puppet/r10k`. Instructions to configure Code Manager in PE either in node groups or via Hiera are available at `https://puppet.com/docs/pe/2021.7/code_mgr_config.html`.

In both cases, as part of these instructions, an SSH key will be generated to allow for communication between `r10k` and any Git repositories you have declared.

An alternative option for Puppet Open Source is to use `g10k` (`https://forge.puppet.com/modules/landcareresearch/g10k`), which is a rewrite of `r10k` in **Go** and has substantial performance improvements.

> **Note**
>
> You can still use `r10k` directly in PE, but this is not an approach Puppet will provide support for.

For open source Puppet, having configured and deployed `r10k`, it is then possible to run a `sudo -H -u puppet r10k deploy production` command to deploy a specific branch or leave off an environment name to deploy all available environments. A Webhook can also be configured using the Sinatra server, as detailed in the `r10k` instructions at `https://forge.puppet.com/modules/puppet/r10k/readme#webhook-support`.

For PE, Puppet Code Manager is a **Clojure** application that exposes an `/code-manager` API using a token generated in the PE **role-based access control** (**RBAC**) system, which will be covered in detail in *Chapter 14*. It can be accessed either directly to the API or by running the `puppet code deploy` command. For example, the following code will generate a token for the currently logged-in user for the next 2 hours and then deploy in the production environment:

```
puppet-access login --lifetime 2h
puppet code deploy production --wait
```

In either version, to see the deployed modules you can use `puppet module --list`, which will also show any dependency issues.

> **Note**
>
> Puppet Code Manager uses `r10k` underneath. To get more detailed debugging information, the following command can be run, which is used for deploying in production:
>
> ```
> runuser -u pe-puppet -- /opt/puppetlabs/puppet/bin/r10k -c /opt/puppetlabs/server/data/code-manager/r10k.yaml deploy environment production --puppetfile --verbose debug2
> ```

For these deployments, it is important to understand caching that can take place. All Puppet code is read and parsed when the environment is loaded—as is the `hiera.yaml` file—and it is not re-read until the environment cache is expired, or the JRuby instance is refreshed. The `environment.conf` file by default sets this to `unlimited`. While Puppet templates and Hiera data are read anew from disk on every function call, they are not cached. This means that if any local edits take place to Hiera data or Puppet templates outside of `r10k`, they will be viewed. It also means that if environments have module paths that look into other environments, a deployment would result in it only seeing the Hiera and template updates. This is why it is strongly recommended to avoid this approach.

When using compilers to synchronize code, open source Puppet,\ has various options depending on your environment as to how to deploy the code: installing and running `r10k` on every compiler node, performing a `rsync` operation from the primary server to compilers, or using a read-only **network file share** (**NFS**) from the primary to all compilers. This choice will be entirely down to what is best for your organization in terms of network configuration and security standards.

On PE, Code Manager has a specific implementation using the file sync client and server, as shown in *Figure 11.3*:

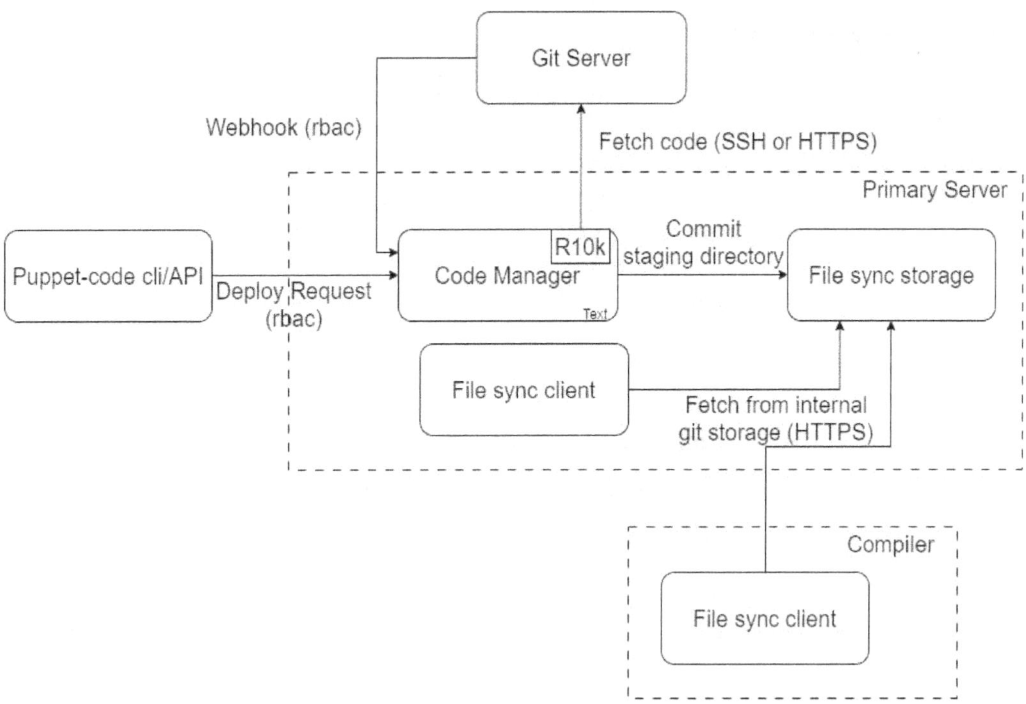

Figure 11.3 – Puppet Code Manager architecture

A code deployment request will come in a request with an RBAC token either via the command line or tooling. This will pull down the code to the commit staging directory on the primary server. The file sync clients for all infrastructure nodes have a polling watcher that sees the deployment and alerts the file sync process. This will result in the file sync process doing one of two things, depending on whether lockless code deploys are enabled (which were introduced in PE 2021.2). If lockless code deploys are not enabled on the relevant server, all JRuby instances will need to be reserved to prevent any catalog runs using inconsistent environments. Remembering how different environment data is cached in the *Puppet runs* section, once reserved, the files will be synced into the environment directory, and the JRuby instances released. This does mean code deployments can be impactful on performance.

If lockless code deployment is enabled, symbolic links or symlinks are used for the environment directories, which means the file sync will synchronize to a folder named after the version commit and, on completion of synchronization, redirect the environment symlink to this new folder. This requires more disk space because more environments will be deployed at once but ensures catalogs can continue to run since they will use the directory the symlink had when they started to run. To enable lockless code deploys, follow the instructions at `https://puppet.com/docs/pe/2021.7/lockless-code-deploys.html`.

Now that we understand how Puppet deploys code to environments, we will look at workflows that can be used to manage the promotion of module code through those environments.

Creating a workflow

There are two common approaches for creating a workflow to deploy code. The first method is to put the control repo as a central gatekeeper of versions. This means that every module declaration on the Puppetfile has specific versions and typically will have the lowest-level environment updated with a specific reference such as `tag`, `commit`, or `branch`. These changes are tested in feature branches and then promoted through environments by merging the changes from one branch to the next, running the code on servers, and confirming expected results. For example, the steps involved in such as process may include the following:

- Creating a feature branch of the control repo and updating the `module1` tag version from 1.1 to 1.2
- Merging the feature branch with the development branch and deploying development
- Merging the development branch with the UAT branch and deploying UAT
- Merging UAT with production and deploying production

This is not a natural Git flow and does not use the main branch. It is very focused on deployment, requiring a lot more management of environments. This approach can be particularly difficult with multiple teams since it will result in the requirement of a gatekeeper such as the Puppet platform team to manage changes to the Puppetfile control repo and manage the schedule of when code deployments are made.

If this approach is taken, it is advisable to have multiple control repos using the prefix configuration settings—this can be useful for teams that want to use different sets of modules, such as Windows and Linux, or want to have isolation and protection around the control repo and have separate ownership of code and servers but want to share infrastructure.

The second approach is to have all modules in the Puppetfile in the control repo set to use a branch of `control_branch` and a default of `main`. Maintenance of the Puppetfile will then only involve the addition and removal of modules. The management of versions will be on the modules themselves, with code changes pushed to the main branch from temporary feature branches before being merged into each static environment branch. Here's an example:

- Create a feature branch on `module1` and control repo testing code changes
- Merge the feature branch of `module1` with `main`
- Merge changes of the module branch from main to development, then deploy and test
- Merge changes of the module branch from development to UAT, then deploy and test
- Merge changes of the module branch from UAT to production, then deploy and test

Using pipelining tools as part of the **pull request** (**PR**) and deployment process is strongly advised. **Continuous Delivery for PE** (**CD4PE**) (discussed in *Chapter 14*) comes with prebuilt checks to make this easier, but various tools exist, such as Jenkins or GitHub, with which you can ensure the pre-commit hook checks and testing we discussed in *Chapter 8* are performed before a PR can be completed.

> **Note**
>
> Some excellent sources of existing pre-commit hooks to copy into place can be found at `https://pre-commit.com/hooks.html`, `https://github.com/pre-commit/pre-commit-hooks`, and `https://github.com/mattiasgeniar/puppet-pre-commit-hook`.

Lab – classifying and deploying code

In this lab, complete the following tasks:

- Create a node definition that assigns the `motd` module to any node with `node` in the certname in the `manifest/site.pp` file of the production environment
    - The `motd` module is already in the `Puppetfile` file in the production control repo
    - The defaults for `motd` should be fine using `include motd`
    - See an example solution at `https://github.com/PacktPublishing/Puppet-8-for-DevOps-Engineers/blob/main/ch11/default.pp`
- Create a node definition to assign the `icinga2` module to all Windows nodes that get created
    - The `icinga2` module is already in the `Puppetfile` file in the production control repo
    - The defaults for `icigna2` should be fine using `include incigna2`
    - Windows nodes will always contain `windows-node` in the `certname`
    - To deploy on the PE web console, run the `peadm code_manager` task from the **Orchestration** task menu, entering the `'deploy production'` action string
    - See an example solution at `https://github.com/PacktPublishing/Puppet-8-for-DevOps-Engineers/blob/main/ch11/node.pp`
- Create and deploy a development environment, adding the `docker` module:
    - On the GitHub web page, go to your control repo and select the arrow next to the production branch, and type `development`
    - Click the text generated below, which should say `create branch: development from 'production'`

    - Add the line `mod 'puppetlabs-docker', latest` to `Puppetfile`, making sure you are on the development branch
    - On the PE web console, run the `peadm code_manager` task from the task menu, entering the `'deploy development'` action string

- Create a node group for a role that includes Docker pinning one of your nodes to it and development, then promote the development branch to production and deploy:
    - Create a node group called `docker` by selecting **node groups** under **inventory** and **add node groups**, ensuring this node group is under **All Nodes**, has the environment set to development, and has your choice of node pinned to it
    - On the **Development environment** node group, pin your choice of node to it
    - Under **Enforcement**, select **jobs**, run **puppet**, select your choice of node, and set it to `apply docker`
    - See sample solutions at `https://github.com/PacktPublishing/Puppet-8-for-DevOps-Engineers/blob/main/ch11/docker_group1.png`, `https://github.com/PacktPublishing/Puppet-8-for-DevOps-Engineers/blob/main/ch11/docker_group2.png`, and `https://github.com/PacktPublishing/Puppet-8-for-DevOps-Engineers/blob/main/ch11/docker_group3.png`

Summary

In this chapter, we discussed how Puppet environments can be used to manage specific versions of modules, classification, and data to apply to groups of Puppet clients. The directory structure and variables to configure this was reviewed.

The options to classify servers into environments and to assign classes and parameters were reviewed, looking at node definitions in manifest files, using Hiera in the node definitions to create more complex data-driven calculations, and then ENC scripts that can access sources such as PuppetDB and return YAML output of classes, environment, and parameters for classification. PE was then shown to build on the ENC approach with its own ENC script used in conjunction with node groups to store data on how to classify servers into environments and assign classes.

It was highlighted that the various methods could be used together but the recommended approach was to keep it simple; for open source Puppet, just use a default node definition to look for `pp_role` trusted facts to classify and to put the environment setting in `puppet.conf`, while for PE, it was recommended to use 1-to-1 matching of node groups with `pp_role` and `pp_environment` trusted facts.

It was then shown how a Puppet catalog request sends data to the Puppet server and how classification files and scripts are used to generate catalogs highlighting how different types of Puppet resources are cached.

The methods of deploying environments were then shown, using a Git-based Puppet control repo to contain the files and directories of environments, with each Git branch representing a particular environment. The Puppetfile was shown as a way to list which modules should be deployed to an environment, specifying the version and location of the module.

It was then discussed how `r10k` and the PE Code Manager implementation on top of `r10k` can deploy code to servers. For servers using compilers, we reviewed various approaches to keep code deployed on all infrastructure, which would depend on local infrastructure and standards. For PE, it was shown that Code Manager contained **File sync**, which kept code synchronized.

Workflow approaches were then viewed, showing the more gated and traditional approach of using a control repo with a Puppetfile at set versions and updating the lowest-level environment such as development before pushing these module version changes up through the environments. The second recommended approach showed the control would rely on the modules themselves and the control repo would look for environment-named branches, allowing teams to work and deploy independently. Highlighting the idea in either of these systems is to use a proper pipelining tool with Webhooks to automate deployment.

Having focused on Puppet infrastructure and language for stateful configuration management in this chapter, the next chapter will look at Bolt and orchestrator to show how procedural tasks can be run either using Bolt as an independent tool or through the PE infrastructure via the PE orchestrator.

12

Bolt for Orchestration

In this chapter, we will cover **Bolt** and Puppet Enterprise's **orchestrator**. We will show how Bolt is Puppet's tool for ad hoc orchestration, allowing work to be done that does not fit into Puppet's state-based enforcement model. We will discuss how to configure it to connect to clients with different transport mechanisms and credentials and run simple commands and upload files. Furthermore, we will show how **tasks** allow single-action scripts in various languages to be run via Bolt, while **plans** allow combinations of tasks to be written using logic and variables in the Puppet or YAML language. The project directory structure will be examined, allowing Bolt content to be stored and shared. This will be compared to how plans and tasks can be stored in a Puppet module using the **Puppet Enterprise Cloud Deployment Module** (**PECDM**) **Bolt project** as an example. We will then show how Bolt can be extended with plugins to dynamically load information from other sources. We will also show how Bolt can directly be used with Puppet to apply manifest blocks, connect to PuppetDB, and use Hiera.

In this chapter, we're going to cover the following main topics:

- Exploring and configuring Bolt
- Understanding the structure of projects
- Introducing tasks and plans
- Plugins

Technical requirements

Clone the control repo, `controlrepo-chapter12`, from `https://github.com/puppetlabs/control-repo` to your GitHub account and update the Puppetfile with the contents of `https://github.com/PacktPublishing/Puppet-8-for-DevOps-Engineers/blob/main/ch12/Puppetfile`

Build a standard cluster with two Unix clients and two Windows clients by downloading the `params.json` file from `https://github.com/PacktPublishing/Puppet-8-for-DevOps-Engineers/blob/main/ch12/params.json` and updating it with the location of your control repo and your SSH key for the control repo. Then, run the following command from your `pecdm` directory:

```
bolt --verbose plan run pecdm::provision --params @params.json
```

Exploring and configuring Bolt

Throughout this book so far, we have focused on Puppet's strengths as a state-based and idempotent configuration management tool. But there are situations where this approach simply doesn't fit, such as service restarts as part of troubleshooting or ordering application deployments with vendor-based install scripts. There is any number of tasks that fit into the wider automation effort that are ad hoc and single use; therefore, Bolt was introduced by Puppet to act as an agentless orchestrator. Bolt is now in its 3.x version, since its release in 2017, and a lot of rapid development has taken place. Over 2022, it stabilized, with far fewer releases and changes to features, but we would strongly advise you to keep Bolt as up to date as possible to avoid any confusion.

Having reviewed the general purpose of Bolt as an ad hoc task runner, the first step is to understand how Bolt can connect to clients with transports and targets.

Connecting to clients with transports and targets

Bolt is a fully open sourced project available at `https://github.com/puppetlabs/bolt`, written in **Ruby** and installed as a single package with a binary, `bolt`. It connects to devices via one of the various **transports** it offers, which is the mechanism/protocol that allows it to establish a connection to multiple platforms, such as virtual machines, network devices, or containers without an agent. The transports available are as follows.

System transports:

- Local, which, as would be expected, just runs commands on a local machine.
- **Secure Shell** (**SSH**), using the `net-ssh` Ruby library or `native ssh`, if selected. Commonly used for Linux and Unix machines.
- **Windows Remote Management** (**WinRM**) for connecting to Microsoft Windows-based machines.

Remote, which is used for API- or web-based devices, for example, network devices such as switches.

Puppet Enterprise transport:

- **Puppet Communication Protocol** (**PCP**), used with the Puppet Enterprise orchestrator service, discussed in *Chapter 14*

Container transports:

- Docker, which is an application container technology developed by Docker Inc
- **Pod Manager** (**Podman**), which is an application container engine developed by Red Hat
- **Linux Container Hypervisor** (**LXD**), which is a system container engine that uses **Linux Containers** (**LXC**), developed by `https://linuxcontainers.org` and sponsored by Canonical

> **Note**
> Bolt fails to connect to targets from Windows using SSH unless `native-ssh` is set to `true` in the transport settings, as per `https://puppet.com/docs/bolt/latest/bolt_known_issues.html#unable-to-authenticate-with-ed25519-keys-over-ssh-transport-on-windows`.

By default, Bolt will use local SSH configuration and at its simplest level can run commands directly on devices that are known in Bolt terminology as **targets**. A simple example command is as follows:

```
bolt command run 'uname' --targets examplehost.example.com
```

Here, the command is within single quotes and the provided target is a resolvable hostname or an IP address. Bolt also has **PowerShell cmdlets**, which provide a more integrated experience for PowerShell users with more flexibility for chaining commands and using structured data for arguments. The same command as previously but as a PowerShell cmdlet would look as follows:

```
Invoke-BoltCommand -Command 'uname' -Targets examplehost.example.com
```

This takes the default settings of the SSH transport using the current user and any saved credentials. To make a choice on the command line, a transport choice can be added ahead of the transport name, `<transport_name>://`, multiple targets are listed, separated with commas (`,`) and additional options are set to configure the transport. For example WinRM requires a username, password and the `no-ssl` option set if SSL is not setup for WinRM connections. Take the following example command:

```
bolt command run 'systeminfo' --targets winrm:// host1.example.
com,winrm://host2.example.com --user windows --password Pupp3tL@
b5P0rtl@nd! --no-ssl
```

This command will run `systeminfo` on the `host1.example.com` and `host2.example.com` targets using `winrm` to connect, and the `windows` and `Pupp3tL@b5P0rtl@nd!` credentials with no SSL check. Bolt runs requests concurrently, by default up to 50 at a time. This can be changed using the concurrent argument, `--concurrent`.

A full list of options available to be used with each transport can be viewed in the documentation: `https://puppet.com/docs/bolt/latest/bolt_transports_reference.html`.

> **Note**
> Bolt 1.3.6 deprecated the `nodes` flag in favor of `targets` and removed it in Bolt 2.0.0.

Running ad hoc commands with Bolt

In this section, we will show how to run ad hoc commands with Bolt, using both Windows PowerShell and Linux Shell command examples. The following table shows how these commands compare across the implementations:

PowerShell Command	Linux Shell Command	Function
`Invoke-BoltCommand`	`bolt command run`	Run an ad hoc command
`Invoke-BoltScript`	bolt script run	Run a script
`Invoke-BoltApply`	bolt apply	Apply Puppet code
`Send-BoltFile`	bolt file upload	Upload a file
`Receive-BoltFile`	bolt file download	Download a file

Figure 12.1 – PowerShell and Linux Bolt commands

To run quoted commands, use double quotes or backslashes (\) to escape. For example, we could run a search for `lang` in `/etc/locale` with `grep -I 'lang'`. To do this, the following command could be run in PowerShell:

```
Invoke-BoltCommand -Command "grep -i 'lang' /etc/locale" -Targets
ssh://examplehost.example.com -User centos -PasswordPrompt -RunAs root
```

In this example, the `password-prompt` option would ask for the password securely on the command line rather than directly entering it into the executed command.

To run multiple commands listed in a file, we are not suggesting running a script but a step-by-step set of commands; for multiple targets in a file, the at symbol (@) can be used with the filename within quotes (`' '`). So, for example, to run a list of commands from a file called `commandlist` on a list of targets in `targetfile`, the following command could be run:

```
bolt command run '@commandlist' --targets '@targetfile'
```

For Unix-based systems, to read input from `stdin` for targets or commands, the minus symbol (`-`) can be used in place of a target or command string. So, to take the same `targetfile` and send the output of the `cat` command to the `bolt` command, the following could be run:

```
cat targetfile | bolt command run '@commandlist' --targets -
```

To run the `uname` and `date` commands on the `hosts1.example.com` and `host2.example.com` targets, the following could be command could be used:

```
echo -e "uname \\ndate" | bolt command run - --targets host1.example.
com, host2.example.com
```

> **Note**
>
> Using both a file and `stdin` with a list of commands will result in a single connection to the target to run all the commands.

To run a script in a file, the `bolt script run` command or `Invoke-BoltScript -Script` PowerShell cmdlet can be used along with any arguments to be passed at the end of the command. For example, on a Unix host, the following command could be used to run an `install.sh` script on the targets in the `application_clients` file with `10.6 no-gui` arguments:

```
bolt script run ./scripts/install.sh --targets @application_clients
10.6 no-gui
```

The `arguments` flag can be used to be clearer on the name of the argument for each passed value. Any argument with spaces can be surrounded with quotes (`''`). For example, on a Windows system running the `dotnet-install.ps1` script on a list of targets on a file with the `-Channel LTS` argument, the command would be as follows:

```
Invoke-BoltScript -Script dotnet-install.ps1 -Targets @targetsfile
'-Channel LTS'
```

In Unix, any script can be executed on a target by including a shebang (`#!`) line at the top of the file specifying the interpreter. For Windows targets, the `.ps1`, `.rb`, and `.pp` files are enabled by default, but further extensions can be enabled in configuration files, which will be discussed in the next section. The scripts can be located from the `modulepath`, this can be of the form `<modulename>/scripts/install.sh`, a relative path from the root of the `bolt` folder, or as an absolute path.

In Unix systems, Puppet manifest files and sections of Puppet code can be applied to a set of targets with the following:

```
bolt apply manifests/exampleapp.pp --targets @targetsfile
```

In PowerShell, this can be achieved with the following command:

```
Invoke-BoltApply -Manifest manifests/exampleapp.pp -Targets @
targetsfile
```

To apply Puppet code, the following command would ensure that a /etc/exampleapp directory exists on the Unix systems:

```
bolt apply --execute "file { '/etc/exampleapp: ensure => present }"
--targets servers
```

For PowerShell cmdlets, the command used would be as follows:

```
Invoke-BoltApply -Execute "file { '/etc/exampleapp': ensure => present
}" -Targets servers
```

This format should seem similar to puppet apply and puppet apply -e '<code>'. Similarly, for code to be applied via Bolt, we must ensure that the code is declared to be included in a catalog and not just defined. When a class or type is defined, it is available to be used in the catalog but it will not have been added to the catalog. In the previous example, if exampleapp.pp contained a class definition with resources, this would result in a warning: Manifest only contains definitions and will result in no changes on the targets. The class itself would need to be included for it to be added to the catalog and applied via Bolt.

There are also commands to upload files from your local machine to the target or to download from the targets to your machine. Some simple examples using both the Unix version and the Windows version are shown in the following commands. The first file listed is the source and the second is the target, regardless of whether you're uploading or downloading:

```
bolt file upload /rpms/cowsay.rpm /tmp/ --targets @targets
Send-BoltFile -Source /installer/installer.exe -Destination /users/
exampleuser/installer.exe -Targets @targets
bolt file download /etc/exampleapp//logfile.log /var/tmp/logfile.log
--targets @targets
Receive-BoltFile -Source /ProgramData/exampleapp/logfile.log\puppet.
log -Destination /user/exampleuser/puppet.log -Targets @targets
```

Now, let's take a look at the output.

Output and debugging

So far, the focus has been on how to run commands and not the output. Bolt by default logs these commands to the bolt-debug.log file in the directory from which the Bolt command was run, as well as to the console. There are six logging levels:

- trace: The most detailed level of logging, which shows the inner workings of Bolt.
- debug: Information about target-specific steps.
- info: This is high-level logging showing the steps taking place in Bolt.
- warn: Warning about deprecations and other harmful scenarios. This is the default console level.

- `error`: Error messages experienced during the execution of Bolt commands.
- `fatal`: Error messages from Puppet code used with Bolt.

A specific log level can be chosen using the `--log-level` flag and the output format can be selected using the `format` flag, which can use `human`, `json`, or `rainbow`. The output from a Bolt command running `uname` on three hosts would look like this in JSON:

```
{ "items": [
{"target":"host1.example.com","action":"command","object":"uname",
"status":"success","value":{"stdout":"Linux\n","stderr":"","merged_
output":"Linux\n","exit_code":0}}
,
{"target":"host1.example.com","action":"command","object":"uname",
"status":"success","value":{"stdout":"Linux\n","stderr":"","merged_
output":"Linux\n","exit_code":0}}
,
{"target":"host1.example.com","action":"command","object":"uname",
"status":"success","value":{"stdout":"Linux\n","stderr":"","merged_
output":"Linux\n","exit_code":0}}
],
"target_count": 3, "elapsed_time": 2 }
```

In comparison, in human-readable format, it would look like this:

```
Started on host1.example.com...
Started on host2.example.com...
Started on host3.example.com...
Finished on host1.example.com:
  Linux
Finished on host2.example.com:
  Linux
Finished on host3.example.com:
  Linux
Successful on 3 targets: host1.example.com, host2.example.com, host3.
example.com
Ran on 3 targets in 2.89 sec
```

The `rainbow` output looks similar to human-readable format but, as the name suggests, it makes the lines multi-colored.

As part of this output, a `.rerun.json` file will be generated. This will list the targets that were processed during the run, indicating which targets failed and which succeeded. For the next Bolt command, we can use a `--rerun` flag with a value of `success`, `failure`, or `all`. This reads the relevant target section from `.rerun.json` to use targets from the previous run. For example, the following command could be run as a result of an `install` task failing and choosing to run a cleanup task on all failures:

```
Invoke-BoltTask -Name install_failure_cleanup -Targets @targets.file
-Rerun failure
```

There are more options for the commands; the full command reference is available at `https://puppet.com/docs/bolt/latest/bolt_command_reference.html` for Unix-based commands and `https://puppet.com/docs/bolt/latest/bolt_cmdlet_reference.html` for PowerShell-based commands.

> **Note**
>
> Bolt has a built-in CLI guide that can be accessed by running `bolt guide` on the Unix or PowerShell command line.

So far, what we have discussed using Bolt for is useful on a very small scale but clearly would work with large numbers of servers and more complex configurations. So, the next area to cover is the project structure and configuration files.

Understanding the structure of projects

A **Bolt project** is a simple directory structure providing configuration and data for Bolt to use. Within this structure, Puppet modules from both Forge and private repositories can be stored along with task plans and policies. Bolt identifies a directory as a Bolt project if a `bolt-project.yaml` file exists in it, and this file contains a name key. To create this file, run `bolt project init` for Unix systems or `New-BoltProject` for PowerShell from within a directory in which you wish to add Bolt project files. This will use the name for the project as the name of the directory, but you can override this by running it with a name using the `bolt project init customname` or `New-BoltProject -Name customname` command, for Unix systems and PowerShell, respectively.

This project name must start with a lowercase letter and can only use lowercase letters, digits, and underscores. This is because Bolt projects are like modules and get loaded into the module path. This is important to note because modules contained within the Bolt project will essentially be overwritten in the module path if the Bolt project has the same name as the module.

In the directory, the `init` command will have created a `bolt-project.yaml`, `inventory.yaml`, and `.git-ignore` file.

Now, let's look at how we can configure a Bolt project.

Configuring a project

`bolt-project.yaml` contains settings to override the default Bolt behavior, a lot of which was discussed in the previous section. The settings to be used with the Bolt command can be set here, as well as project configuration such as paths to configuration files and data. Largely, the defaults for these settings will not need to be changed and the core settings that will be configured include the `modules` attribute, which defines modules to manage in the Bolt project, and the `plans`, `policies`, and `tasks` attributes, which limit the visibility of each item by providing a list that will be visible to project users. A sample `bolt-project.yaml` file containing some modules and choosing plans, policies, and tasks to be publicly visible could look as follows:

```
name: packtproject
modules:
- name: puppetlabs-stdlib
- name: puppetlabs-peadm
  version_requirement: 3.9.0
- name: puppetlabs/bolt_shim
- git: https://github.com/binford2k/binford2k-rockstar
  ref: 0.1.0
plans:
- packproject
- peadm::provision
policies:
- packproject::lab
tasks:
- bolt_shim::command
```

The full list of settings can be found at `https://puppet.com/docs/bolt/latest/bolt_project_reference.html`.

The `module` attribute has multiple ways to be updated. When adding items from Forge, this can be updated via the `bolt module add` Unix command or the `Add-BoltModule` PowerShell cmdlet. For example, in Unix systems, `bolt module add puppetlabs/apt` will update the `modules` parameter to contain `- name:puppetlabs-apt` in `bolt-project.yaml`.

Then, the `bolt module install` Unix command or the `Install-BoltModule` PowerShell cmdlet can be used, which will automatically do several things:

- Find dependencies on all Forge modules
- Find compatible versions
- Update the Puppetfile
- Install the modules into the Bolt project

The modules can also be added at project creation time using the following command in Unix systems:

```
bolt project init example_project --modules puppetlabs-
apache,puppetlabs-mysql
```

In PowerShell, this can be done using the following command:

```
New-BoltProject -Name example_project -Modules puppetlabs-
apache,puppetlabs-mysql
```

If you need modules pinned at a specific version or Git modules added, you will need to add these manually to the Bolt project file and run the `Force` flag with the following Bolt module installation command: `Install-BoltModule -Force` on Windows or `bolt module install --force` on Unix systems.

These modules allow us to use Puppet code within plans, as well as bring in plans and tasks from modules, which will be shown in detail in the *Introducing tasks and plans* section.

Configuring transports

The `inventory.yaml` file contains configuration information about targets, creating groups of targets with details about how Bolt connects to them. The inventory contains a top level that includes settings that act as defaults for all targets, group objects that allow targets to be grouped based on common settings, such as all Windows nodes using certain WinRM settings, and target objects, which are individual settings. For each setting, there are common fields that can be used:

- **Alias**: An alias to use instead of the **Uniform Resource Identifier** (**URI**), which can be shorter and more human readable
- **Config**: A map of transport configuration options for the target
- **Facts**: A map of facts for the target(s)
- **Features**: An array of features to be enabled (features will be discussed later in the chapter)
- **Name**: Used with groups to give a human-readable name
- **Plugin hooks**: A map of plugin configurations (plugins will be discussed in the *Plugins* section of this chapter)
- **URI**: A target's URI
- **Vars**: A map of variables

A sample inventory file could look as follows:

```
config:
  transport: ssh
  ssh:
```

```
    host-key-check: false
    run-as: root
    native-ssh: true
    ssh-command: 'ssh'
groups:
  - name: agents
 groups:
  - name: linux_agents
    targets:
      - 20.117.165.119
  -name: windows_agents
    targets:
     - 20.117.165.218
     config:
      winrm:
        user: windowsuser
        password: Pupp3tL@b5P0rtl@nd!
        ssl: false
targets:
  - name: primary:
  - 20.117.166.6
```

This would provide default settings for SSH transport. It should be noted in this example it was shown how to create groups within groups in any inventory to ease management and settings for groups. In this case, we have an agents group, which contains a `linux_agents` group and a `windows_agents` group. The `windows_agents` group contains WinRM transport configuration. This allows us to run Bolt against all agents but with different transports set. There is then a single target called `Primary` outside of these groups.

The full `inventory.yaml` configuration documentation is available at `https://puppet.com/docs/bolt/latest/bolt_inventory_reference.html`, while the transport configuration is available at `https://puppet.com/docs/bolt/latest/bolt_transports_reference.html`.

To return the contents of the `inventory.yaml` file, the `bolt inventory show` Unix command or the `Get-BoltInventory` PowerShell cmdlet can be used. Specific targets can be viewed with the `targets` flag.

As was discussed in the previous section, for Windows scripts, additional extensions can be allowed using the inventory file, so in a `config` section, the following could be added to allow `.py` and `.pl` scripts to be run:

```
config:
  winrm:
    extensions:
```

```
    - .py
    - .pl
```

Having reviewed how to configure settings at a project level in Bolt, it is now important to also know how system-level settings can be set in Bolt and also how previous legacy versions of Bolt projects may be configured differently.

System level and legacy

In addition to the project settings, system-level settings can be set in the `/etc/puppetlabs/bolt/bolt-defaults.yaml` file on Unix-based systems and the `%PROGRAMDATA%\PuppetLabs\bolt\etc\bolt-defaults.yaml` file on Windows systems. User-level settings can be set in `.puppetlabs/etc/bolt/bolt-defaults.yaml` in the user's home directory.

Bolt will choose which project to use with its commands based on the following order of priority:

1. The project location set in the `BOLT_PROJECT` environment variable
2. The `project` flag on a Bolt command with the project location set (`--project /tmp/myproject`)
3. By traversing from the current directory up until a `bolt-project.yaml` or `boltdir` directory is found
4. The `.puppetlabs/bolt/` folder in the home directory of the user

> **Note**
>
> In the Unix environment, Bolt does not load a world-writeable Bolt project directory.

If you want to ship Bolt within an application project but the base Bolt project files would clutter the application, it is possible to embed a Bolt project by creating a `boltdir` directory within the application directory. Bolt could still be run from the parent directory as it would recognize `boltdir` as containing the project.

If you have used older versions of Bolt before 2.36, you will note that projects used to create a single `bolt.yaml` file instead of `bolt-project.yaml` and `inventory.yaml`. Support for v1 `bolt.yaml` projects was removed in v3.0.0 of Bolt. Additionally, Bolt-managed modules changed with the deprecation of manual editing of the Puppetfile in v2.42 and the removal of manual editing in v3.0.0. This also changed the module path from containing the `site-modules` and `site` modules to the modern version of `modules` and `.modules`. Previously, managed modules had existed in `modules` and unmanaged modules in `site` and `site-modules`. This has now been changed to managed modules in `.modules` and unmanaged modules in `modules`. To migrate an old-style Bolt project to the new style, the `bolt project migrate` Unix command or `Update-BoltProject` PowerShell command can be run. As with all automated conversions, ensure your pre-migration configuration is backed up in revision control. The full details of changes made during the migration process can be found at `https://puppet.com/docs/bolt/latest/projects.html#migrate-a-bolt-project`.

Having reviewed the structure created for Bolt configuration and target transport, it is now time to look at more structured ways of running Bolt via tasks and plans.

Introducing tasks and plans

Tasks and **plans** are more forms of scripts and allow users to manage parameters, logic, and flow between actions. Unlike normal Puppet code, plans and tasks run through the script in sequential order, even for Puppet plans that compile a catalog.

Creating tasks

Tasks are single-action scripts that can be in any language that will run on a target machine. The key differences between the normal scripts we have run with Bolt previously and a task are as follows:

- Tasks are paired with a JSON file to provide metadata such as parameters, which allow them to be shared and reused more easily
- Tasks can handle structure/typed input and output
- Tasks can handle multiple implementations to make them cross-platform

They can be stored in the task directory of a Bolt project or the task directory in a Puppet module. Task implementations should contain their extension in the name. The name can include digits, underscores, and upper and lowercase letters

When calling these tasks, a namespace is created that is made up of the name of the Bolt project or module containing the task and the task name, except if the task has been named `init`, in which case it will be referred to only by the Bolt project or module name.

For example, the task to install an agent with the **Puppet Enterprise Administration Module (peadm)** is named `peadm::agent_install`.

> **Note**
>
> The `.json` and `.md` extensions are reserved and cannot be used for tasks.

For Unix shell systems the script part must contain a shebang (`#!`) line at the top of the file specifying the interpreter.

An example of task implementation is when the PEADM module is used to configure the labs using the following code on Unix systems under the `agent_install.sh` task name:

```
#!/bin/bash,
set -e
if [ -x "/opt/puppetlabs/bin/puppet" ]; then
echo "ERROR: Puppet agent is already installed. Re-install, re-
configuration, or upgrade not supported. Please uninstall the agent
before running this task."
exit 1
fi
flags=$(echo $PT_install_flags | sed -e 's/^\["*//' -e 's/"*\]$//' -e
's/", *"/ /g')
curl -k "https://${PT_server}:8140/packages/current/install.bash" |
bash -s -- $flags
```

Parameters are passed based on variables starting with `$PT_`.

With PowerShell, which has a built-in argument handler, this can be done without `$PT_` using the `param` function in a task called `agent_install.ps1`:

```
param(
  $install_flags
  $server
)
if (Test-Path "C:\Program Files\Puppet Labs\Puppet\puppet\bin\puppet")
{
Write-Host "ERROR: Puppet agent is already installed. Re-install, re-
configuration, or upgrade not supported. Please uninstall the agent
before running this task."
Exit 1
}
$flags=$install_flags -replace '^\["*','' -replace 's/"*\]$',''
-replace '/", *"',' '
[Net.ServicePointManager]::ServerCertificateValidationCallback =
{$true}; $webClient = New-Object System.Net.WebClient; $webClient.
DownloadFile("https://${server}:8140/packages/current/install.ps1",
'install.ps1'); .\install.ps1 $flags
```

To make these files visible to Bolt commands and allow callers to pass the parameters, a JSON file is written with the same name as the task. For the `agent_install` example, it looks like this:

```
{
  "description": "Install the Puppet agent from a master",
  "parameters": {
    "server": {
      "type": "String",
      "description": "The resolvable name of the Puppet server to
```

```
install from"
    },
    "install_flags": {
      "type": "Array[String]",
      "description": "Positional arguments to pass to the shell
installer",
      "default": []
    }
  },
  "implementations": [
    {"name": "agent_install.sh", "requirements": ["shell"]},
    {"name": "agent_install.ps1", "requirements": ["powershell"]}
  ]
}
```

The metadata provides a description of the task, which is displayed when listing tasks. In addition, the metadata includes a list of parameters with names that must start with a lowercase letter and only include lowercase letters, underscores, and digits. The parameter type, which may match any Puppet type that can be represented in JSON format, and default values for the parameter can also be specified.

Ensuring the type is an enum or more specific type, such as an integer within a specified size range, can make the task much more secure, limiting the input and therefore the attack vector. Also, within tasks, you should ensure that the parameters for the implementation you are working on are properly separated and do not allow strings to be called. Precise examples can be seen at `https://puppet.com/docs/bolt/latest/writing_tasks.html#secure-coding-practices-for-tasks`.

The `implementations` parameter allows us to define what scripts are used in what environments. In this case, ensure the `.sh` implementation is run on the Unix shell and `.ps1` on PowerShell.

With this file in place, the `bolt task show` Unix command or `Get-BoltTask` PowerShell cmdlet will show all modules available in the module path, and specific tasks can be viewed with `bolt task show <name of task>` or `Get-BoltTask -Name <name of task>`.

Setting the `private` parameter to `true` prevents the task from being listed and can be useful for hiding tasks that are under development, although as we showed in the *Configuring a project* section, the same could be achieved at the Bolt project level.

A parameter can be marked as `sensitive` by setting the parameter value to `true`, and variables can be set to `sensitive` within code to ensure they will be redacted in logs and output.

A parameter of `supports_noop` in the metadata allows users to pass a `noop` argument to the task and will result in the `_noop` parameter being `true` or `false`. It is then possible to use this parameter in your task code to logically check whether changes should be made or just tested.

If the `remote` parameter is set to `true`, the task will only be able to run on remote transport to prevent tasks from being run on incompatible transports.

For a task with lots of options or that returns a lot of information, it may be better to use structured input and output rather than just simple parameters.

Bolt passes task parameters as a single JSON object on `STDIN`, as well as environment variables by default. These can then be read in as parameters by a Ruby script with the following line: `params = JSON.parse(STDIN.read)`.

For complex output, it should be ensured that the task prints a single JSON object to `stdout` in the task. This can be useful if you want to use the result within another task. For example, in Python, the following code snippet would dump the JSON of two value sets to stdout, using `json.dump` to convert the result string into JSON and passing it to the `sys.stdout` method Python uses to print to stdout:

```
result = { "example1": "value1 , "example2": "value2" }
json.dump(result, sys.stdout)
```

To return error messages from tasks, an `Error` object can be returned. In structured output, the `_error` key is expected and the `msg` key is available as a human-readable message for the UI, `kind` as the string for script handling, and `details` with structured data about the task failure, such as exit code tails. Take the following example:

```
{ "_error": { "msg": "Task exit code 1", "kind": "puppetlabs.tasks/
task-error", "details": { "exitcode": 1 } } }
```

If the `_error` key is not present, Bolt generates a generic error instead.

> **Note**
>
> Within a module, `pdk new task <taskname>` can be run to generate a `<taskname>.json` file and a `<taskname>.sh` file in the task folder.

To run these tasks, the `bolt task run` Unix command or `Invoke-BoltTask` PowerShell cmdlet can be used with the parameters either passed as arguments or using the `@` symbol via a string of JSON or a filename with the `.json` extension. For example, the first task would install a Puppet agent on targets in the agents group with the server and `install_flags` parameters set:

```
bolt task run peadm::install_agent --targets agents server=primary.
example.com install_flags= ["--puppet-service-ensure","stopped","agent
:certname=node.example.com"]
```

The second task would run the `package` task and take a JSON string with the `params` flag to check the status of the `apache2` package:

```
Invoke-BoltTask -Name package -Targets @targetservers -Params
'{action="status";name="apache2"}'
```

Having seen how to create and run tasks, it is now time to review plans, which allow for greater structure, logic,and flow to be applied in managing tasks and the ability to use Puppet code.

Creating Puppet plans

Plans are written in Puppet code or YAML and allow multiple tasks and commands to be brought together and to apply logic and control of flow and data between them.

A Puppet plan is written in a manifest and in a similar format to a Puppet class. It starts with the `plan` keyword, then the name of the plan, attributes within brackets `()`, and code between curly braces `{}`. So, for example, a sample plan in a sample project contained in the plan directory would look as follows:

```
plan exampleproject::exampleplan(
  TargetSpec $nodes,
  Enum ['true', 'false'] $manage_user,
) {
  <code>
}
```

Plans are named similarly to tasks, with the first segment the name of the module or project and the second segment and all following segments named with lowercase letters, digits, and underscores.

They must not use a reserved word or have the same as a Puppet data type.

The `init.pp` class, as with tasks and modules, is different. It would skip the need for the task to be named directly. However, it can only be used at the base level but not in any subdirectory.

To create a new plan, the following commands can be used for Unix systems and PowerShell, respectively:

```
bolt plan new <PLAN NAME> --pp
New-BoltPlan -Name <PLAN NAME> -Pp
```

Having reviewed how to create a plan, we will now see how plans receive their target and transport information via the `TargetSpec` type.

Constructing targets

In addition to the normal attribute data types, plans use the `TargetSpec` type, which allows for strings exactly like were used with Bolt command targets in the *Connecting to clients with transports and targets* section, such as `ssh://examplehost.com`, arrays of `Target` types, and recursively, an array of `TargetSpec` types.

The `Target` type represents a target and its specific connections in such a way that they can be added to an inventory file.

Within a plan, the `get_targets` function can be used to return targets from a `TargetSpec`. The following is a simple example of how this is used:

```
plan restart_apache_servers(
TargetSpec $apache_servers,
){
 get_targets($apache_servers).each |Target $apache_server | {
 run_task('apache', $target_node, 'action' => 'reload')
 }
}
```

This plan takes a `TargetSpec` object of `apache_servers`, which is passed to the `get_targets` function. The Apache `reload` task is then run on each individual target server, with the `action` parameter set to `reload`.

Target objects can also be constructed and changed within a plan manifest using functions beginning with `set_` or `add_` for the various parts of the inventory config, such as the `set_config`, `set_var`, `add_facts`, and `add_to_group` functions. For example, a new target could be assembled like so:

```
$example_server = Target.new('name'; => 'exampleserver')
$example_server.set_config('transport', 'ssh')
$example_server.set_config(['ssh', 'password', 's3cur3!')
$example_server.add_facts({'application' => 'example'})
```

It is possible to access parts of the target, such as `$example_server.config['ssh']`, but the targets will only last in memory for as long as the plan is running.

Now that we understand how to connect to clients using plans, we will show how functions can be used in the Puppet code block of a plan to use features of Bolt and the Puppet core language.

Using plan functions

As was shown in the *Constructing targets* section, using `run_task`, Bolt plan functions can be used within the Puppet code block itself, many of which are the same types of commands that were run directly in Bolt, such as `run_command`, `run_script`, and `run_task`. The full list is available at `https://puppet.com/docs/bolt/latest/plan_functions.html`.

It is also possible to run a plan from within a plan using the `run_plan` function. This can be useful to ensure no plan gets too large and they can be more easily reused. A pattern that can be observed in the PEADM module is the use of the `subplan` folder for plans we only expect to be used within plans, reducing the size and complexity of the catalogs.

It should be noted that most Puppet language features, such as functions, the sensitive type, and lambdas, can be used within this code, but other features, such as deferred functions, cannot since the catalog is not being sent to the node to be applied. The differences are fully documented at `https://puppet.com/docs/bolt/latest/writing_plans.html#puppet-and-ruby-functions-in-plans`.

For example, within PEADM, the following `run_command` function stops Puppet on all the targets stored in the `$all_targets` variable and then runs a `modify_certificate` plan on the targets in the `covert_target` variable, passing in a primary `add` parameter and the extensions to be added:

```
run_command('systemctl stop puppet', $all_targets)
run_plan('peadm::modify_certificate', $convert_targets,
  primary_host => $primary_target,
  add_extensions => {
    'pp_auth_role' => 'pe_compiler',
  },
)
```

Puppet code can also be applied via the `apply` function similar to how a `puppet apply` command would run. For example, PEADM uses the following code to create node groups:

```
apply($primary_target) {
class { 'peadm::setup::node_manager_yaml':
  primary_host => $primary_target.peadm::certname(),
}
```

This applies the `node_manager_yaml` class, passing a `primary_host` parameter. It should be noted that if Puppet libraries are needed in advance of applying Puppet code, the `apply_prep` function can be used to ensure they are available before using the `apply` function.

Logging and results

To add logging to plans, the `out::message` and `out::verbose` functions are used, with message logging on every run and verbose message output only if Bolt is run in `verbose` mode. Take the following example:

```
out::message('Error')
out::verbose("Heres the error: $detailed_output")
```

`Error` would be printed on every Bolt run, but only when the `-verbose` flag is used would the second message be displayed.

Each function returns an object type of `ResultSet` with each target containing its own `Result` object type, except the `apply` function, whose `ResultSet` contains `ApplyResult` objects. A plan returns a `PlanResult` type as output, which can contain all these data types and just about any Puppet data type.

These objects can be assigned to variables and then functions used to expose data. There are two common functions used in all of these object types. `ok` is a function that returns a simple Boolean confirming whether there were any errors and the `value` function returns the output of the run.

Further type-specific functions can be viewed in the documentation at `https://puppet.com/docs/bolt/latest/bolt_types_reference.html`.

To return output from the plan, the return function should be used with any appropriate data type; this could be the direct output from a task or as simple as a string. If no return function is used, the output will be `undef`. For example, the following code will run the task `error_check_task` and only if that is successful will it return the `ResultSet` type output from the task `output_task`; otherwise, it will return the string `OH NO`:

```
plan return_result( $targets )
$did_this_work = run_task('error_check_task', $targets)
If $did_this_work.ok {
out::message('It worked')
return run_task('output_task', $targets)
}else{
Return "OH NO"
}
```

Now, let's look at how we can handle errors.

Handling errors

To perform a simple check and fail a plan as a result, the `fail_plan` function can be used. For example, the following code would check whether the `$targets` variable only contained a single target:

```
unless get_targets($targets).size == 1 {
    fail_plan('This plan only accepts one target.')
  }
```

If Bolt functions fail and `_catch_errors` is not set to `true`, then the plan will fail. If `_catch_errors` is used, this allows the plan to continue and the error can be handled:

```
$install_agent_results = run_task('agent_install', $agents , '_catch_
errors' => true)
$ install_agent_results.each |$agent_result| {
$target = $agent_result.target.name
if $result.ok
```

```
  { notice("${target} installed correctly ${result.value}")
 } else {
  notice("${target} failed install with error: ${result.error.
 message}")
  }
 }
```

Alternatively, the `catch_result` function can be used to catch specific types of errors, as follows:

```
$install_agent_results = catch_error(agent_install/connection_error)
|| { run_task('agent_install', $agents , '_catch_errors' => true)
}
```

With an understanding of logging and error handling in plans, we can now look at how external data can be used in plans. Since Bolt uses Puppet as a library, it can use Hiera to access external data. As was covered in *Chapter 9*, this can ensure we separate code and data into plans as we do with Puppet code.

Managing data sources

Facts can be collected from the hosts using the built-in facts plan or from PuppetDB using `puppetdb_facts`, assuming PuppetDB is set up in a Bolt configuration already. Using either plan would cause the targets to query PuppetDB to automatically have their in-memory inventory updated with the facts. The following example would run `facts` on `targets`, and those targets for which the `os.name` fact is equal to `Windows` are assigned to the `windows_targets` variable:

```
run_plan('facts', 'targets' => $targets)
$windows_targets = get_targets($targets).filter |$target| { $target.
facts['os']['name'] == 'Windows' }
```

PuppetDB can also have general queries run against it using the `puppetdb_query` function. To return all the `certnames` fact values of `windows` hosts listed in PuppetDB, use the following code:

```
$windows_targets = get_targets (puppetdb_query('inventory[certname] {
facts.os.name = "windows" }'))
```

Hiera can be used with plans by using either modules or Bolt project-level Hiera and having an appropriate `hiera.yaml`. The `lookup` function can then be used either inside `apply` functions or simply in the plan. If `lookup` is used within an `apply` function, and assuming the `apply_prep` function is run, we can gather all the facts and Hiera will function as expected. When using it within a plan, the important differences to note are that Bolt has no automatic parameter lookup capabilities like normal Puppet code with classes and the Bolt hierarchy can't use top scope variables or facts. When Hiera is used within Bolt, it uses two levels of hierarchy, the project and module levels, with the project level being higher in precedence.

An example of the Bolt hierarchy would be a `hiera.yaml` project containing a hierarchy with Node data and the `plan_hierarchy` key without Node data:

```
Hierarchy: -
- name: "Nodes" path: "targets/%{trusted.certname}.yaml"
- name: "Org" path: "%{org}.yaml"
plan_hierarchy:
- name: "Org" path: "%{org}.yaml"
```

A `lookup` function in the plan could do the following and with an `application` variable be able to look up the `dns_server_name` variable in the org level of the plan hierarchy:

```
plan exampleproject::exampleplan(
TargetSpec $nodes,
String $application
){
$dns_server_name = lookup('dns_server_name)
}
```

In the following section, we will look at how comments can be used to document metadata.

Documenting plan metadata

Unlike tasks, since plans do not have a `metadata.json` file, it is necessary to document via comments so that when `puppet plans show <plan name>` is run, a description is provided. The first comment line is taken as the description, or an `@summary` tag can be used. Using `@param <param name>` on a comment will indicate it is the description of a parameter and using `@api private` will mark the plan as private. An example of using all of these fields is as follows:

```
# @summary This plan is just for example
# @api private
# @param example_servers The targets to run this plan on
# @param manage_user Whether the user account should be managed
plan exampleproject::exampleplan(
TargetSpec $example_servers,
Enum ['true', 'false'] $manage_user
){
```

The data type details are picked up automatically by the `bolt plan show` command.

> **Note**
>
> It can be useful to add plans and tasks to the control repo, but it should be noted that when using PDK validate, PDK cannot validate plans and will only ignore plans in the default bottom-level plan directory. If you have a structure that puts plans at a lower level, you will have to run `pdk` to ignore these lower level directories of plans, such as `pdk set config project.validate.ignore subdir1/subdir2/plan`.

Plan testing

Testing Puppet plans is beyond the scope of the book. This is because plan testing is currently not fully implemented and difficult compared to the normal RSpec testing we saw in *Chapter 8*. Certain things are simply not implemented, such as mocking uploading a file or custom functions, which makes it difficult to carry out meaningful and complete testing compared to module testing. The testing functions currently available can be viewed at `https://puppet.com/docs/bolt/latest/testing_plans.html`.

Introducing YAML plans

YAML plans will be summarized here due to their much lower level of usage than Puppet plans. They are named similarly to Puppet-based plans but end with an extension of `.yaml` (not `.yml`). However, there is no command to create them. YAML plans contain the following:

- `Description`: What will be displayed in the `show` command
- `Parameters`: A hash of parameters that can be passed to the plan
- `Private`: A Boolean stating whether the plan is visible to the `show` command
- `Return`: An array, Boolean, hash, number, or string to return from the plan
- `Steps`: An array of steps to be run

The steps essentially represent the action to be performed in that step and the variables the step needs. There is a similarity between the options available in Bolt and the actions in Puppet plans, such as commands, tasks, scripts, file downloads, and file uploads. As with Puppet plans, YAML plans can call other plans with a plan step.

The following example task plan, which uses the Docker `puppetlabs` module from Forge `https://forge.puppet.com/modules/puppetlabs/docker` to create and join an additional manager node to a Docker swarm, shows some of these features in use:

```
description: configure docker swarm
paramters:
  firstnode
    type: TargetSpec
```

```
    Othernodes
      Type: Targetspec
  - name: init
      task: docker::swarm_init
      targets: $firstnode
    - name: token
      task: docker::swarm_token
      targets: $firstnode
    - name:facts
      Fact:
      targets: $firstnode
    - name: managersjoin
      task: docker::join_swarm
      targets: $othernodes
      parameters:
        token: $token.map |$token_result| { $token_result['stdout'] }
         manager_ip: $facts.map |$facts_result| { $facts_
  result['stdout']['networking']['interfaces']['ip'] }

  return $managersjoin.map | $managersjoin_result| {$managersjoin_
  result['stdout']}
```

This task takes the `firstnode` and `othernodes` variables of the `TargetSpec` type to provide the servers to the target. It uses the `swarm_init` task to initialize on the first node and runs the `swarm_token` task on this node. The `Fact` task is then run on `firstnode`, and in the final step, the `join_swarm` task is run on `othernodes`. It can be seen calling a variable with the name of previous step allows us to access the output created by that step. So we can take the output of the token step and map out the taskspec type returned to use the `stdout` as the token. For the `manager_ip` parameter, we perform a similar action, but this time, as there is more content in `stdout`, we must find the `networking.interface.ip` address fact we wish to pass. The plan then sets the return key to take the `stdout` output of the `join` step to confirm the result of the plan.

It is also possible to use the `eval` step to calculate values, and both Puppet and Bolt functions can be used with this. The `message` and `verbose` steps are available for output just as they were in Puppet plans, while string interpolation follows the normal Puppet principle of single quotes (`''`) having no interpolation, just printing the text, double quotes (`""`) performing interpolation, and also using a pipe (`|`) with a new line to allow for expressions of a block of Puppet code to be displayed on the next line.

To show some of this, the following plan takes an array of strings to install as packages:

```
parameters:
  packages:
    type: Array[String]
  servers:
    type: Targetspec
```

```
Steps:
  -name: unique_packages
  eval: $packages.unique
  -name: numer_of_packages
  eval: $unique_packages.size
  - verbose: 'Installing ${number_of_packages} packages'
  - name: install
    task: example::install_packages
    parameters:
      packages:  $unique_packages
      Targets: $servers
Return: $install.map | $install_result| {$install_result['stdout']}
```

We can see that the `unique_packages eval` step uses the `unique` function to find only unique values in the array and the `numer_of_packages eval` step uses the `size` function, the result of which is passed to the `verbose` output and interpolated into a string which shows the number of packages. The `example::install_packages` task is run with the output of the `unique_packages` eval step before its output is used in the returned value.

This has been just a summary of using YAML plans. The full options available on each step and more are available in the documentation: `https://puppet.com/docs/bolt/latest/writing_yaml_plans.html`.

In the following section, let's look at some examples of plugins that are commonly used with Bolt.

Plugins

Plugins allow Bolt to dynamically load data during a Bolt run. Plugins are essentially just modules containing tasks with a `bolt_plugin.json` file identifying which tasks are plugins and what type of plugins they are. Some are built into Bolt, while others can be added to extend the functionality.

There are three types of Bolt plugins:

- **Reference**: Used to fetch data from external sources, such as loading information into the inventory file
- **Secret**: Used to create keys to encrypt text and decrypt cipher text
- **Puppet library**: Used to install Puppet libraries when the `apply_prep` function is called on a target

We will look at these in detail in the following subsections.

Reference plugins

Reference plugins can be used in configuration files such as `inventory.yaml` or `bolt-project.yaml` using a `_plugin` key with the plugin name as the value and followed by parameters associated with the plugin. For example, to use the `puppetdb` plugin and query and select all the window nodes in PuppetDB, we could add the following group to `inventory.yaml`:

```
groups:
  - name: windows
    targets:
      - _plugin: puppetdb
        query: 'inventory[certname] { facts.kernel = "Windows" }'
```

This is assuming the PuppetDB connection configuration details are set in one of the configuration files.

> **Note**
>
> With the PuppetDB plugin configured a one-time query like this can be used to query PuppetDB: `bolt task run 'inventory[certname] { facts.kernel = "Windows" }'`.

Another approach to reference plugins can be with passwords where the `prompt` plugin will result in user input from the command line that sets a password. For example, the following will ensure that when running against `target1.example.com`, Bolt will connect on `winrm` with the user `bill` and the password prompted by the message `Enter your password`:

```
targets:
  - target1.example.com
  config:
  winrm:
    user: bill
    password:
      _plugin: prompt
      message: Enter your password
```

Plugins can also be used in plans via the `resolve_references` function. The following example shows a subsection of the `pecdm` module using plugins via the `resolve_references` function:

```
$inventory = ['server', 'psql', 'compiler', 'node', 'windows_node'
].reduce({}) |Hash $memo, String $i| {,
$memo + { $i => resolve_references( {
'_plugin' => 'terraform',
'dir' => $tf_dir,
```

In the preceding code block, it essentially iterates through each group name and builds an array of target entries read in from the `terraform` directory set by the `tf_dir` variable. To see a further example, look at the contents of the `inventory.yaml` file for your lab setup, which uses the `terraform` plugin.

Secret plugins

Secret plugins allow for the creation of keys and secrets, and encrypted values to be passed in. Currently, `pckcs7` is Bolt's default and only secret plugin. To create the encryption keys, run the `bolt secret createkeys --force` Unix command or the `New-BoltSecretKey -Force` PowerShell cmdlet. This will create the keys in the `keys` folder of your project. Cipher text can be generated via the `bolt secret encrypt 'N33dt0kn0wba515!' --plugin pckcs7` Unix command or the `Protect-BoltSecret -Text 'N33dt0kn0wba515! ' -Plugin pckcs7` PowerShell cmdlet.

The cipher text from this command can then be used in places such as `inventory.yaml` using the `pkcs7` reference plugin, for example:

```
targets:
  - uri: target1.example.com
    config:
      ssh:
        password:
          _plugin: pkcs7
          encrypted_value: |
            ENC[PKCS7,MIIBiQYJK]
```

Note the previous encrypted string was shortened and has a default key size of `2048`. This can be changed by configuring the plugin in `bolt-project.yaml` or the default and user configuration.

Puppet library

Puppet library plugins install Puppet libraries on a target when the `apply_prep` function is called in a plan. Each target the plugin is run on must be able to use the scripting language used by the plugin. Currently, only `puppet-agent` exists as a Puppet library plugin and it is configured to be available by default. But any future libraries or custom written libraries would be added to the Bolt, user, or default configuration in a similar fashion to this example:

```
plugin-hooks:
  puppet_library:
    plugin: task
    task: package
    parameters:
      name: puppet-agent
      action: install
```

The full list of supported and built-in plugins can be viewed at `https://puppet.com/docs/bolt/latest/supported_plugins.html`. Writing plugins is beyond the scope of this book, but the documentation at `https://puppet.com/docs/bolt/latest/writing_plugins.html` advises further.

Having covered Bolt in detail, we will now practice creating and using a bolt project in the following lab.

Lab – creating and using a Bolt project

In this lab, we will create a Bolt project. We will create a task that runs the `facter` command on a Windows and Linux node.

The steps are as follows:

- Create a Bolt project with the following line of code:

```
bolt project init packtlab
```

- Create an `inventory.yaml` file by performing a lookup of Windows and Linux clients from the PECDM Bolt project and copying the output:

```
bolt inventory show --targets agent_nodes --detail
bolt inventory show --targets windows_agent_nodes --detail
```

- Write a task to cover both Windows and Linux that runs the `facter` command, taking a single argument if only a single fact should be returned.
- Write a plan that uses `run_command` to run `facter` and return the results of the plan.
- Run the task and plan on your Windows and Linux clients.
- You can find example solutions at `https://github.com/PacktPublishing/Puppet-8-for-DevOps-Engineers/tree/main/ch12`.

Summary

In this chapter, we showed how Bolt complements Puppet's state-based management by providing a capability to run ad hoc actions for anything that doesn't fit the declarative enforcement methods of Puppet. We also showed how transports provide the ability for Bolt to connect to targets. We saw how, using the Bolt commands via Unix or PowerShell, we could run commands, scripts, Puppet code, and manifests on targets, as well as uploading and downloading files. We reviewed how Bolt logs to `bolt-debug.log` and how to configure logging to get more logs for different issues.

We then showed how Bolt projects provide a directory structure to contain the configuration and data for Bolt. Bolt projects provide the `inventory.yaml` file to contain target and transport configuration and the `bolt-project.yaml` file to contain project-level configuration settings for Bolt and to allow module dependencies to be downloaded into the project. We discussed how the Bolt project is loaded into the module path along with any modules it had downloaded. We then highlighted how the project format has changed over different versions of Bolt and how the `bolt migrate` command can convert older projects to the new format.

We then discussed how tasks are single-action scripts that can be in any language that will run on a target machine, paired with a JSON file to provide metadata such as parameters. We also showed how a task can list multiple implementations depending on the target. We looked at how sensitive parameters allow passwords and other secrets to be used by tasks without logging in the APIs. The `noop` option was introduced as a standard way to pass a parameter to a task and run in no-execute mode. We also showed how remote tasks contain the `remote` parameter, set to `true`, and a remote transport to allow web access services to use tasks despite being unable to log on in the traditional way.

Then, we discussed how tasks are capable of sharing scripts in implementations and referring to other modules. Some security practices were discussed to ensure parameters are safely passed to tasks.

Plans were then discussed as a way of running multiple tasks together and providing logic and control flow. We saw how plans could be written in either the Puppet language or YAML and how targets can be created using the `targetspec` data type and functions. We also saw how structured results can be returned after running a plan.

We then discussed how Bolt plugins provide ways to dynamically load data into Bolt runs using reference plugins to fetch data and store it, such as to fill the inventory with data from Terraform. We can also use secret plugins to provide keys for encrypting and decrypting values in Bolt runs. The third type of plugin we looked at was Puppet library plugins, which has only been implemented currently for the installation of Puppet agent via Bolt.

In this chapter, we saw how Bolt can be paired with Puppet to get the best of both declarative and stateful language approaches to allow for flexibility in Puppet configurations.

Having reviewed how to use Bolt and Puppet Enterprise, in the next chapter, we will look at more advanced topics on how to monitor and scale Puppet infrastructure, review performance issues, and use the **Puppet Data Service** to implement the external data pattern and allow users to enter data into Puppet setups with self-service APIs.

13

Taking Puppet Server Further

This chapter will look at how you can monitor, tune, and integrate your Puppet infrastructure with third-party sources. You will understand how to find the logs of the various services we have discussed in previous chapters and how to find the current status APIs. You will then learn how these logs and statuses can be integrated into services such as **logstash** to provide greater visibility and alerting options. Then, we'll review the metrics provided by Puppet, along with how these can be integrated with dashboarding tools such as Splunk and Grafana to provide **monitoring** and **observability** for Puppet's infrastructure. We will set up a lab for both **Splunk** and **Grafana** as part of the Puppet Operational Dashboard to show these dashboards. Using these metrics, you will learn how the various components of Puppet's infrastructure can be tuned and scaled to deal with common issues and problems as Puppet grows. After, you'll learn how the external provider pattern can allow for facts, classification, and Hiera data to be fed from external data sources into Puppet and to allow Puppet platform teams to provide self-service with Puppet data without requiring full knowledge of Puppet or the environment release procedures. Various third-party implementations, including **ServiceNow** and **1Password**, will be shown. The **Puppet Data Service** (**PDS**) will be implemented in this chapter's lab to demonstrate this pattern.

In this chapter, we're going to cover the following main topics:

- Logging and status
- Metrics, tuning, and scaling
- Identifying and avoiding common issues
- External data sources

Technical requirements

Clone the control repository from `https://github.com/puppetlabs/control-repo` to your GitHub account (`controlrepo-chapter13`) and update the following files in this repository:

- `Puppetfile` with `https://github.com/PacktPublishing/Puppet-8-for-DevOps-Engineers/blob/main/ch13/Puppetfile`
- `hiera.yaml` with `https://github.com/PacktPublishing/Puppet-8-for-DevOps-Engineers/blob/main/ch13/hiera.yaml`
- `manifests/site.pp` with `https://github.com/PacktPublishing/Puppet-8-for-DevOps-Engineers/blob/main/ch13/site.pp`

Build a large cluster with three compilers and three clients by downloading the `params.json` file from `https://github.com/PacktPublishing/Puppet-8-for-DevOps-Engineers/blob/main/ch13/params.json` and update it with the location of your control repository and your SSH key for the control repository. Then, run the following command from your `pecdm` directory:

```
bolt --verbose plan run pecdm::provision --params @params.json
```

First, we will look at where to find the logs and current status of the Puppet services and infrastructure. This will be fundamental to how you will need to tune and troubleshoot Puppet.

Logging and status

When we discussed different Puppet components previously in this book, we listed logging directories, but it is useful to have a single reference point for these logs.

Exploring log locations

This section provides a list of these logs, titled with the core function, the containing directory, and the list of logs in that directory:

- **Primary server logs**:
  - `/var/log/puppetlabs/puppetserver/`: The primary server logging directory
  - `puppetserver.log`: The primary server which logs its activity
  - `puppetserver-access.log`: Requests to access endpoints
  - `puppetserver-daemon.log`: Crash reports and fatal errors
  - `puppetserver-status.log`: Debug status logging for the service

- **Database logs**:
    - `/var/log/puppetlabs/postgresql/<version>`: PostgreSQL logging directory
    - `pgstartup.log`: Start-up logs
    - `postgresql-<Mon - Sun>.log`: Daily debugging logs
    - `/var/log/puppetlabs/puppetdb/`: PuppetDB logging directory
    - `puppetdb.log`: The PuppetDB service activity log
    - `puppetdb-access.log`: Requests to access endpoints
    - `puppetdb-status.log`: Debug status logging for the service
- **Primary server logs** (Puppet Enterprise only):
    - `/var/log/puppetlabs/puppetserver/`: The primary server logging directory
    - `code-manager-access.log`: Requests to access endpoints of the code manager
    - `file-sync-access.log`: Requests to access endpoints of file sync
    - `pcp-broker.log`: Puppet Communications Protocol brokers on compilers
- **Console and console services logs** (Puppet Enterprise only):
    - `/var/log/puppetlabs/console-services/`: Puppet Enterprise console service logging directory
    - `console-services.log`: Console service activity logs
    - `console-services-api-access.log`: Requests to access the console service API endpoints
    - `console-services-access.log`: Requests to access the console service endpoint
    - `console-services-daemon.log`: Crash reports and fatal errors logged
    - `/var/log/puppetlabs/nginx/`: nginx logging directory
    - `access.log`: Requests to nginx endpoints
    - `error.log`: nginx errors and general console errors

- **Agent logs:**

  The agent output that you see on your screen when you run Puppet manually is logged to a location based on the `logdest` and `logdir` settings in the `puppet.conf` file. The `logdest` parameter can be set to `syslog` (to be sent to the POSIX syslog service), `eventlog` (to be sent to the Windows event log), `console` (for logs to be sent to the console), or a filename so that they're outputted to a file of this name in the location set by `logdest`. `syslog` is the default for Unix-based systems, while `eventlog` is the default for Windows. The defaults for `logdest` are `/var/log/puppetlabs/puppet` for Unix and `C:\ProgramData\PuppetLabs\puppet\var\log` for Windows.

> **Note**
>
> It is possible to turn on server profiling, which can generate detailed catalog logging information. This can then be graphed to show in-depth debugging information about the catalog compilation. It is beyond the scope of this book to dive into this. More information can be found in Puppet's documentation at `https://github.com/puppetlabs/puppet/blob/main/docs/profiling.md`.

Having examined the various locations of logging, it becomes clear that it would be useful to forward these server logs to specialized tools so that they can be indexed and processed.

Forwarding server logs

As the number of Puppet clients grows, the log tracking exercise, which we completed in *Chapter 10*, simply becomes impractical. In this scenario, more specialized **log tooling** should be used to filter and view events. Previously, we saw that Puppet uses the **Logback** library, `http://logback.qos.ch/`, for logging Java services on servers. This can be configured to output logging in JSON format in `logback.xml`, which can be sent to a logging backend such as Elastic's Logstash or Grafana's Loki. The `logback.xml` file contains `appender` definitions, which are the Logback components for writing logs.

By observing the `appender` configuration for `puppetserver.log`, we will see the current configuration:

```
    <appender name="F1" class="ch.qos.logback.core.rolling.
RollingFileAppender">
<file>/var/log/puppetlabs/puppetserver/puppetserver.log</file>
        <append>true</append>
        <rollingPolicy class="ch.qos.logback.core.rolling.
SizeAndTimeBasedRollingPolicy">
<fileNamePattern>/var/log/puppetlabs/puppetserver/puppetserver-
%d{yyyy-MM-dd}.%i.log.gz</fileNamePattern>
            <!-- each file should be at most 200MB, keep 90 days worth
of history, but at most 1GB total -->
```

```
            <maxFileSize>200MB</maxFileSize>
            <maxHistory>90</maxHistory>
            <totalSizeCap>1GB</totalSizeCap>
        </rollingPolicy>
        <encoder>
            <pattern>%d{yyyy-MM-dd'T'HH:mm:ss.SSSXXX} %-5p [%t]
[%c{2}] %m%n</pattern>
        </encoder>
    </appender>
```

Here, we can see how it appends to the log, the filename pattern it will use, along with dates for rolling the log, and that the file size will not exceed 200 MB per file, no more than 1 GB, and that logs will not be retained for more than 90 days. The encoder shows how the log entries should be formed.

To add a JSON version of the log, we could make a similar entry that consists of just 5 days of logging. Here, the encoder is the Logstash encoder to output in JSON. Here's the code for this `appender`:

```
<appender name="server_JSON" class="ch.qos.logback.core.rolling.
RollingFileAppender">
    <file>/var/log/puppetlabs/puppetserver/puppetserver.log.json</
file>
    <rollingPolicy class="ch.qos.logback.core.rolling.
TimeBasedRollingPolicy">
        <fileNamePattern>/var/log/puppetlabs/puppetserver/
puppetserver.log.json.%d{yyyy-MM-dd}</fileNamePattern>
        <maxHistory>5</maxHistory>
    </rollingPolicy>
    <encoder class="net.logstash.logback.encoder.LogstashEncoder"/>
</appender>
```

To enable this `appender` toward the bottom of the `logback.xml` file, add the following definitions:

```
    <root level="info">
        <!--<appender-ref ref="STDOUT"/>-->
        <appender-ref ref="${logappender:-DUMMY}" />
        <appender-ref ref="F1"/>
    </root>
```

Adding `<appender-ref ref="server_JSON"/>` within this root section would enable our JSON `appender`. Restarting the `puppetserver` service would enable the new `appender`.

To set up `server-access.log`, you can use the code at `https://github.com/PacktPublishing/Puppet-8-for-DevOps-Engineers/blob/main/ch13/appender_example.xml`, which will add an `appender` that will configure the JSON output with an appropriate pattern.

You will need to consider disk space in such cases. It's possible to run this with JSON logging. Logback is a powerful library, but it's beyond the scope of this book to go through the full options. These can be reviewed at `http://logback.qos.ch/manual/configuration.html` and `https://logback.qos.ch/manual/appenders.html`.

Now that logfiles exist in JSON, a tool such as Grafana's **Promtail** or Elastic's **Filebeat** could be configured to forward the log files to a service such as Elastic's Logstash or Grafana's Loki. The Ruby logs managed by **Logrotate** can also be gathered, but it would require more work to put suitable patterns on them to be processed.

> **Note**
> Puppet Enterprise services, `console-services`, PuppetDB, and orchestration services all use Logback and can have their logs forwarded like this.

Having reviewed how logs can be sent to external services to be processed, we will now see how the reports that are generated from applying Puppet catalogs can also be sent to external tools using report processors.

Report processors

As well as server logging, as shown in *Chapter 10*, every catalog run generates reports. In *Chapter 10*, you saw how this was configured by `peadm` to be stored in `puppetdb` using `reports = puppetdb` in the `master/server` section of `puppet.conf`. Setting this `reports` value told the server to use a report processor, which is a Ruby script that's run when Puppet Server receives a report. The script then performs actions to pass it on to a target. In the case of PuppetDB, this is to send reports to be stored in PuppetDB. There are three built-in report processors: `http`, `log`, and `store`. `http`, `which` sends the report to an HTTP address set by the `reporturl` setting in `puppet.conf` in YAML format. `log` sends the report output to the logging file specified in `logdest` and `logdir` in `puppet.conf`, and `store` puts the report's output into files specified by the `reportdir` setting that's set in `puppet.conf`. Other custom report processors are available in Puppet Forge, including the Splunk integration module, as described at `https://forge.puppet.com/modules/puppetlabs/splunk_hec`, and the Datadog agent module, as described at `https://forge.puppet.com/modules/datadog/datadog_agent`, which allows report data to be viewed in those third-party services. The instructions vary, depending on the module, but normally, the minimum actions required to add a report processor is for Puppet Server to have the module deployed in an environment and for the reports to have the name of the forwarder set. Writing custom report processors is beyond the scope of this book but details can be found at `https://puppet.com/docs/puppet/latest/reporting_write_processors.html`.

In addition to logs and reports, we can see what condition the current Puppet Infrastructure is by calling the status APIs.

Accessing status APIs

Puppet provides a status endpoint that can be called at `GET /status/v1/services`. This endpoint returns the status of all known services on the server. This access is controlled by `auth.conf` and can be accessed locally via the Puppet CA certificates, as follows:

```
cert="$(puppet config print hostcert)"
cacert="$(puppet config print localcacert)"
key="$(puppet config print hostprivkey)"
uri="https://$(puppet config print server):8140/status/v1/services"
curl --cert "$cert" --cacert "$cacert" --key "$key" "$uri" | jq
```

The final pipe to JQ (a command-line JSON processor) is optional but makes the output more readable. Here's an example of the output Puppet Server's status would provide:


```
{
  "server": {
    "service_version": "7.6.0",
    "service_status_version": 1,
    "detail_level": "info",
    "state": "running",
    "status": {},
    "active_alerts": []
  },
```


It is possible to target an individual service by adding the service name to the URL – for example, `GET/status/v1/services/server`. For PE installations that have additional services, the specific port for each service should also be called. PE services will be discussed in *Chapter 14*.

The return code from these calls will be `200` for all services running, `404` when a service is not found, or `503` when a service state is in any state other than running.

The server state can be `running` when all services are running, `error` if any service reports an error, `starting` or `stopping` if any services are in those states, and `unknown` if any service reports an unknown state.

Puppet Enterprise has an extra command-line option to call APIs via the `puppet infrastructure status` command, which produces an output similar to the following:

```
Puppet Server: Running, checked via https://pe-server-davidsand-
0-cffe02.tq2kpafq5bsehkpub4ur5a35ya.xx.internal.cloudapp.net:8140/
status/v1/services
  PuppetDB: Running, checked via https://pe-server-davidsand-0-cffe02.
tq2kpafq5bsehkpub4ur5a35ya.xx.internal.cloudapp.net:8081/status/v1/
services
```

In the web console, you can find the Puppet Service's status by clicking the **Puppet Services status** button, as per the following figure:

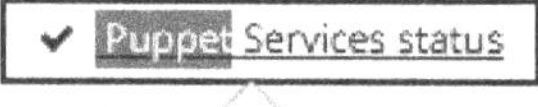

Puppet Services are all accepting requests.

	Service	Message
✔	Activity Service https://pe-server-davidsand-0-cffe02.tq2kpafq5bsehkpub4ur5a35ya.xx.internal.cloudapp.net:4433	Operational
✔	Classifier Service https://pe-server-davidsand-0-cffe02.tq2kpafq5bsehkpub4ur5a35ya.xx.internal.cloudapp.net:4433	Operational
✔	Code Manager Service https://pe-server-davidsand-0-cffe02.tq2kpafq5bsehkpub4ur5a35ya.xx.internal.cloudapp.net:8140	Operational
✔	Orchestrator Service https://pe-server-davidsand-0-cffe02.tq2kpafq5bsehkpub4ur5a35ya.xx.internal.cloudapp.net:8143	Operational
✔	Puppet Server https://pe-server-davidsand-0-cffe02.tq2kpafq5bsehkpub4ur5a35ya.xx.internal.cloudapp.net:8140	Operational
✔	PuppetDB Service https://pe-server-davidsand-0-cffe02.tq2kpafq5bsehkpub4ur5a35ya.xx.internal.cloudapp.net:8081	Operational
✔	RBAC Service https://pe-server-davidsand-0-cffe02.tq2kpafq5bsehkpub4ur5a35ya.xx.internal.cloudapp.net:4433	Operational

Figure 13.1 – Puppet Services status in the web console

> **Note**
>
> The `puppet status` command, which was depreciated in Puppet 5, was removed in Puppet 7. It didn't use API endpoints.

The logs, reports, and statuses we have viewed so far allow us to observe what the Puppet infrastructure and clients are doing, but they don't tell us about the overall performance of the infrastructure and its clients. Next, we will look at the metrics that Puppet supplies and how they can be used to monitor the performance and capacity of infrastructure.

Metrics, tuning, and scaling

To provide more detailed data on the performance and health of Puppet services via the services status API, the `level` flag can be set to `debug`; this will return metrics. For example, to return the metrics for Puppet Server and filter them using JQ, the following commands can be run:

```
cert="$(puppet config print hostcert)"
cacert="$(puppet config print localcacert)"
key="$(puppet config print hostprivkey)"
uri="https://$(puppet config print server):8140/status/v1/services/
server?level=debug"
curl --cert "$cert" --cacert "$cacert" --key "$key" "$uri" | jq
".status.experimental"
```

This would output data such as the following metric for the `puppet-v3-catalog` endpoint:

```
{
  "http-metrics": [
    {
      "route-id": "puppet-v3-catalog-/*/",
      "count": 41,
      "mean": 4459,
      "aggregate": 182819
    },
```

This gives us a `count` of how many calls have been made to the endpoint since the service last restarted. `mean` is the average response time over 5 minutes, while `aggregate` is the total time spent since the service started.

There are many metrics across all the different services. To find the definitions of these metrics, the API services can be viewed in the documentation (for example, `https://puppet.com/docs/pe/2021.7/status_api.html`). However, overall, they are poorly documented and may take some exploration or you asking questions on Puppet's Slack channels and support channels if you have a contract. Do not be concerned by most of the metrics having *experimental* in their title – most of the metrics have been available for years; they just haven't had the experimental tag removed by Puppet.

> **Note**
>
> In-depth details explaining how the underlying metrics library works for Puppet are available at `https://www.youtube.com/watch?v=czes-oa0yik&t=0s`, provided by the author of the metrics library.

Now, let's take a look at the dashboards that are used to display metrics data.

Exploring metrics dashboards

Puppet provides three implementations that automate the process of gathering and displaying metrics data:

- **Puppet Operational Dashboards**, available at `https://forge.puppet.com/modules/puppetlabs/puppet_operational_dashboards`, is a supported Puppet module from Puppet Forge that implements Telegraf, InfluxDB, and Grafana to provide mechanisms to send the API metric data, store it in a time series database, and visualize the data in preconfigured dashboards. Operational Dashboards supports both Puppet Enterprise and open source Puppet. An example of such a dashboard can be seen in the following figure:

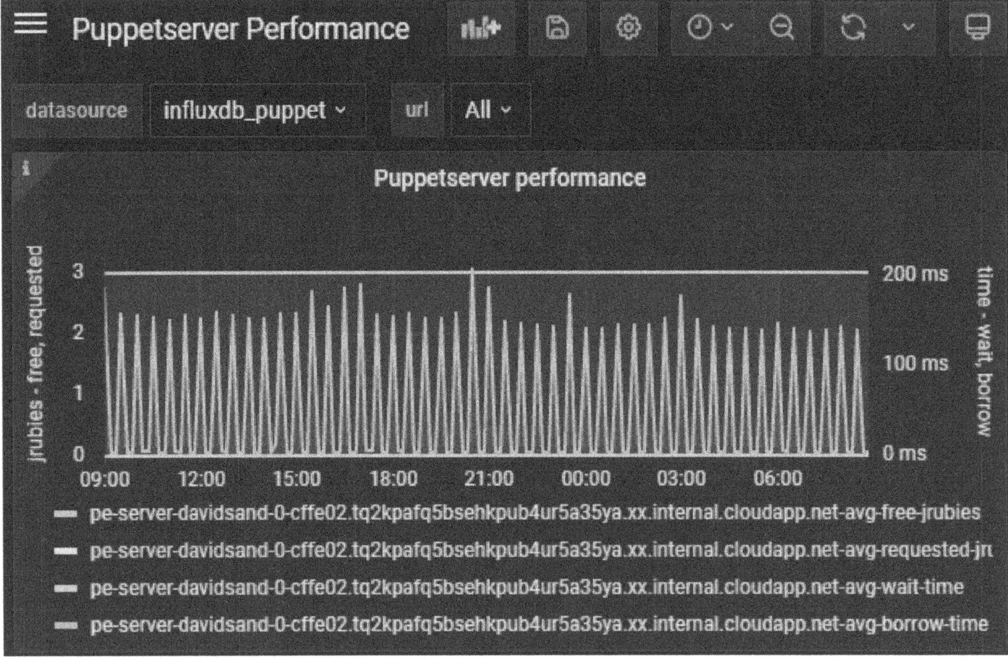

Figure 13.2 – Grafana Puppetserver performance dashboard

- The **Splunk Plugin**, an app on the Splunk store, available at `https://splunkbase.splunk.com/app/4413/#/overview`, can be added to your Splunk setup to provide preconfigured dashboards. The Splunk **HTTP event collector** (**HEC**) module can be found at `https://forge.puppet.com/modules/puppetlabs/splunk_hec`. `https://forge.puppet.com/modules/puppetlabs/pe_event_forwarding` can be combined with this module to send the metrics over HTTP to the HEC module. An example of a Splunk dashboard is shown in the following figure, with Puppet Server Memory graphed:

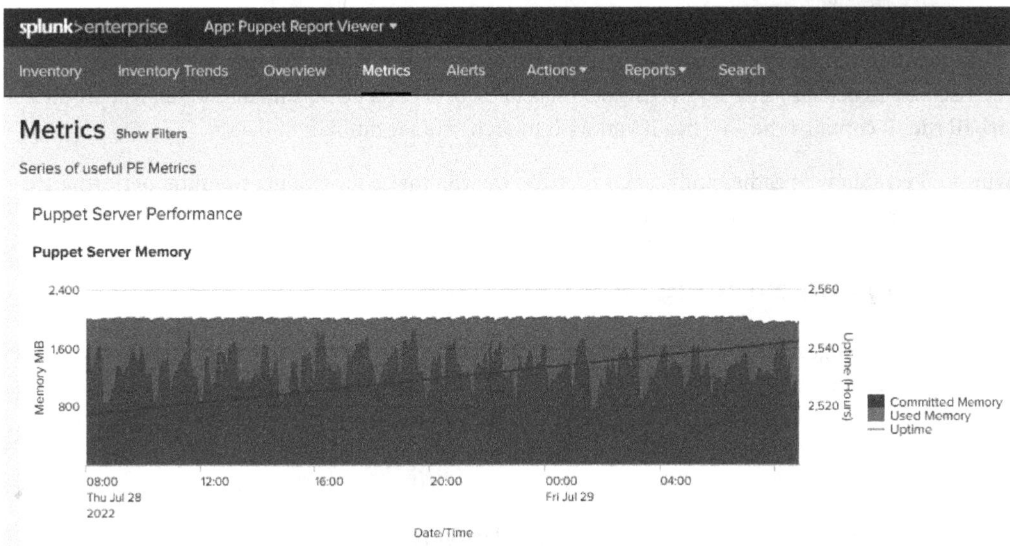

Figure 13.3 – Splunk Puppet Server Memory dashboard

Puppet teams work together to keep the Splunk and Grafana dashboards consistent.

For Puppet Enterprise, there is also the **Puppet Metrics collector** module (`https://forge.puppet.com/modules/puppetlabs/puppet_metrics_collector`) installed by default on Puppet Enterprise 2019.8.7 onwards. This module gathers metrics from the APIs and outputs them to JSON files in the `/opt/puppetlabs/puppet-metrics-collector` directory. These JSON files can then be searched using commands such as `grep` or `JQ` (assuming the terminal was in the metric collector directory). Two common queries, which will be explained in detail in the **Identifying and avoiding common issues** section, are `average-free-jrubies` and `queue_depth`. These can be added like this:

```
grep -oP '"average-free-jrubies.*?,' puppetserver/primary.
example.com/*.json puppetserver/pe-server-davidsand-0-cffe02.
tq2kpafq5bsehkpub4ur5a35ya.xx.internal.cloudapp.net/*.json

"puppetserver/pe-server-davidsand-0-cffe02.tq2kpafq5bsehkpub4ur5a35ya.
xx.internal.cloudapp.net/20220731T220502Z.json":"average-free-
jrubies":3
```

```
jq '.. |."queue_depth "? | select(. != null)| input_filename , .'
-- puppetdb/pe-server-davidsand-0-cffe02.tq2kpafq5bsehkpub4ur5a35ya.
xx.internal.cloudapp.net/*.json

"puppetdb/pe-server-davidsand-0-cffe02.tq2kpafq5bsehkpub4ur5a35ya.
xx.internal.cloudapp.net/20220731T221001Z.json"
0
```

To make it easier to share this data, it can be archived by running the `/opt/puppetlabs/puppet-metrics-collector/scripts/create-metrics-archive` command, which will produce a **tarball file**. It contains the `-r` flag if you wish to archive a set number of days.

Having looked at how to gather and display metrics, we will now discuss some common performance and capacity issues and how to manage them using metrics.

Identifying and avoiding common issues

Having set up logging, statuses, and metrics, we need to consider what to look for and how to examine our Puppet infrastructure. Normal monitoring of CPU, memory, and disk usage should be in place but there are some key functional areas to focus on. We will discuss these in the following sections.

Catalog compilation

In *Chapter 10*, we learned how each **catalog compilation** required a JRuby instance to compile a catalog for each Puppet request and that the Puppet Primary or Compiler Servers provided this JRuby capacity. To calculate the necessary number of JRuby instances to handle the load for infrastructure, we can take the **run interval** (how often servers will check in) and divide this by the average length of compilation. This will sum up how many JRuby instances are required per server. We can take the number of Puppet Clients we expect the infrastructure to have and divide this by the previous figure to provide an estimate of the total required JRuby instances:

Add this here: *Total JRuby Instances = Number of Puppet clients / (run interval / average compilation length*

Choosing the sizing for your infrastructure can be complicated. The number of JRuby instances on a primary server or compiler can be set by running `max-active-instances` in the `puppetserver.conf` file; this defaults to the number of CPUs – 1 for a range of 1 to 4. Each JRuby instance will require memory in the JVM stack. For Puppet Enterprise, this file is controlled by setting the `hiera` value to `puppet_enterprise::master::puppetserver::jruby_max_active_instances:`.

The total JVM stack memory is allocated by the Puppet Server startup script, which depending on the operating system, will be at `/etc/sysconfig/puppetserver` or `/etc/defaults/puppetserver`. This is set by the `xmx` argument, which can be calculated as each JRuby instance requiring 512 MB of memory by default and leaving 512 MB of headroom for other Java tasks:

Add this here: *Total stack heap size required = 512mb + maximum active instances * 512MB*

It is recommended that you never exceed 32 GB of stack size for JVM. As seen from various field experiments, the maximum effective number of JRuby instances appears to be between 11 and 13. These maximum figures tend to be for much larger estates and concern should be given to allow for compiler failures. In this case, it would be unwise to focus entirely on horizontal scaling; instead, it should be balanced with **vertical scaling** (having more compiler servers).

Sizing recommendations can be difficult when you're just starting – it can be very unclear what your average compilation time will be and seconds in compilation can have a big impact, so it is wise to monitor both JRuby usage and catalog compilation time as the estate grows and look for outliers as they appear. Puppet has some guidelines for Puppet Enterprise sizing at `https://puppet.com/docs/pe/2021.7/tuning_infrastructure.html`.

When monitoring catalog performance, we have some key concerns. The `jruby.num-jrubies` and `jruby.num-free-jrubies` metrics show how many JRuby instances are on a server and how many are free. When looking at these metrics, the average used capacity of the infrastructure should be calculated. It is recommended that you avoid going beyond 80% usage as performance tends not to scale beyond this. You should also confirm that there are no issues with load balancers and that the free JRuby instance usage is even across compilers. One issue that can occur is known as **Thundering Herd**, where many servers request catalog compilations at the same time. This can be seen in the metrics as large spikes in JRuby instance usage. If you experience this, you can use the **puppet run scheduler** module at `https://forge.puppet.com/modules/reidmv/puppet_run_scheduler` to distribute the scheduling of Puppet agent runs.

If the capacity of the JRuby pool is exceeded, then requests will queue and timeout after a default of 10 seconds. The `borrow-timeout-count` metric provides a count of the number of requests that have timed out while waiting for a JRuby instance to become available.

Catalog runtimes

As highlighted previously, the catalog's runtime has a huge impact on the number of JRuby instances that are required in the infrastructure. Looking at the `metrics-time-total` metric, which shows the compile time for report events sent, we can look at the average time to compile to help with our capacity calculations. We can also look at the distribution of these figures to see if we have any extreme outliers we would want to investigate in that catalog and its Puppet code.

Some key areas to check can be seen in the following table:

Metric Definition	Metric Name
Catalog compilation time	`Metrics-time-config_retrieval`
Time to apply the catalog	`Metrics-time.catalog_application`
Number of resources in the catalog	`Metrics-resources-total`
Time to generate facts	`Metrics-changes-total`

Table 12.1 – Catalog metrics

Within each of these measures, you should ensure that there was a reason for it to be an outlier. If the code or fact is complex or any particular resource is known to be slow, this may be normal, or it may be inefficient code that can be reviewed before it is applied to more servers, thus saving infrastructure capacity.

PuppetDB and PostgreSQL tuning

For PuppetD, the best metrics to monitor are `jvm-metrics.heap-memory.committed` and `jvm-metrics.heap-memory.used`. If the used memory's size is regularly approaching the committed memory's size, then it's best to increase the stack's size. Similar to compilers, this involves updating the `puppetdb` or `pe-puppetdb` config file at `/etc/sysconfig/` or `/etc/default/puppetdb`, respectively, depending on your operating system, and updating the `JAVA_ARGS` argument. For example, if you found that `jvm-metrics.heap-memory.committed` was set to 512 MB but `jvm-metrics.heap-memory.used` was approaching this limit regularly, the maximum heap size could be updated to **1 GB** by changing `JAVA_ARGS="-Xmx512m"` to `JAVA_ARGS="-Xmx1g"` in the config file. After doing this, you would need to restart the PuppetDB service. However, note that all jobs that have been queued due to them running out of memory would just continue after a restart. For Puppet Enterprise, this file can be controlled by setting the `hiera` value to `puppet_enterprise::profile::puppetdb::java_args:`.

Another good indication of performance is the queue depth, which is represented by the `puppetdb-status.status.queue_depth` metric. If this is high and there are free CPUs, it would be beneficial to increase the number of CPU threads available to PuppetDB. This can be done in the PuppetDB configuration file at `/etc/puppetlabs/puppetdb/conf.d`. If PuppetDB has been installed by a package in the `[command-processing]` section with the `threads` key or if the Puppet PuppetDB module has been used, as will be the case in Puppet Enterprise, the class should be adjusted using the module's settings. In Puppet Enterprise, this can be done in the **node groups** section in the web console. Any changes that are made to threads will require you to restart the PuppetDB service.

The reverse scenario, where CPU usage is high and throttling but the PuppetDB queue is low, should allow threads to be released to improve the throughput of other services.

Tuning sizing

To assist in getting the right server settings based on available hardware, Puppet Enterprise has the `puppet infrastructure tune` command.

This calculates the optimal settings to apply for your servers. The following example output extracts only the suggested Hiera settings printed by the command:

```
puppet_enterprise::profile::database::shared_buffers: 3176MB
puppet_enterprise::puppetdb::command_processing_threads: 1
puppet_enterprise::profile::puppetdb::java_args:
  Xms: 1588m
  Xmx: 1588m
puppet_enterprise::profile::orchestrator::jruby_max_active_instances:
2
puppet_enterprise::profile::orchestrator::java_args:
  Xms: 1588m
  Xmx: 1588m
puppet_enterprise::profile::console::java_args:
  Xms: 1024m
  Xmx: 1024m
puppet_enterprise::master::puppetserver::jruby_max_active_instances: 2
puppet_enterprise::profile::master::java_args:
  Xms: 1536m
  Xmx: 1536m
puppet_enterprise::master::puppetserver::reserved_code_cache: 192m
```

This extracted data could be put in a Hiera file and classified against the primary server. Note that if the RAM is high enough, it will recommend configurations for heap sizes above 32 GB. This is sub-optimal, as we discussed when we looked at compilation sizing and issues.

The following is general advice:

It may seem difficult to process the number of metrics, especially given the lack of clear definitions, but if the Splunk plugin or Operational dashboard is used, this can give you a view that's consistent with what Puppet Support teams use and monitor. Learning how your estate normally behaves in these values and looking for spikes in the graph and relating them to others can go some way to finding issues.

Using Puppet's knowledge base, which has been open to any user since April 2021, can help you search for issues. Looking at collections of articles such as `https://support.puppet.com/hc/en-us/sections/360000926413-Performance-tuning` can assist you in gaining a deeper understanding of any issues that you experience.

Lab – configuring metric dashboards

Having discussed metrics and Puppet's two options for viewing them, we can configure the Splunk dashboard and the Puppet Operational dashboard to see the dashboards provided. Using the issues that were described in the previous section around PuppetDB and compiler capacity, find the graphs that would assist in your investigation.

Configure the Puppet operational dashboard:

1. Choose one of your nodes to host the Puppet Operational dashboard and classify `puppet_operational_dashboards` on the web console as a node group.
2. In the node group's PE Infrastructure Agent, add `puppet_operational_dashboards::enterprise_infrastructure` to the list of classes on the **Classes** tab.
3. Run `puppet` on all the nodes until the servers are showing clean.
4. Log in at `https://<public_ip_of_operational_dashboard_node>:3000 user=admin password=admin`.

Configure Splunk:

1. Sign up for a Splunk account at `https://www.splunk.com/en_us/sign-up.html?301=/page/sign_up` (this is free).
2. Choose a different node to host Splunk Enterprise by classifying `splunk::enterprise` as a node group on the web console and pinning the chosen node.
3. Install the Puppet report viewer:

    I. Log in and download `https://splunkbase.splunk.com/app/4413/`.

    II. Log in to `https://<public_ip_of_splunk_server>:8000 username=admin` and `password=changeme`.

    III. Select the cog to the top left of the **Apps** bar.

    IV. On the next screen, select **upload app from file** at the top right.

4. Set the license to free in Splunk:

    I. Click **Settings** at the top right.

    II. In the dropdown under system, select **Licensing**.

    III. Select **change license group**.

    IV. Select **free license** and click **Save**.

5. Create an HEC token in Splunk:

   I. Navigate to **Settings | Data Input** in your Splunk console.

   II. Add a new HTTP Event Collector with a name of your choice.

   III. Ensure **Indexer acknowledgement** is not enabled.

   IV. Click **Next** and set **source type** to **Automatic**.

   V. Ensure **App Context** is set to **Puppet Report Viewer**.

   VI. Add the main index.

   VII. Set **Default Index** to **main**.

   VIII. Click **Review** and then **Submit**.

6. Classify `puppet_metrics_collector` with the `metrics_server_type` parameter set to `splunk_hec` on the PE Infrastructure node group.
7. Classify `splunk_hec` on the **PE HA Master** node group with the following parameters:

   - `enable_reports` set to `true`
   - `events_reporting_enabled` set to `true`
   - `manage_routes` set to `true`
   - `token` set to `<token number of generated in step 5>`
   - `url` set to `https://<public_ip_of_splunk_server>:8088/services/collector`

8. Log in at `https://<public_ip_of_splunk_server>:8000` and run `index=* sourcetype=puppet:summary` to ensure data is being gathered (it may take some time to start).

Now that both the Puppet operational dashboard and Splunk have been configured, tour the various graphs and panels to find the graphs that are relevant to catalog capacity and PuppetDB performance.

If possible, leave Splunk set up for the next section as the inventory and inventory trend views are the dashboard views of the Facter terminus output.

Now that you know how to integrate Puppet's status, logging, and metrics, we can look at a pattern that allows us to integrate Puppet with other services and provide self-service to Puppet users.

External data provider pattern

Puppet will not be the only source of configuration and information in your estate. There are likely to be numerous sources for **Configuration Management Databases (CMDBs)** such as ServiceNow or internally developed systems that are used by application teams to store their information. Several of your colleagues and internal customers will want to be able to create exceptions and customizations without having to understand Puppet code and the workflow for deployment. There will also be demands to be able to feed Puppet data back into external systems. The external data provider pattern allows this to happen by allowing you to do the following:

- Make changes in the classification
- Add and change trusted facts
- Feed existing data in as a fact or Hiera data
- Send Facter data to external sources

Having introduced the core concept of the external data provider pattern, we will now look at each of the technical components used within it.

Understanding external data provider components

The underlying components of this pattern are shown in the following figure:

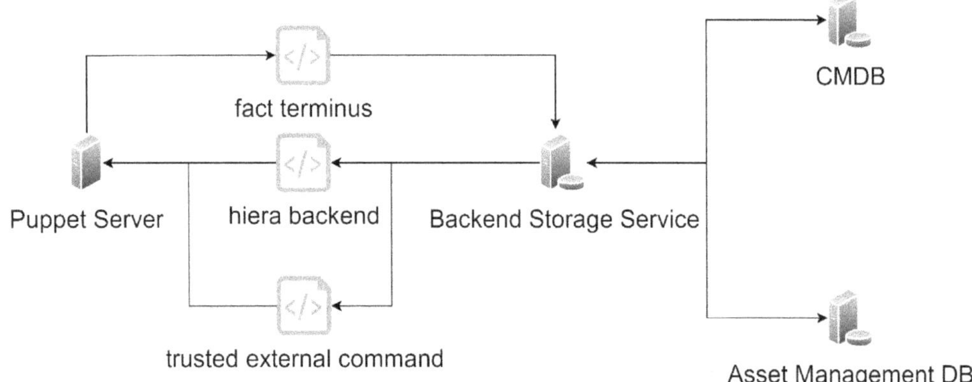

Figure 13.4 – Core components of the external data provider pattern

We will run through each to show how this pattern works. It is beyond the scope of this book to show you how to write each of these components, but we will detail where documentation exists for this. In the following section, sample implementations will be referenced:

A **backend storage service** (**BSS**) allows you to store data for consumption. The technical solution for the BSS is not important, but it must be resilient and provide high throughput on reads.

This throughput can be calculated at *2 + <number of Hiera levels>* reads per Puppet agent run. To show how this would be calculated if an estate of 10,000 servers used the default agent run time of 30 minutes and had a 5-level Hiera setup, this would be calculated as *(2+5) * 10,000 / 1,800 = 39 queries per second* (rounded up).

Tools such as CMDB or internal applications can be directly queried and act as the BSS, but the tool can deal with the workload.

The `trusted` external command, which was introduced in Puppet 6.11 and 7.0, allows a script to be run during Puppet runs and gather facts and classification information from external sources. This script should take the `certname` property of the client as its argument, return a JSON hash of facts, and exit with an error code for any unknown `certname`. This script can be configured by using the `trusted_external_command` setting in the `master/server` section of `puppet.conf` on each primary and compiler server. The facts that are returned by this command will be contained under `trusted.external.basename`, where `basename` is the name of the script. Since Puppet 6.17 and 7.0, it is also possible to use multiple trusted external commands by setting `trusted_external_command` to be a directory containing multiple scripts. This can be useful for querying multiple sources. Each source would then get a different base name. In the external data provider pattern, it is used to query the BSS.

The Hiera backend uses functions written in Ruby or Puppet to query APIs or other sources when Hiera lookups are performed. In the external data provider pattern, the backend queries the BSS for values. Documentation on this is available at `https://puppet.com/docs/puppet/latest/hiera_custom_backends.html`.

Puppet allows pluggable backends known as **termini** and uses indirectors, as discussed in *Chapter 10*, to allow Ruby scripts to access key-value pairs at endpoints. Fact termini are Ruby scripts that access the fact endpoints and allow the data to be sent on to other external systems. Further details on this are available at `https://puppet.com/docs/puppet/latest/indirection.html` and `https://puppet.com/docs/puppet/latest/man/facts.html`.

External data provider implementations

At the time of writing, no single implementation of the external data provider pattern implements all parts, but they can be used together to integrate multiple systems and purposes. The list of examples in this section is not meant to be exhaustive but should show the breadth of integrations that can be investigated. We will also provide documentation examples, which can be expanded on if necessary.

Satellite

Red Hat Satellite can receive reports from Puppet Server through a report processor while using the module available at `https://forge.puppet.com/modules/puppetlabs/satellite_pe_tools`. However, it is also possible to use the `puppetserver_foreman` module to configure a trusted external command to gather the various configuration data from Satellite, such as smart parameters and organization as facts. With Puppet being removed as the default configuration management choice and instead used as an optional plugin, and Puppet versions not keeping up with development in the Satellite platform, as per `https://www.redhat.com/en/blog/upcoming-changes-puppet-functionality-red-hat-satellite`, the use of this trusted external command allows Puppet

Server's functionality to be migrated as a separate Puppet infrastructure while the configuration is maintained in the Foreman component of Satellite. See the files/Satellite at `https://github.com/theforeman/puppet-puppetserver_foreman` to see the trusted external command.

ServiceNow

Several ServiceNow integrations have been developed for use with Puppet; the CMDB integration allows a trusted command provided by the module available at `https://forge.puppet.com/modules/puppetlabs/servicenow_cmdb_integration`.

ServiceNow should only be used with BSS on a smaller scale since using a large number of nodes could overwhelm ServiceNow with queries. It provides a useful example of using the trusted command.

A better approach has been developed to ensure scaling where the ServiceNow graph connector connects to the Puppet API and gathers the necessary data: `https://store.servicenow.com/sn_appstore_store.do#!/store/application/42ae987a1b832c10fa34a8233a4bcb0b`.

Azure Key Vault

Azure Key Vault's integration is a function that calls `azure_key_vault::secret` in Puppet code. This can be used with a Hiera backend to access Azure Key Vault secrets. It is an approved module (`https://forge.puppet.com/modules/tragiccode/azure_key_vault`).

1Password

The 1Password integration is a Hiera backend that allows lookup calls to be made for secrets in the 1Password setup: `https://forge.puppet.com/modules/bryxxit/onepassword_lookup`.

Vault

The two Vault solutions (server-side and client-side) were discussed and demonstrated in *Chapter 9*, but to recap, it is the server-side Vault lookup that implements a Hiera backend lookup function called `hiera_vault` in the module. As discussed at `https://forge.puppet.com/modules/petems/hiera_vault`, this allows secrets from Vault to be called via Hiera and compiled into code.

Puppet Data Service

Puppet Data Service (**PDS**) provides one of the most complete implementations of the external data provider pattern, except it implements a fact terminus. The following components are a part of PDS:

- A REST API and CLI that allow user and application interaction
- A pluggable backend database to provide a BSS (at the time of writing, only PostgreSQL is supported)
- A Hiera backend to query the BSS
- A trusted external command to query the BSS

PDS was designed to be less focused on a particular set of integrations and allow Puppet platform teams to leverage the external data provider pattern to provide self-service and reduce operational burdens. PDS has an install module (`https://github.com/puppetlabs/puppetlabs-puppet_data_service`). The code that makes up the application and API (`https://github.com/puppetlabs/puppet-data-service`) is packaged in `deb` and `rpm`, which are used by the `install` module. At the time of writing, both modules are only designed for a Puppet Enterprise installation, but nothing within the underlying setup limits the application to Puppet Enterprise, which means it can be adapted to open source Puppet.

Splunk

In the *Logging and status* section of this chapter, we discussed and demonstrated how the `splunk_hec` module provided communication from Puppet Server to the Splunk Hec URL on a Splunk server using the same modules fact terminus (`https://github.com/puppetlabs/puppetlabs-splunk_hec/blob/main/lib/puppet/indirector/facts/splunk_hec.rb`).

Facts from Puppet runs can be sent to Splunk, which can then be viewed in the Splunk app (`https://splunkbase.splunk.com/app/4413/`) in terms of inventory and inventory trends.

Lab – hands-on with Splunk and Puppet Data Service

Having discussed several integrations, if you left the Splunk installation or reinstallation as per the previous lab, you can log into Splunk and view the **inventory** and **inventory trend** tabs. Here, you will see the Facter terminus output and can experiment with viewing the data from your nodes.

In this part of the lab, you will see how PDS can be used to classify nodes, update Hiera data, and add trusted facts.

To install PDS, you will need to perform the following tasks:

1. Observe the `hiera.yaml` and `site.pp` file in the control repository you cloned and see how PDS will use them.
2. Configure the two required application roles.
3. For the database server, do the following:

   I. Add a new node group from the PE console:

   ```
   Parent name: PE Infrastructure
   Group name: PDS Database
   Environment: production
   ```

   II. Add the `puppet_data_service::database` class to the PDS database group you created in the previous step.

   III. Add your existing primary server to the group using the Rules tab node before following these steps.
   IV. Commit your changes.

4. For the PDS API servers, do the following
   I. Select the **PE Master** node group.
   II. Select the **classes** tab.
   III. Add the new puppet_data_service::server class
   IV. Include the database_host: <FQDN of your primary server> parameter
   V. Select the **Configuration** data tab
   VI. Configure the sensitive pds_token parameter. You can use https://www.uuidgenerator.net/ to generate a token
   VII. Commit your changes.
5. Run Puppet on all nodes until reports show unchanged.
6. Create an SSH session to the primary server and one of the nodes in separate terminal windows.
7. On the primary server, run the pds-cli node upsert <fqdn_of_node> -c motd -e production command.
8. On the node, run puppet agent -t. In the output of the command, you will see that motd has been applied with default settings.
9. On the primary server, run pds-cli hiera upsert nodes/<fqdn_of_node> motd::content -v '"Hello world its PDS\n"'.
10. On the node, run puppet agent -t. In the output of the command, you will see that motd has been applied with the Hiera override we set.
11. On the primary, run pds-cli node upsert <fqdn_of_node> -c motd -d '{"status": "Testing"}' -e production and pds-cli hiera upsert nodes/<fqdn_of_name> motd::content -v '"Hello world, I am a PDS %{trusted.external.pds.data.status} Server\n"'.
12. On the node, run puppet agent -t. Observe that it applies the new motd with the Hiera override and value testing set for the trusted fact.
13. Log into the console and look at the node and its facts to see if it has a trusted fact, pds.data.status, set to testing.

Summary

In this chapter, we summarized the various log locations and showed you how logs could be turned into JSON and exported so that they can be handled in logging toolsets such as Elastic or Grafana, which can better index them for viewing and analysis. We learned how report processors can be used on Puppet Server to allow the reports to be generated by applying catalogs on clients. This allows them to be sent to tools such as Splunk and allows for advanced visualizations and searches. The available status APIs were discussed, indicating how an API call could be made to find the status of all running services or a particular service. Puppet Enterprise was shown to have a command line (`Puppet Infrastructure status`) and web console option to call this API. Using these mechanisms, you learned how to access critical logging and metrics to understand the current state of the system.

To use this information and understand the performance of the services in depth, you learned how Puppet metrics become available upon using the `debug` flag of the status API and how tools such as the Puppet Operational Dashboard and the Puppet plugin for Splunk could be used to gather this data and visualize it. Puppet Enterprise was noted as having the Metrics Collector module, which gathers metrics locally in JSON files, which can be viewed manually or exported.

To better understand how these metrics and dashboards can be used, we reviewed some common issues, looking at how to size the infrastructure for catalog compilation and avoid issues such as Thundering Herd as servers squeeze demand and how PuppetDB could be adjusted as demand increases or decreases. Various infrastructure tuning tools were shown to be an option in PE to optimize settings for deployed hardware.

Then, we covered the external data provider pattern, which provides mechanisms for self-service and access to Puppet data on external services so that it can be integrated better. The core components of a backend storage service were shown to provide a store for data that could cope with the level of queries Puppet would make while trusted external commands and the Hiera backend were shown as ways to query that data. Fact termini were shown to be ways to export data from the BSS to external services.

Various implementations of these components were shown when using various Hiera backends, with 1Password, Azure Key Vault, and Vault being shown as ways to access external secret managers, while Satellite and ServiceNow were shown to have trusted commands that allowed data within those applications to be fed into Puppet code.

Puppet Data Service was shown to be one of the most complete implementations of the pattern and provides a solid design to allow for self-service of internal customers who would be able to access suitably exposed Puppet options without requiring full knowledge of the Git flow and Puppet language.

This coverage of the external data provider pattern showed you how powerful integrations can be made with Puppet Enterprise to feed data into and out of different tools and work toward building a platform with Puppet as a vital component.

Having covered the components of Puppet Server how to monitor performance at scale and integrate it, the next chapter will look at Puppet Enterprise-specific services and their components. It will describe what Puppet Enterprise is, how it differs from the open source version, what extra services it provides, and the reference architectures provided by Puppet to allow for easier scaling and tooling to automate a deployment and its status. Projects and integrations specifically for Puppet Enterprise will also be discussed.

Part 4 – Puppet Enterprise and Approaches to the Adoption of Puppet

This part will look at Puppet Enterprise and how it differs from open source. It will review some Puppet-related products that can extend Puppet Enterprise and some specific integrations for Puppet Enterprise. We will then discuss approaches that can help organizations successfully adopt Puppet. We will look at correctly scoping use cases to benefit from regular delivery, and how Puppet can work within platform engineering as well as with heritage estates, and even in highly regulated, change-managed estates.

This part has the following chapters:

- *Chapter 14, A Brief Overview of Puppet Enterprise*
- *Chapter 15, Approaches to Adoption*

14
A Brief Overview of Puppet Enterprise

This chapter will give an overview of **Puppet Enterprise**, what it is, and what it provides compared to **Open Source Puppet**. Although the author of this book is a Puppet employee, this is not intended as a hard sell but to present where and how to use Puppet Enterprise well. It will cover the extra Enterprise console services in the Puppet platform, showing how code deployment, orchestrator service, RBAC, web console, and various other services are automatically configured and work with each other. This will assist in understanding how Puppet Enterprise differs from Open Source Puppet and the preconfigured and built-in features that would need to be manually created in Open Source Puppet. Supported architectural patterns will be highlighted that help to understand how to deploy and scale Puppet infrastructure using Puppet Enterprise packaging and modules to automatically deploy these patterns. Some related projects and integrations will be discussed, along with how they fit into the Puppet Enterprise environment.

In this chapter, we're going to cover the following main topics:

- What is Puppet Enterprise?
- Exploring the Puppet Enterprise console and services
- Using Bolt with Puppet Enterprise
- Automating deployment and reference architectures
- Puppet Enterprise-related projects and tooling
- Lab—Puppet Enterprise extensions and configuration

Technical requirements

Clone the control repo from `https://github.com/puppetlabs/control-repo` to your `controlrepo-chapter14` GitHub account and update the `Puppetfile` file in this repo: `https://github.com/PacktPublishing/Puppet-8-for-DevOps-Engineers/blob/main/ch14/Puppetfile`.

Build a large cluster with a replica with three compilers and three clients by downloading the `params.json` file from `https://github.com/PacktPublishing/Puppet-8-for-DevOps-Engineers/blob/main/ch14/params.json` and updating it with the location of your control repo and your SSH key for the control repo. Then, run the following command from your `pecdm` directory:

```
bolt --verbose plan run pecdm::provision --params @params.json
```

What is Puppet Enterprise?

A common misconception when discussing Puppet Enterprise is that features of the product are held back and not available for open source users. The aim of Puppet Enterprise isn't to limit what is available to open source users but to instead provide value to customers who want to consume Puppet easily and focus on gaining the value of configuration management by putting less of their own development and automation work into the platform itself.

Puppet achieves this by ensuring that in Enterprise, the packing of components is versioned and tested together with an automated installation script and module, reducing the effort required by users managing the infrastructure. Puppet Enterprise works on two different types of releases. Puppet Enterprise, which works on an *xxxx.y* pattern, is normally updated every 3 months, which at the time of writing would be `2023.0`. This version is planned to upgrade to Puppet 8.x versions in 2023.3 and will receive new features throughout its lifetime. This release is recommended for users who want to access the latest features and fixes and will require a regular update pattern. The other type of release is the **long-term support** (**LTS**) version; this follows an *xxxx.y.z* pattern. This branch is normally updated every 3 months, but the updates would only include fixes and not new features. The LTS versions last 2 years and have an overlap of 6 months with the next major *xxxx* release, so the current 2021.7.z LTS will end mainstream support on August 31, 2024, at which point overlap support will continue until February 28, 2025, after which users should migrate to whichever version of Puppet they require. 2023.y becomes the new LTS release to continue to have support from Puppet. The two running Puppet Enterprise versions generally mirror two Open Source Puppet versions in active development. The release of 2023.0 retired Puppet 6 and 2023 should move to Puppet 8 in version 2023.3 or shortly after.

The most obvious feature of an Enterprise license is support, with access to raise support cases with teams who can review infrastructure problems and assist with any issues or features required for supported modules.

Puppet also provides various professional services, such as on-site engagements to provide hands-on training and advice. This can lead to architecture reviews to understand how best to implement in your environment and to feed into processes that develop products and solutions such as the **Puppet Data Service (PDS)** and **Puppet Enterprise Administration Module (peadm)**. Further, **technical account managers (TAMs)** are assigned to give you a regular point of contact and champion you in Puppet, supporting you in creating a success plan for your organization and focusing your deployment to achieve its goals.

Puppet provides reference architectures and patterns of Puppet products to show how to work at different scales and implementation types. Additional applications built on top of the Puppet Server service allow for access control, server classification, code deployment, visualization, and searching of data to be completed in a standard way from the console. We will look at these in greater detail in the following section.

Exploring the Puppet Enterprise console and services

There are several additional services built into a **Puppet Enterprise primary server**, as shown in the following diagram:

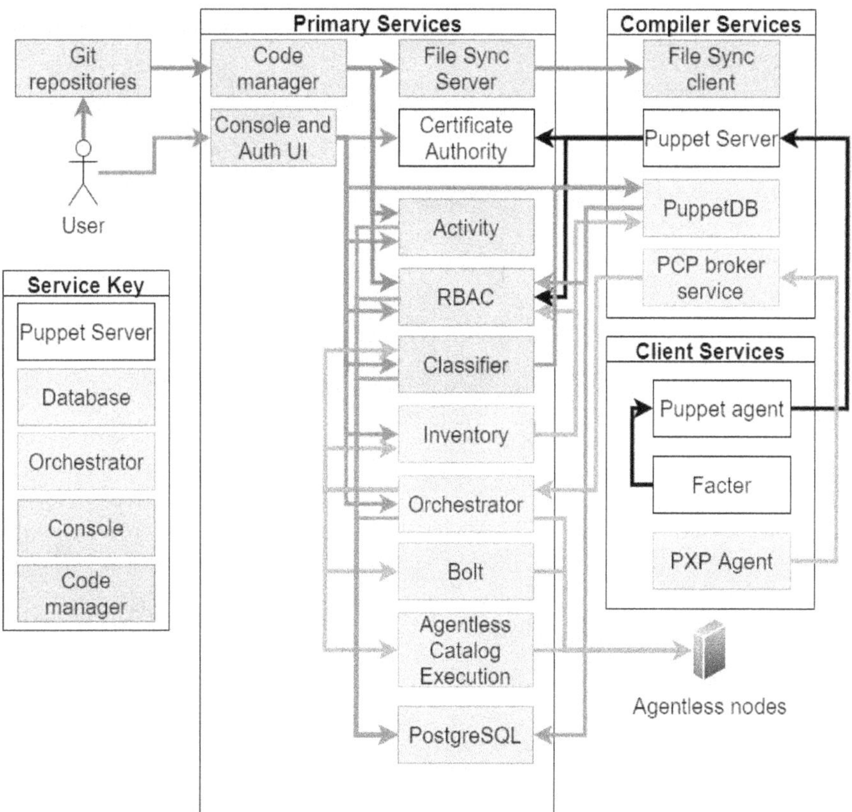

Figure 14.1 – Puppet Enterprise components

Puppet Server

The **Puppet** service is the same as discussed in *Chapter 10*, with the **certificate authority** (**CA**) providing a certificate signing process to secure communication and the Puppet agent contacting a compiler's Puppet Server service to request a catalog compilation. **Facter** is used to provide a server profile. In *Figure 14.1*, we opt not to show that the primary server itself has a Puppet Server service, and both the compiler and primary server have Puppet agents, which both request catalog compilations from the primary server's Puppet server.

Introducing Puppet web console components

The most obvious immediate difference of the Puppet Enterprise server is the **web console**, which provides the login view we have been using in our lab throughout this book. Several services combine to make up the console services

The console is a web frontend Jetty-based Clojure service with an NGINX server that acts as a reverse proxy. The NGINX server listens on port HTTPS `443` and redirects HTTP 80 to HTTPS. The console UI provides an aggregation and translation Jetty-based Clojure service to generate the correct pages and access other console services.

The authentication UI generates login and resets password content pages. The simplest way to show this is to use an example of the communication required when logging in, as shown in the following diagram:

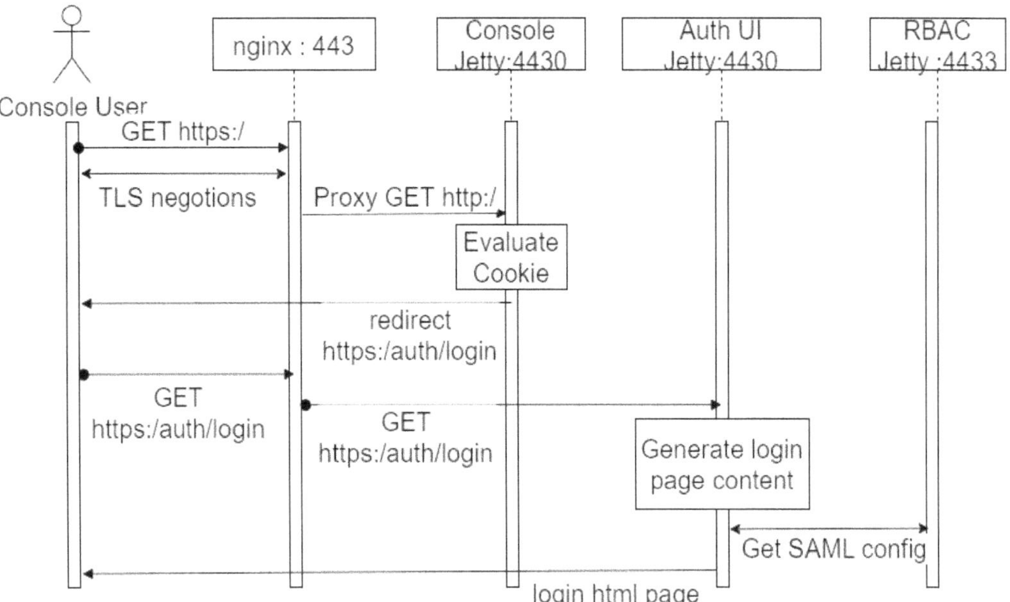

Figure 14.2 – Steps to generate the login page

It can be seen from the diagram that as a first step, the NGINX server receives a `GET` request and, after performing TLS negotiation, redirects to the console Jetty page. This page evaluates cookies, establishes the user is not logged in, and redirects to the auth/login page, which is requested from NGINX and redirected to the authentication UI. The authentication UI generates a login page and gets **Security Assertion Markup Language** (**SAML**) configuration from the RBAC Jetty page, and this login page is then passed back to the user.

The **RBAC service** has users and roles to construct access policies. It allows for both local and remote users in Puppet Enterprise, with integration possible to **Lightweight Directory Access Protocol** (**LDAP**) and SAML services. All users by default are denied permission to create, edit, or view any part of Puppet Enterprise, and permissions are then granted via roles.

By default, there will be a local administrative user who acts as a superuser for the Puppet service and an API user for authentication for Puppet services to communicate within Puppet Enterprise. It cannot be used for login and only authenticates with certificate authentication. There is an allow list that has the `certname` values of certificates that can be used with the API user.

Roles allow the grouping of permissions to give users permission to perform actions. A permission is made up of a **type**, a **permission**, and an **object**. The type is what the permission will allow actions on, such as users or node groups. A permission is a level of access from create, edit, or view, and an object is a specific instance of the type such as the Puppet Enterprise infrastructure node group for a node group type.

There are five roles provided by default:

- **Administrators**: All permissions
- **Operators**: Permission to create and modify node groups, deploy code, run Puppet, run sign certificates, and view the console
- **Viewers**: Permission to view the console, node groups, and jobs
- **Code deployers**: Permission to deploy code with Code Manager
- **Project deployers**: Permission to deploy projects, run tasks and plans from projects, and start, stop, and view jobs in orchestrator

Custom roles can be created with the consultation of `https://puppet.com/docs/pe/2021.7/rbac_permissions_intro.html#user_permissions` to find the right granularity of user permissions.

LDAP solutions such as **Active Directory** (**AD**) can map user groups to roles, while **SAML solutions** such as **Okta** can do similar user group mapping with attributes to match to roles. Both LDAP and SAML are configured in the web console on the **Access control** tab by selecting the corresponding option.

Tokens are used for all web sessions, and instead of logging in with a password whenever running commands, tokens can be generated for multiple uses. The tokens are alphanumeric values between `0` and `2^256 - 1` and stored in the database and, depending on the argument, a local file location. Tokens are generated by either the token API endpoint, the web console in the **My account** tab, the **Tokens** tab, or by running the `puppet access login` command on the CLI. The token is against the user credentials provided, so will have the permissions set to that user. By default, the `puppet access login` command will write the token to `~/.puppetlabs/token` with a lifetime of 30 minutes unless the `--lifetime` is used to set a lifetime such as `5h` for 5 hours.. The `--print` flag will cause the token to only be printed and not stored, which is appropriate for service-based API access.

The classifier service was discussed in *Chapter 11*, where we looked at how it used **node groups** to classify servers in Puppet, but to reiterate the key points, node groups are used to classify classes to servers either by using facts or directly pinning named servers. Node groups are inheritance based, so each child of a node group will inherit everything above it.

Code Manager was discussed in *Chapter 11*, showing how the **Code Manager** service used `r10k` to download modules based on a Puppet file in a control repo from a named Git repository, and the `filesync` server and `filesync` clients then kept this copy of code in sync across the services.

The **activity service** is used to log all activities that have taken place through the console service and can be viewed by the API endpoint and on the web console in various places, such as the **activity** tab on any user and role.

The database components of Puppet Enterprise, **PuppetDB** with **PostgreSQL**, are the same as was discussed in *Chapter 10*, but several of the services need additional databases to store their state and records. So, the following databases are created:

- `pe-activity`: All auditable activities of console services
- `pe-classifier`: All node group information
- `pe-inventory`: Agentless client details and their access method for orchestrator
- `pe-orchestrator`: Job runs, job results, users, and node
- `pe-postgres`: Postgres databases for templating and general access. See `https://www.postgresql.org/docs/current/manage-ag-templatedbs.html` to understand template databases further.
- `pe-puppetdb`: Reports, node information, and last run catalog
- `pe-rbac`: Users, roles, groups, and AD/LDAP information

> **Note**
>
> All PostgreSQL communication is done using certificates, including communications to replicas.

The one component which was not covered in *Figure 14.1* was the orchestrator services. We will now cover how orchestrator provides the capability of using Bolt plans and tasks within Puppet Enterprise.

Using Bolt with Puppet Enterprise

In *Chapter 12*, it was seen how Bolt was run using the `bolt` binary within a Bolt project, but it can be used integrated with Puppet Enterprise via the **orchestrator service**, allowing plans and tasks to be run as part of Puppet Enterprise.

The key difference is that currently, only Puppet modules containing tasks and plans can be deployed (including adding them to a control repo); there is no current method of deploying bolt projects to Puppet Enterprise directly.

> **Note**
>
> With plans and tasks deployed through modules, this means the same plan or task can have multiple versions, depending on the environment it is run from.

It is also important to realize not all of the features available to Bolt natively will be available within Puppet Enterprise.

The following list highlights the key differences when running plans and tasks in Puppet Enterprise in orchestrator instead of in native Bolt:

- Various Bolt functions for plans, such as `prompt`, `parallelize`, and `file.upload`, have not been implemented
- `puppet apply` blocks can only be applied to nodes with a Puppet agent
- Targets and the localhost target are unavailable
- File sources must be module based and cannot be absolute paths

Most of these limitations reflect not running Bolt from a local machine and a lack of a prompt to run them. Full details can be viewed in Puppet's documentation at `https://puppet.com/docs/pe/2021.7/plans_limitations.html`.

Puppet Enterprise handles three types of nodes with plans and tasks:

- Nodes with a Puppet agent installed, using the **Puppet Communications Protocol** (**PCP**) and the **PCP Execution Protocol** (**PXP**)
- Agentless nodes via the **Windows Remote Management** (**WinRM**) and **Secure Shell** (**SSH**) transports
- Agentless devices such as switches or firewalls via transports such as F5 and **Palo Alto Netorks Operating system** (**PAN-OS**) or transports provided via the resource API

Having highlighted what orchestrator is capable of running for plans and tasks, we will now look at the components that make up orchestrator, highlighting the purpose of these services and key details such as log locations and configuration files.

Orchestrator services

The orchestrator application is a **Clojure** application made up of the services shown in the following diagram:

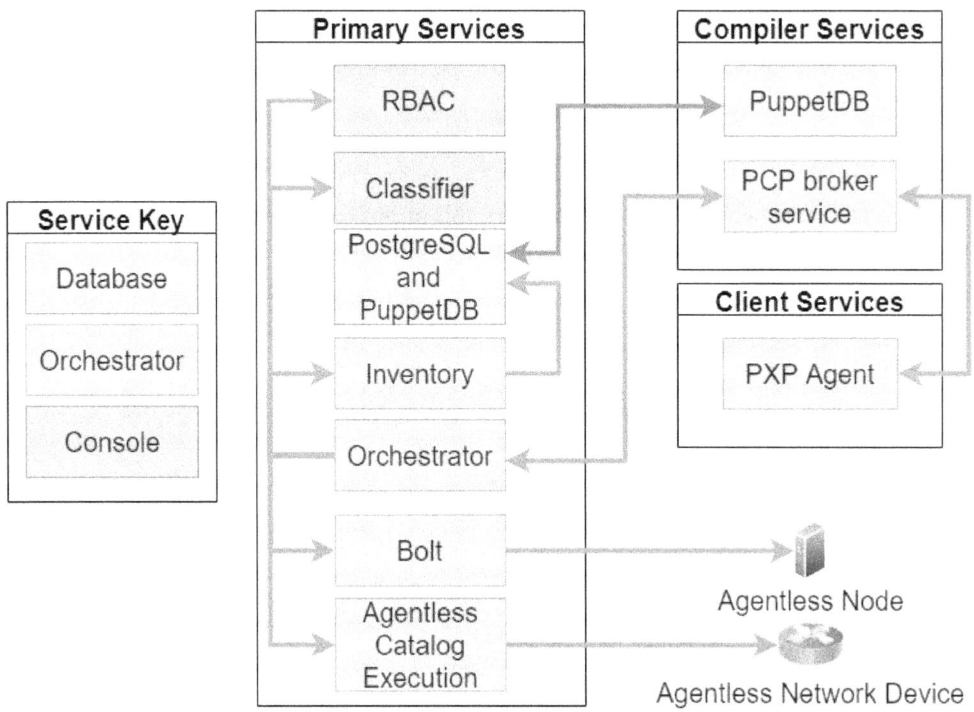

Figure 14.3 – Components of the Puppet orchestrator service

Let's have an overview of these components and their relevant services and log files:

- **Orchestrator service**: The orchestrator service is the core service that requests for jobs, tasks, and plans are made to. Users must first be authenticated, which verifies their user and permission profile, as managed in the RBAC service. For nodes with agents, it contacts PuppetDB to retrieve facts about nodes or inventory for agentless nodes. It will update the PostgreSQL database orchestrator with details of job requests and direct jobs to service based on their transport. The orchestrator service runs under `pe-orchestration-services.service` and logs to `/var/log/puppetlabs/orchestration-services/orchestration-services.log`.

- **Inventory service**: A register of agentless clients and their access method added via the web console inventory page or via the POST `/command/create-connection` inventory API call (`https://puppet.com/docs/pe/2021.7/node-inventory-v1-command-endpoints.html#node-inventory-v1-command-endpoints`). These entries are encrypted by a secret key, by default placed at `/etc/puppetlabs/orchestration-services/conf.d/secrets/keys.json`, and although listed separately, the inventory service runs within `pe-orchestration-services`. It stores its data in the PostgreSQL inventory database.

> Note
>
> Agentless nodes added to the inventory are counted within the overall licensed number of nodes for Puppet Enterprise.

- **Bolt service**: A Ruby service that allows actions such as tasks and commands to be run over SSH and WinRM tasks to agentless nodes. It will also compute module metadata content. It runs under `pe-bolt-server.service` and logs to `/var/log/puppetlabs/bolt-server/bolt-server.log`.
- **Ace service**: A Ruby service that can run tasks, plans, and Puppet runs on agentless targets remotely, such as network switches and firewall devices using transports such as PAN-OS, F5, or any other transport defined with the resource API. The Ace service runs under `pe-ace-server.service` and logs to `/var/log/puppetlabs/ace-server/ace-server.log`.
- **PCP broker**: A Clojure application running on a **Java Virtual Machine** (**JVM**) acting as a broker service on compiler servers that routes PXP messages using PCP, which routes PXP messages to an agent and returns them to the orchestrator. API requests are logged to `/var/log/puppetlabs/orchestration-services/pcp-broker-access.log` and general service logs are logged to `/var/log/puppetlabs/orchestration-services/pcp-broker.log`.
- **PXP agent**: The agent that allows requests for tasks to be run on Puppet clients via PXP, which sends requests for task plans etc to be applied and returns results. It runs under `pxp-agent.service` and logs to `/var/log/puppetlabs/pxp-agent/pxp-agent.log`.

The orchestrator service will verify via RBAC that the Puppet Enterprise console user has the correct permissions. For plans, it is only possible to specify users or groups and which plans they can run with no limit of nodes or which environment the plan will come from. For tasks, task targets allow a list of tasks and either a **Puppet Query Language** (**PQL**) query of nodes or groups of nodes to be specified, which the tasks can be run against. This can be done either via the API call, as shown at `https://www.puppet.com/docs/pe/2021.7/orchestrator_api_commands_endpoint.html#orchestrator_api_post_command_task_target`, or in the RBAC GUI, as shown in the following screenshot:

Permissions

Add a permission

Tasks | Run Tasks | adhoc::win_example

Permitted nodes

PQL Query

inventory[certname] { facts.os.name = "windows" }

Figure 14.4 – Creating a task target on the web console

In the next section, we will learn how to run tasks, plans, or Puppet runs through orchestrator.

Running jobs

When tasks, plans, or Puppet runs are run through orchestrator, they become known as **jobs**. There are three ways to run jobs, as follows:

- The first way to do this is via the GUI by selecting the relevant menu on the left bar.
- The second is via the CLI on the primary server with largely the same syntax as the `puppet task run` and `puppet plan run` Bolt commands. The key differences compared to Bolt are that the `--nodes` flag is used instead of `targets` (reflecting the fact you will be just providing a node name, for which orchestrator will lookup transport information) and extra flags are available, such as the `--node-groups` flag, for choosing a node group to run against. Here's an example:

```
puppet task run examplemodule::exampletask paramter1=value1
paramter2=value2 --node-group <node group id>
puppet plan examplemodule::exampleplan
parameter1=value1  --nodes examplehost.com,examplehost2.com
puppet job run --query 'inventory { facts.os.name = "windows" }'
```

- The third way is via the APIs documented at `https://puppet.com/docs/pe/2021.7/orchestrator_api_commands_endpoint.html`, with the key calls listed here:
  - `POST /command/deploy`: Run Puppet on demand
  - `POST /command/plan_run`: Run a plan
  - `POST /command/task`: Run a task on a set of nodes

Jobs in progress can be stopped by pressing *Ctrl* + *C* on the CLI, selecting **Stop job** on the GUI, or by the `POST /command/stop` API command. Although we should be careful to note a stopped jobs underlying process may run to completion regardless.

An API command was introduced in PE 2021.7.1 `POST /command/stop_plan` to allow for plans to be stopped.

It is also possible to schedule jobs in orchestrator via the GUI or by `API POST /scheduled_jobs/environment_jobs`, but great care should be taken to be aware of the system load of using the scheduler. Orchestrator has limitations with how it scales since there is no way to horizontally scale, and the queuing system for tasks and plans can be easily blocked by certain types of requests.

Configuring performance settings

The settings discussed in this section can all be configured in the Puppet Enterprise orchestrator infrastructure node group on the web console or as code in Hiera.

orchestrator can run a maximum number of tasks concurrently; this maximum number of concurrent tasks is configured with the `puppet_enterprise::profile::orchestrator::task_concurrency` parameter (default: `250`), along with `puppet_enterprise::profile::bolt_server::concurrency` (default: `100`) and `puppet_enterprise::profile::ace_server::concurrency` (default: `100`), which limit Ace and Bolt directly (they should not be greater than the `orchestrator::task_concurrency` total). Their sizes are mainly limited by orchestrator memory, which will reserve approximately ± 1 MB of RAM for each instance of capacity you add. Tasks are dealt with in the order they are received until they are completed; this means long-running tasks and tasks with large numbers of targets can potentially block other tasks from running and monopolize resources. Taking the case of running tasks taking 10 minutes to complete on 1,000, servers this would result in the task using the queue capacity of 250 four times and taking a total executing time of 40 minutes to run the tasks on all targets, during which time all other tasks would need to queue until it was complete. It is strongly recommended that a task should take no longer than 5 minutes and that careful management should take place to run tasks in smaller batches. It should also be noted there is no limit in the task queue and it risks running **out-of-memory (OOM)** resources if too many requests are sent. Another effect can be the time a task takes to time out. Every 12 seconds, orchestrator will request the status of a task and after a default of 35 attempts will time out, meaning a timeout after 7 minutes. This number of attempts can be adjusted by setting the `puppet_enterprise::profile::orchestrator::allowed_pcp_status_requests` parameter. It is important to understand this does not mean the task has failed but simply that orchestrator cannot get a status for it within the timeout. The task itself may have completed after this time.

For plans, orchestrator is similar to Puppet Server in requiring JRuby instances to compile plans. This capacity is set by `puppet_enterprise::profile::orchestrator::jruby_max_active_instances`, with heap memory for the JVM set at `puppet_enterprise::profile::orchestrator::java_args`.

Having discussed the core components and services of Puppet Enterprise, we will now look at how these components can be deployed using automated tools, deploying to Puppet-advised reference architectures to ensure that infrastructure will scale to user requirements.

Automating deployment and reference architectures

Puppet Enterprise focuses on creating standard architectures and configurations and the automation to deploy them. This ensures that less design effort is required from Puppet Enterprise customers who can find the right standard architecture and pattern and deploy it using provided tooling.

Understanding supported architectures

Puppet documents three supported architectures for Puppet Enterprise, as follows:

- The **standard installation** is just a standalone primary server and supports up to 2,500 clients
- The **large installation** is a primary server with compile servers behind a load balancer and supports up to 20,000 clients
- **Extra-large installations** are a primary server, a separate server with PuppetDB, and compile servers behind a load balancer supporting over 20,000 servers

These are illustrated in the following diagram:

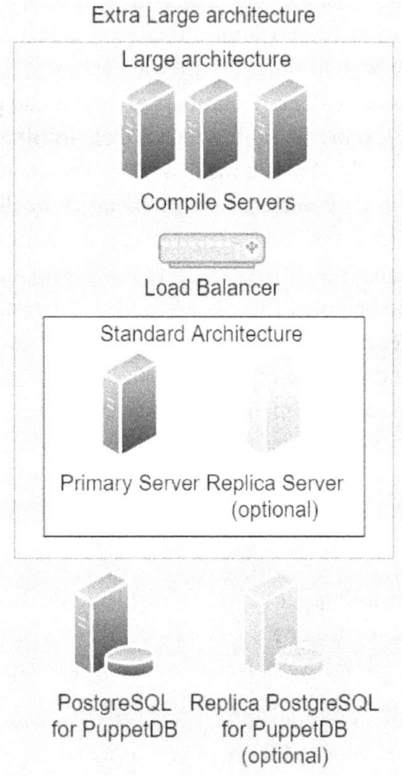

Figure 14.5 – Standard architectures

The standard architecture is limited by how many clients a primary server can run catalogs for by itself, up to 2,500 nodes. Over this level, the large architecture allows horizontal scaling using compiler nodes but reaches limits of how much load a single primary server can take running all the services together. So, at 25,000 nodes, the extra-large architecture recommends separating out PuppetDB as one of the heaviest services to its own server.

In all these architectures, it is possible to provide a replica server to the primary server and a separate PostgreSQL server, through a method named **disaster recovery** (**DR**). In the event of loss of the primary or PostgreSQL server, DR gives the ability to perform failover actions and recover services with an expected loss of some services, as listed in the following tabular breakdown of services:

Service name	**Replication type**	**Failover approach**
Puppet Server	None	Active / Active
Console services UI	None	Read-only until manual promotion
ACE service	None	Read-only until manual promotion
Bolt service	None	Read-only until manual promotion
CA	One-way replication	Read-only until manual promotion
RBAC	One-way replication	Read-only until manual promotion
Classifier	One-way replication	Read-only until manual promotion
Activity	One-way replication	Read-only until manual promotion
Orchestration	One-way replication	Read-only until manual promotion
File sync	One-way replication	Read-only until manual promotion
PuppetDB	Bi-directional	Active – Active

Table 14.1 – Service replication and failover approach for DR

PuppetDB is unique in its synchronization within Puppet Enterprise; it performs a read-write synchronization between primary and replica, which is why it is the only service in the previous list that synchronizes and is available on promotion. The other services that use PostgreSQL rely on a `PGLogical` synchronization from primary to replica, making the data read-only on the replica.

What can be seen from this list is during the failure of a primary server, the replica will only be able to take over and compile catalogs of servers already registered, queries and reports from PuppetDB, and queries of node classification via the API. This means no new servers can be registered or removed, no new code can be deployed, the web console cannot be used, classification cannot be changed, and most of the CLI tools will be non-functional until manual promotion actions are taken via the `puppet infrastructure promote replica` command on the replica.

This is an irreversible action, and the original failed primary server must be redeployed as a replica before it can be used again. Therefore, for many users attempting to fix the original primary server, this is less time-consuming than going through the DR process.

DR should not be confused with **high availability** (**HA**), which would be expected for continuous service in the event of the loss of a server, and that is not possible in any current Puppet architecture.

> **Note**
>
> When using DR, peadm ensures that the compilers are split and configured into two groups and PuppetDB requests are distributed across the two sides of the PuppetDB replication to maximize capacity. If you choose not to use pecdm, ensure you follow this optimization, which can be seen in code at `https://github.com/puppetlabs/puppetlabs-peadm/blob/main/manifests/setup/node_manager.pp`, with the A and B groups setting parameters for databases.

The Puppet architecture also defines a set of multi-region patterns for how to deploy across regions both public and private cloud, where a region is defined by cloud vendors as data centers with regional low-latency connections. Full details are available at `https://puppet.com/docs/patterns-and-tactics/latest/reference-architectures/pe-multi-region-reference-architectures.html`. Best practice requires compilers to have low-latency connections, and these are therefore best placed in the same region as primary and replica servers; similarly, the connection between primary and replica must be low latency. The best practice is, therefore, to use a centralized deployment where all Puppet infrastructure is in a management region that all regions can communicate with, as shown in the following diagram:

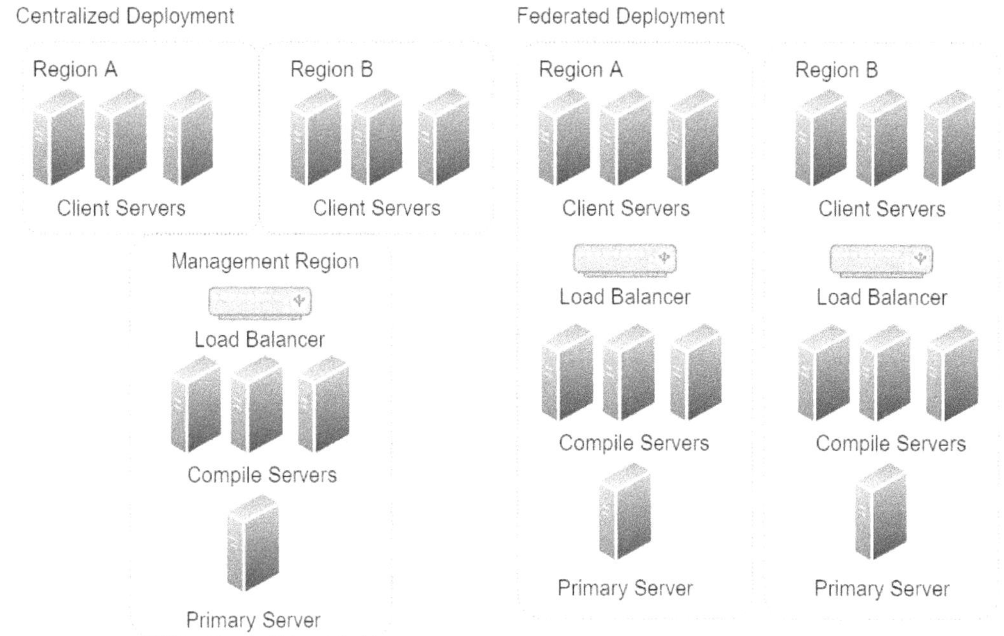

Figure 14.6 – Centralized and federated deployments

Alternatively, a federated model can be used whereby Puppet infrastructure is placed in each region, with the downside that no single console views the whole estate.

Having discussed the architectures and patterns in full, it is time to see which tooling is available to deploy these patterns.

Deployment and configuration

Puppet automates the deployment of its server infrastructure in several layers. The first layer uses the Puppet Enterprise installer, a tarball file that is downloaded from Puppet containing all the necessary packages and scripts to install Puppet Enterprise. Once downloaded on a target server and untarred, the basic install can be done by running `./puppet-enterprise-installer`. It is possible to add custom configurations by creating a **Human-Optimized Config Object Notation** (**HOCON**)-formatted file and adding a `-c` flag to its location, following the guidance at `https://puppet.com/docs/pe/2021.7/installing_pe.html`. Once a Puppet server is configured, the install scripts can be used to automate adding agents; a Bash script for Unix-based systems and a PowerShell script for Windows are hosted on a file server on the primary, which ensures the correct agent package is installed:

```
uri='https://<PRIMARY_HOST>:8140/packages/current/install.bash' curl
--insecure "$uri" | sudo bash -s -- --puppet-service-ensure stopped
agent:environment=production

[Net.ServicePointManager]::ServerCertificateValidationCallback =
{$true}; $webClient = New-Object System.Net.WebClient; $webClient.
DownloadFile('https://<PRIMARY_HOST>:8140/packages/current/install.
ps1', 'install.ps1'); .\install.ps1 -PuppetServiceEnsure stopped
agent:environment=production
```

In the example, the options set `environment` to `production` in the `puppet.conf` file and ensure the service is not running. The full range of options is available and documented at `https://puppet.com/docs/pe/2021.7/installing_agents.html`.

This `install` script would only have installed the primary server and would require further manual steps to add compilers and replicas depending on the architecture we wanted. To deploy the next layer instead of using the Enterprise installer directly, we use the peadm module (`https://forge.puppet.com/modules/puppetlabs/peadm`), a supported Puppet module that provides an automated way to run the Puppet Enterprise installer script and configure it to one of the supported architectures automatically. This module assumes the infrastructure required for the requested configuration is available and it is possible to go to another level and automatically provision in public cloud environments using the pecdm module (`https://github.com/puppetlabs/puppetlabs-pecdm`). An example of usage of these modules was discussed in detail in *Chapter 12* and is what we have been using throughout this book to deploy labs.

The `peadm` module itself goes beyond simple deployment and has plans and tasks to show the status of the server and allow the performance of version upgrades via its tasks and plans.

Puppet Enterprise combines modules installed in the `Enterprise` folder and configured either in the classifier or Hiera data with other file locations to place customizations. The console has a number of configurations that can be set either in the classification in the web console or via Hiera, such as failed login attempts set by `puppet_enterprise::profile::console::rbac_failed_attempts_lockout` and password complexity rules such as minimum password length, set by `puppet_enterprise::profile::console::password_minimum_length`. A full list of console customizations can be found at `https://puppet.com/docs/pe/2021.7/config_console.html#configure_the_pe_console_and_console_services`.

In addition, files can be placed for the console, by placing a file at the path specified by `puppet_enterprise::profile::console::disclaimer_content_path`, which defaults to `/etc/puppetlabs/console-services`. You can create a message to display when logging in to the console, such as a legal warning your organization may have.

Additionally in the console, it is possible to search for nodes based on PQL with predefined PQL examples selectable. It is possible to add your own PQL examples to the web console by simply placing a file at `/etc/puppetlabs/console-services/custom_pql_queries.json` using `/etc/puppetlabs/console-services/custom_pql_queries.json.example` as a template. The web console itself uses a self-signed CA by default, and this can be replaced with one signed by your organization's CA system by placing the generated certificate at `/etc/puppetlabs/puppet/ssl/certs/console-cert.pem` and `/etc/puppetlabs/puppet/ssl/private_keys/console-cert.pem`. One last key file to consider is the license key, which is issued to you by Puppet and placed at `/etc/puppetlabs/license.key` with `644 root:root` permissions. You can view the details of licensing under the **License** tab on the web console. A Puppet agent run should be made for these changes and the console service restarted.

Some areas of Puppet Enterprise are not currently definable through native code such as RBAC, classification, and LDAP, but there are APIs and Puppet modules that take advantage of those APIs, which can allow for storing configuration. For classification, there is an API to view the classification and configure node groups; this can also be done via the **node_manager** module (`https://forge.puppet.com/modules/WhatsARanjit/node_manager`), which is used by peadm. For RBAC and LDAP, the RBAC API (`https://puppet.com/docs/pe/2021.7/rbac-api.htm`) has endpoints that can be used to manage groups, roles, and users. A Puppet module has been developed to use these APIs (`https://forge.puppet.com/modules/pltraining/rbac`) and it has an LDAP endpoint that has similarly had a module developed to use the APIs (`https://forge.puppet.com/modules/abuxton/puppet_ds`).

Having reviewed the architecture and deployment recommendations, we will discuss other supporting tools and products to work with Puppet in the following section.

Puppet Enterprise-related projects and tooling

Puppet Enterprise has several modules and tools developed by Puppet to ease the management and support of Puppet infrastructure. The most direct is the built-in support script; this command gathers logs and system information and compresses it allowing users to send detailed status information to cases with Puppet's support teams. The simple version of the command is shown here: `/opt/puppetlabs/bin/puppet enterprise support`.

Various options can be found in the documentation at `https://puppet.com/docs/pe/2021.7/getting_support_for_pe.html#pe_support_script` that allow for selecting services to be collected, to directly **Secure File Transfer Protocol** (**SFTP**) upload the archive as part of the command, and to encrypt the archive, assuming **GNU Privacy Guard** (**GPG**) keys are available.

> Note
>
> It is possible to use **SOScleaner** to remove hostnames and IP addresses from the support script contents. Visit `https://support.puppet.com/hc/en-us/articles/115003312887` for details on how to install and run it.

Having seen how to deploy Puppet infrastructure, it is important for you to understand how to monitor and troubleshoot any issues found, so let's look at that next.

Monitoring and troubleshooting Puppet Enterprise infrastructure

The **Puppet Enterprise status_check module** (`https://forge.puppet.com/modules/puppetlabs/pe_status_check`) performs checks on both Puppet infrastructure servers and Puppet agents based on commonly found issues in support cases, such as confirming services are running, disk space is free, and certificates are not expiring. These checks can be run as tasks, Puppet code that will notify issues into reports, or as facts—the Splunk plugin shown in *Chapter 13* has a dashboard for displaying the fact output. Using these checks means if you do experience any issues when you raise your support case with Puppet, you can reference the check number.

The **support_tasks module** (`https://forge.puppet.com/modules/puppetlabs/support_tasks/tasks`) provides tasks that perform actions set out in knowledge base articles such as regenerating certificates, running the support script, and printing Puppet database table sizes.

Some extra console views can be configured to be visible and usable in the console; value reporting simply needs values entered in the **Value report** tab for how much time is to be reclaimed by using tasks, plans, corrective changes, and intentional changes, and it will also generate statistics.

Puppet Enterprise can gather additional information about packages including unmanaged packages; this information is made visible in the **Packages** tab. This will show which packages are installed on each server, what type of package they are, their version, and if they are managed by Puppet. It is enabled by adding the `puppet_enterprise::profile::agent` class to a node group covering nodes you wish to collect from and by setting the `package_inventory_enabled` parameter to `true`.

The final extra that can be enabled allows the monitoring and management of patching. In the **Patches** tab, it will create a view of nodes managed, patches available, and an option to run a task to patch. This is enabled by creating a node group under the **PE Patch Management** group that contains the `pe_patch` class.

In addition to the core Puppet Enterprise infrastructure, there are additional Puppet products allowing management of pipelines for code deployment onto Puppet Enterprise and for compliance scans to be run on Puppet nodes.

Managing deployments and ensuring compliance

There are two additional Puppet products to consider using with Puppet Enterprise, **Continuous Delivery for Puppet Enterprise** (**CD4PE**) is a pipelining product for Puppet built on the acquired Distelli pipelining product that looks to automate the process of managing deployment of Puppet code. It can watch for events such as **pull requests** (**PRs**) or commits to control or module repositories and then runs through pipelines that can automatically perform checks such as **Puppet Development Kit** (**PDK**) or **Onceover** checks and bring its own check of impact analysis. If checks whether the pipelines pass or are approved, and can then deploy and apply the code in various patterns. Impact analysis uses the `v4` catalog API to compile a new catalog with the new code and compare it with the current code's catalog, displaying the difference to ensure the impact is as the developer expected. These pipelines can be made in the web console for CD4PE or created as code in YAML files inserted into modules and control repos to be deployed.

Puppet Comply is a compliance tool based on the **Centre for Internet Security** (**CIS**) benchmarks. It builds automation around the Java scanner developed by CIS, CIS-CAT Pro accessor (`https://www.cisecurity.org/cybersecurity-tools/cis-cat-pro`). This allows hosts to be accessed against the CIS benchmarks, using orchestrator in Puppet Enterprise to automate and schedule runs of the scanner via tasks and producing dashboards of their compliance in a separate Puppet Comply console. An example of the home screen of Comply is shown in the following screenshot:

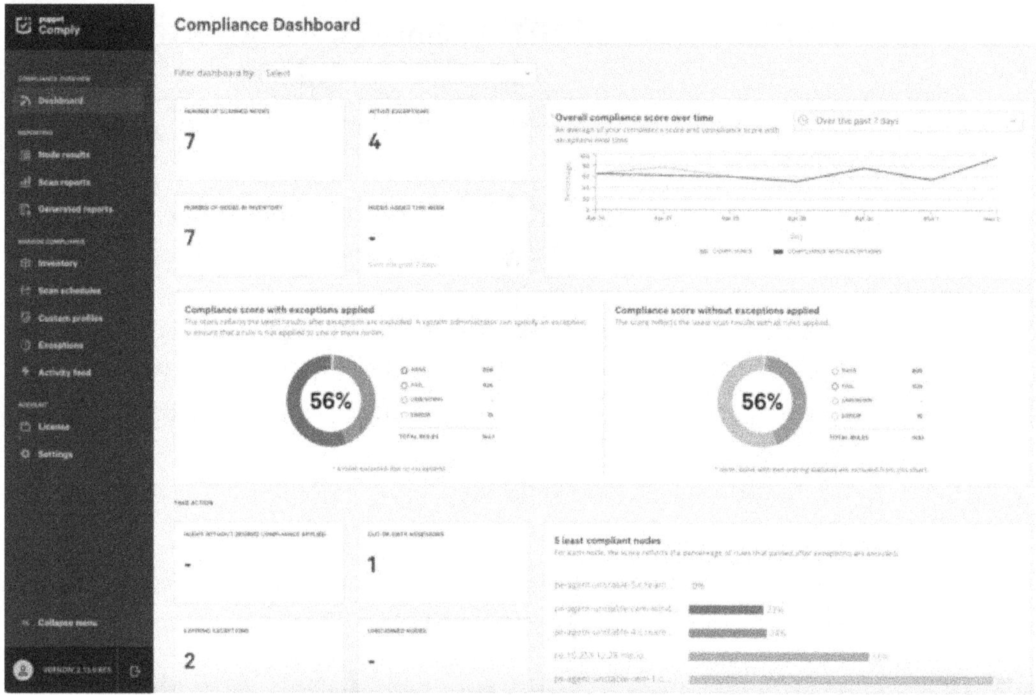

Figure 14.7 – Puppet Comply home dashboard

It can be seen from the dashboard how many of the nodes are achieving compliance, how many nodes have a compliance profile set, and a list of node results listing which profile is assigned and compliance scores in a particular scan.

It also comes with the premium **compliance enforcement modules** (CEM) of `cem_linux` (`https://forge.puppet.com/modules/puppetlabs/cem_linux`) and `cem_windows` (`https://forge.puppet.com/modules/puppetlabs/cem_windows`) to speed up your adoption of Puppet, allowing base security configuration to be taken based on CIS benchmarks via pre-made Puppet modules. These modules are maintained and supported by Puppet, ensuring the enforcement code is up to date with the latest CIS benchmarks.

Both products run in the framework known as **Puppet Application Manager** (**PAM**), a Kubernetes-based tool for managing Puppet applications.

Lab – Puppet Enterprise extensions and configuration

Executing the bolt command in the *technical requirements* section deploys a large deployment of Puppet Enterprise 2021.5. With this infrstructure setup, we will try various extensions and configurations we have discussed, as follows:

1. Examine the code in peadm and the node groups that set up the A and B groups. Note `https://github.com/puppetlabs/puppetlabs-peadm/blob/main/documentation/classification.md` provides an explanation of the groups.
2. Create a personal user with permission to view the console and create node groups and view the activity log of the administrator user (try to log in without the view console permissions).
3. Enable package management on the web console for all nodes, log in as your personal user, and view the activity log of this.
4. Enable patch management for the nodes by applying code using the `node_manager` module.
5. Customize the login message.
6. Perform an upgrade to 2021.6 using the peadm upgrade plan. *Note*: Since pecdm includes peadm, this can be performed from your development environment.

Sample solutions are provided at `https://github.com/PacktPublishing/Puppet-8-for-DevOps-Engineers/tree/main/ch14`.

Summary

This chapter has reviewed how Puppet Enterprise builds on top of the open source tooling, providing the services necessary to secure and automate the deployment of Puppet. It was discussed how Puppet Enterprise bundled the open source packages into consistent versions, with support offerings and services from Puppet architecture and services teams.

We also discussed the additional services of Puppet Enterprise that secure user and API access via RBAC, giving a web frontend and additional APIs in the console services the ability to deploy code from Code Manager.

Puppet orchestrator was then seen, to show how tasks and plans could be run in Puppet Enterprise with the orchestrator service running tasks and plans via PCP using PXP brokers to direct communication from PXP agents on nodes. The agentless clients could be added to the inventory service storing their transport details, and tasks or plans to run on them would ego via the Bolt server for nodes connected by WinRM or SSH, while other transports' particular network devices such as switches or firewalls used the ACE server. We saw how orchestrator would store all the job details updating the activity service. RBAC access was discussed, showing how you could only limit which plans were available to a user but could set tasks to particular users and particular groups of nodes using target sets. Performance and capacity aspects of orchestrator were discussed, as well as how to run tasks or plans via the web console GUI or the CLI interface.

The supported architectures customers could take off the shelf to implement Puppet at scale and regional requirements for their estate were reviewed, showing the modules and scripts that wrap up to automatically deploy these architectures, the pecdm module deploying infrastructure in the public cloud, peadm automating the various steps of install and maintenance, and using the installer script.

Extras services that could be enabled in the web console to help report on the value Puppet delivers, patch management, and packaging reporting were reviewed, along with customizations and methods to automate configuration within the console covering customization of the console message, the certificate used on the web console, and the license key. Several modules were then discussed that could assist in reporting the status of the infrastructure and running standard tasks in the `support_task` and `status_check` modules.

Two further Puppet products that integrate with Puppet Enterprise were then discussed: CD4PE, which provides a pipeline to assist in automating the deployment of code, and Puppet Comply, which gives pre-written modules and dashboards to allow for reporting on CIS benchmarks.

While all of the architecture, tooling, packaging, and general automation could be achieved with Open Source Puppet, it would require development and support work from your own teams. So, Puppet Enterprise should be seen as a decision about the skills and people available in your team, tooling already invested in the organization and money available for tooling, and where your organization wants to focus its work.

Now, having fully reviewed the language, the platform, and how Puppet Enterprise can provide preconfigured infrastructure to reduce the operational burden and design required, in the final chapter, we will discuss approaches to adopting and using Puppet, focusing on getting the best use in your organization, since understanding the technology is only part of the battle while understanding how to integrate with people and processes is often the greater challenge.

15

Approaches to Adoption

Having discussed the Puppet language and platform in detail, this chapter will now look at approaches to adoption and implementation. This chapter does make certain assumptions about the most likely adopters of Puppet and their viewpoints. As a result, some of this advice will appear from a Puppet platform team's point of view but it will look to discuss how all the implementation teams, from application to OS, should work together to boost adoption.

Too often, the view taken by a project or modernization program is that technology alone can solve all the problems of an organization, and existing teams and processes are just in the way and will need to be worked around to deliver the future. The most successful adoptions work with the current teams and embed themselves in their processes. This chapter will cover this by discussing how to choose the right scope and focus to make sure that the implementation can achieve its goals, delivering on a regular interval, and showing value to encourage the adoption. We will discuss how to work with other teams and stakeholders to ensure that Puppet as a technology is not an island that battles for space but can be a platform among many tools that can integrate and maximize benefits. While it is often more practical to start with greenfield newly provisioned servers, we will discuss how to safely and progressively reach the heritage brownfield estate, where understanding the level of configuration drift that has happened and developing automation to remediate can have huge benefits in reducing costly auditing processes. Using Puppet in regulated environments will be discussed in detail as it is often assumed that a tool that commits regular change and has elevated access simply cannot be used. We will see how to present the processes and testing to not only make Puppet safe and secure but also to show it is an integral part of enforcing the requirements of any regulated environment. Finally, we will see where Puppet fits into the cloud, its appropriate uses, and how to avoid mistakes made by public cloud migrations and not leave behind the benefits gained by Puppet in the private data center.

In this chapter, we're going to cover the following main topics:

- Scope and focus
- A platform engineering approach
- Managing heritage estates with no-op mode
- Adoption in regulated environments
- Moving to the cloud

Scope and focus

The pressure on scope and focus will depend on why your organization has started using Puppet. If it is an exercise for an individual team, such as the Oracle team, to automate its deployment, the pressure will be less than a transformation program that has bought a large Puppet Enterprise contract. In big transformation programs, it can be tempting to pursue big goals quickly to earn this cost back. This is dangerous because configuration problems are complex, and technologists are prone to optimism about how quickly solutions can be created. Additional pressure can come from sales teams and decision-makers who may have oversold how quickly change can be implemented to get the necessary funding. This is not advocating against having a vision or a Jim Collins-style big hairy audacious goal. The future vision is needed but it has to be shown that it will be an incremental journey of improvement to get there, and these increments will deliver immediate and recurring value. This will develop trust and belief from supporting teams and customers to invest in your platform because you reliably deliver something tangible and not just a distant hope.

The best approach to this delivery is to follow good sprint practices, having epics such as delivering a core OS role or an Oracle role, which can then be broken down to have a small number of focused objectives for each sprint. Each task within an epic should be small enough to be completed in a regular sprint cycle, typically 2 weeks. At the end of each sprint, these features can then be demoed to stakeholders to show progress, benefit, and receive feedback.

> **Note**
> This book does not advocate for any agile methodology; there is a vast collection of books and advice on how to implement agile working practices. What works for your organization will depend on local culture and your team. So, this book's recommendation would be to research various approaches but remain flexible and find what is comfortable and works well for your team and not just try to fully mimic anyone else's system. Using techniques such as retrospectives at the end of sprints can help ensure that how you are working is still effective and that actions are taken on issues.

If this approach is ignored and the team is split among many objectives, it can easily result in developers working in isolation. When developers work in isolation, other team members cannot help or provide meaningful reviews because they do not have an understanding of the work or why decisions have been made.

If the work is too large and complex, it will result in development problems, which are hard to test and break down to understand. This can lead to frustration between management and developers as nothing will be visible in terms of delivery. The pressure to deliver something can then lead to developer exhaustion and this combined with the difficulties of reviewing and testing large complex work can lead to something risky or incomplete being delivered, simply to deliver something. This erodes confidence and morale in a vicious cycle of pressure and mounting issues to fix.

In the *Adoption in regulated environments* section, we will discuss how critical it is to demonstrate an ability to reliably test and deliver to win confidence from change and risk teams in your processes and platform.

With this warning said, if the team does stick to a focus and scope, then an understanding among the team can be built about the ongoing development work and strengthen the review, testing, and learning processes. This produces opportunities for developers to pair on interesting or challenging sections of work and use team breakout sessions to jointly make decisions on coding approaches. As discussed in *Chapter 1*, keeping a Puppet best practices document updated with these decisions helps spread the knowledge further. Most importantly, on review of the submission of code, it becomes something the team has been actively discussing and working on together, not just something that may have only been heard in brief morning meeting updates and something the team has to take the developers' word on. All of this works toward a better understanding of what the code is intended to do and why the approach has been chosen.

To illustrate this approach, it is common for a base OS to need a security profile for the core build. This security profile will contain various aspects such as core OS user accounts, SSH configuration, kernel settings, and various other important settings. Making this profile the focus of a sprint could result in developers pairing and working on the component modules that make up the profile. With the pairs of developers focusing on elements such as user accounts and building up the profile piece by piece, practical progress will be made and knowledge shared. Depending on the size of the profile and the number of developers, it may make more sense to work on multiple profiles but the aim should still be to limit the scope and focus.

This is not to say any development team should expect to get everything their own way and that external pressures will not result in having to split this focus, but it should be strongly represented that it will slow down work and create risk.

The next key thing to understand after the focus and scope of Puppet code are the minimum acceptance criteria for code to be delivered to production. The phrase **minimum viable product** (**MVP**) has been tainted as an excuse to release something that is clearly not fit for production with items such as testing to be added later. It is a simple reality this will not happen since there are always new features to develop, creating an operational burden in the future as the code develops further. So, in your organization and platform, the Puppet best practices should lay out what tests the code should pass. An example standard could contain the following:

- Code must be clean in PDK validation with accepted listed exceptions
- RSpec tests provide 100% coverage of code
- ServerSpec tests for the module and passes core Serverspec tests

Another challenge can be scope creep, which can dilute the higher-level scope and focus of how Puppet is used in your organization. When investing in a tool, it is tempting to maximize the return on investment by expanding use cases, and as the implementation becomes successful, other teams will want to attach to that success and try to use a provided tool. Therefore, it needs to be clear what the use case of Puppet is; inappropriate uses such as the distribution of binaries or large-scale synchronization of files need to be called out as inappropriate in platform documentation. In this example, it would put a lot of burden on the infrastructure, as was discussed throughout this book. Also, in this regard of focus, this book would strongly recommend against any policy encouraging mandatory rewriting/re-platforming strategies to Puppet unless the current implementation has maintenance issues or cannot be developed as needed. This sort of rewrite provides little value and, unless the original implementation is well understood, can lead to mistakes in the translation, particularly for declarative code since only the method is visible not the intended final state.

Having discussed the scope and focus for Puppet, we will now look at how to manage this approach on heritage servers and handle the history and complexities of brownfield sites.

Managing heritage estates with no-op mode

Heritage can be more daunting for implementation; the extent of configuration drift can make it hard to know where to start. Your organization could have been part of several mergers and acquisitions, which have led to not only multiple configuration standards for the core organization itself but also for anything that has been onboarded.

A common pattern of adoption is to progressively build up the automation levels in heritage servers to build confidence, and we will step through a common approach to this.

Installing an agent on all nodes to gather facts is a common starting point, and having this data stored in PuppetDB to create a valuable CMDB source. Then, as was discussed in *Chapter 13*, this data can be sent to services such as ServiceNow to integrate with central CMDB services. In Puppet Enterprise, this gives us access to the package view and the ability to manage patching, as was demonstrated in *Chapter 14*. This rollout gives immediate capability and a better understanding of the estate without even having written any code.

The next step is to consider orchestration. It is likely there are common scripts and tasks performed manually or semi-automatically by various teams on the heritage estate. Taking these scripts and wrapping them in Bolt projects or Puppet modules and using Bolt or Orchestrator to run these scripts and tasks can deliver greater control and process with these scripts without having to perform rework.

The simplest case is if you are using Puppet Enterprise and have rolled out agents in the first step, in which case, Orchestrator can simply take advantage of the presence of the agent being deployed and use the PCP transport to communicate and take advantage of Puppet Enterprise RBAC and logging systems with Orchestrator. For open source Puppet or Puppet Enterprise users who do not want to buy licenses for heritage, a Bolt server can be used to set up a golden host with SSH keys and WinRM. There is an in-between option of using agentless Puppet Enterprise licenses but allowing the Puppet Enterprise host to be used and to still have the RBAC and access logs. It was not discussed in previous chapters but the advantages of agent-based servers are that they are more integrated and can perform more actions and gather more data, and their approach to keys and security is managed by Puppet as part of the product. Agentless approaches can be added without the issue of having to request an agent be installed, which may not be compatible with all servers. Agentless also avoids the potential issues of vulnerabilities and updates of Puppet agent code versions but does have the issue of separate access management, such as deployment and management of SSH keys.

The next step is exactly what was discussed in the *Scope and focus* section: looking at a baseline configuration and ideally finding something non-negotiable to start, which must be enforced on your estate. For example, root logins must be turned off or application agents need versions to be upgraded and managed to avoid vulnerabilities. Once these straightforward configurations are managed, it is time to look at server configurations that may have historical exceptions. The difference for heritage servers is even if the pre-existing configuration of the server does not follow current security and build standards, it should first be flagged as an issue before being remediated to avoid causing potential service issues. To flag configuration issues without immediate remediation, a no-op flag pattern at a profile or module level can be used, as discussed in *Chapter 8*. The configuration drift can then be understood and either accepted as exceptions, which are recorded in Hiera data, or remediated with Puppet by switching from no-op mode to execution mode to apply the configuration.

Once base profiles are complete, this leads to having all the tools available for automated audit reporting and compliance remediation in our heritage estate.

This approach can then be repeated by engaging with application teams to find their needs for configuration and auditing and following the same pattern to build out their own roles and profiles specific to their applications.

Having mentioned different teams involved in the development of Puppet code, it is important to directly address the best approaches to cross-team working in Puppet.

A platform engineering approach

As will be clear from the first two sections of this chapter, a common adoption start of Puppet is for core base OS configuration to be created and then to reach out to application teams. This can often lead to a setup where Puppet is a tool of the Linux/Unix operating system team, who dominate the code base and are gatekeepers for the whole platform. To ensure effective cross-team working, what is required is a platform engineering approach.

> **Note**
>
> More in-depth knowledge about how to run a platform team can be found in books and training such as `https://teamtopologies.com/`, and platform engineering has been popularized via communities such as `https://platformengineering.org/`.

The core concept of platform engineering is to have a platform team who are responsible for managing the tooling, workflows, and development of a self-service platform. This platform should be treated as a product, with its users treated as customers, ensuring their needs are met and that the platform is evangelized throughout the organization. As was discussed in *Chapter 1*, Puppet is likely to be part of a platform along with various other DevOps tools and workflows. *Figure 15.1* shows a common toolset choice:

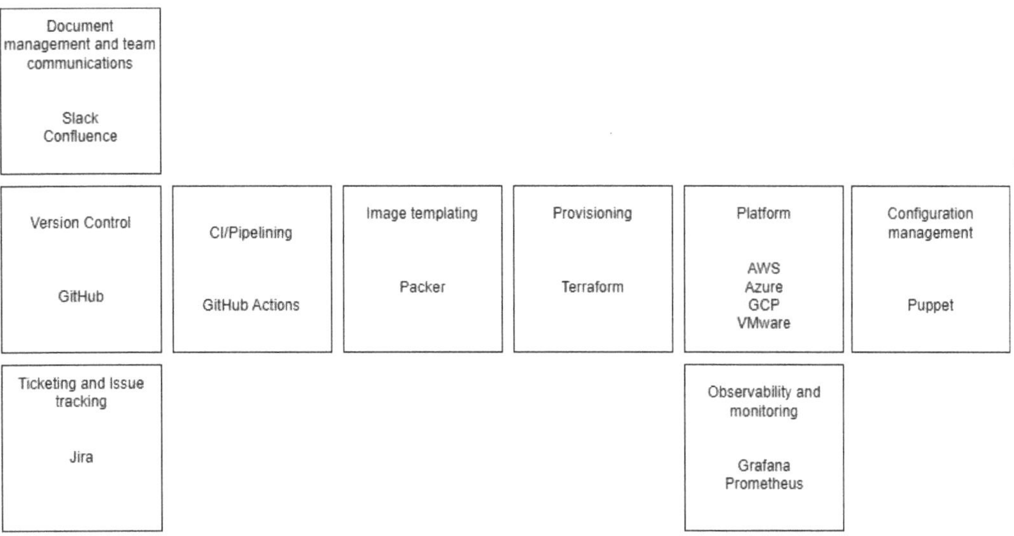

Figure 15.1 – A common DevOps toolset

Looking exactly where Puppet would fit, this would likely work in a day 0, day 1, and day 2 approach, as shown in *Figure 15.2*, whereby provisioning would be done by a specialist tool such as Terraform to create the infrastructure on day 0. Then, on day 1, Puppet code would be applied to the client to configure the OS to build and security standards on the infrastructure. The day 2 Puppet role would be to continue to enforce the configuration to avoid configuration drift as unintentional external changes take place or intentional changes to standards result in code changes.

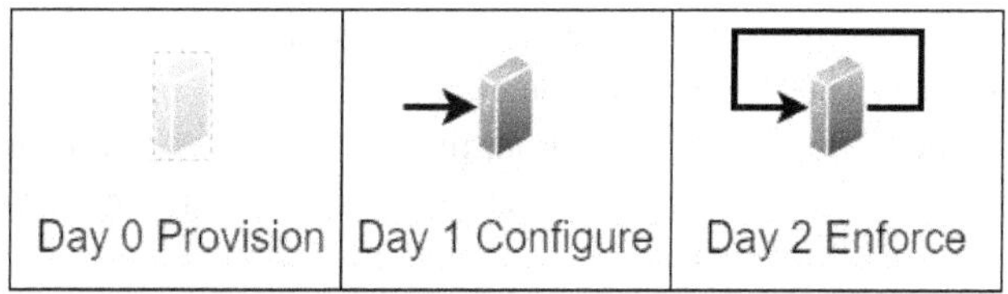

Figure 15.2 – A day 0, day 1, and day 2 approach

The key point to think about Puppet code in these platforms is that, ideally, the responsibility for running the Puppet infrastructure should be part of the platform team's role. This allows separate teams to develop their own code and roles, which they have a clear path to deploy via the self-service platform.

This may not always be possible, and often, the Linux team remains running both their own code base and the Puppet infrastructure. In this case, it would be best to still see this as two separate roles and to not prioritize only the needs of the Linux team's code base over other consumers. The platform team should not try to be a gatekeeper for everyone's Puppet code, as this blocks developers from using Puppet as a self-service platform. The processes of your organization should cover responsibility and escalation, which will be discussed further in the next section, *Adoption in regulated environments*.

It should also be ensured that the team responsible for managing the heritage estate takes ownership of automation efforts. Bringing in a new team to automate systems without a full understanding of them can be challenging. It may take more time to train and involve the heritage teams, but having them lead the integration efforts can result in a more thorough understanding of the systems and their processes.

While each team is responsible for its own code, it is important to collaboratively develop standards and best practices to ensure that teams have the knowledge to appropriately test and pipeline their tools.

Cross-team collaboration is not just limited to using Puppet but also includes other integration points. It is neither practical nor desirable to rewrite and run everything in Puppet. Creating communities of practice, where various teams across departments can meet, discuss, and showcase their approaches and progress toward automation can foster the exchange of ideas. In some cases, it may even be possible to reuse what others have already developed within your organization. This should not be seen as a competition but as an opportunity for mutual benefit and to exchange skills and ideas.

Evangelism at all levels is crucial. Attending various team meetings, lunch sessions, management meetings, and external vendor or trade body events can help spread the news about your platform and create enthusiasm for further development. External vendor events are often seen as legally complicated, but with careful consideration and consultation with your legal and marketing teams, you can increase the visibility of your platform within your organization and attract external talent by generating interest in your work. Moreover, these external events, such as technical advisory boards, are excellent opportunities to exchange best practices with like-minded organizations.

Although already mentioned earlier in the *Scope and focus* section, it is worth emphasizing that you should not try to solve every problem brought to you. People will become enthusiastic if you evangelize well but it is essential to be completely honest about the capabilities and fit of your platform. You should clearly communicate what you can realistically deliver and what they can expect to have to commit if they want to onboard or engage with the Puppet platform.

With a scope and focus set and an understanding of collaborative working, the next major thought should be around how regulation and process can affect these ways of working.

Adoption in regulated environments

Working in highly regulated environments can be challenging but it is often where Puppet can have the most significant impact. Implementing automation may be more difficult in regulated environments but it is even more challenging to perform large-scale manual actions, making the potential returns on investment significant. The worst approach when trying to adopt new technology is to believe that "the processes just need to change." This attitude sets up the team for failure later in the process and can lead to a reputation for being sloppy and neglecting process work, resulting in a setup that will not work in production.

The best approach is to engage with change, risk, audit, and other teams involved in the management of processes in your organization before implementing Puppet. Often, despite regular complaints about processes in the organization, no one has engaged with these teams, and they may have their own programs to modernize to which you can align your adoption. Discussing what Puppet is and how you plan to use it in production can provide credible feedback. Even if this feedback requires scaling back your initial ambitions, it is better than treating these teams as gatekeepers, who end up with a limited understanding of your adoption and have to reject things they haven't had a chance to understand the consequences of or influence the approach.

> **Note**
>
> Invite your process team to the community of practice sessions and demos; you are not on different sides and will find you have far more challenges and objectives in common as you try to deliver value for the organization.

It's important to frame the discussion around what Puppet can do and how your approach to development, testing, and release will work, as well as what scope it will cover. Puppet is a powerful tool that operates at the administrative/root level, so it's crucial to demonstrate that you are in control and that any risks associated with your processes are understood.

To win the trust of all stakeholders, including process teams, you can show them the possibilities of Puppet and discuss how it can benefit the organization. As discussed in the section on focus and scope, iterative improvements to processes can be made, especially if they are discussed in communities of practice sessions. This way, multiple teams and departments can agree on improvements that benefit the organization as a whole without compromising security and risk.

This approach may not seem revolutionary, but in regulated environments, change cannot happen quickly. Therefore, it is important to focus on what can be done within the current constraints, show how your solution fits into this, and work with stakeholders to modernize or improve processes. This requires patience and consistency to win teams over. After completing the view of a traditional private data center environment, it is important to consider how this approach differs in the cloud.

Moving to the cloud

The move to the public cloud has huge opportunities, particularly in terms of flexibility, with opportunities to use cloud-specific technologies to reduce the operational burden on your organization. For example, the ease of using availability zones for compilers to reduce the risk of data center failures is a complex feature to implement in private data centers.

Unfortunately, there are two commonly seen anti-patterns for the cloud adoption approach. The first is a wholesale copy of all infrastructure, processes, and components as they work in the private data centers to the public cloud. This often happens with "cloud-first" programs, which tend to be a result of **Chief Information Officers' (CIOs)** disappointment in the take up of public cloud resources. This forces deployments into the public cloud before organizations are ready and understand what is a suitable fit. This results in surprise bills as the infrastructure deployed is not planned to be flexible and ignores the rental nature of the public cloud, and many of the solutions that make sense in a private data center are far better implemented in cloud-native solutions in the public cloud.

The second is where everything is left behind, which can be seen with application teams or departments that are frustrated with internal processes and time to delivery. They may have justification for their frustrations but rarely have the experience; sadly, the lessons hard won in private data centers are lost and good practices in audit, configuration, and testing must be rebuilt as auditors find issues with the new fractured setup.

You should consider what is really being done in the public cloud; when looking at how to deploy Puppet infrastructure, the multi-region patterns and tactics mentioned in *Chapter 13* show the options. Simply, we can have public cloud servers managed by Puppet infrastructure in a private data center, or the Puppet infrastructure could be migrated to the public cloud and manage both private data centers and the public cloud, or have separate Puppet infrastructure for private data centers and the public cloud.

This choice depends on implementation aims. Is the public cloud going to be used to provide flexible capacity for the private data centers, for example, by providing an alternate site that can be built in disaster recovery? Or is the public cloud being used to start a new way of working with a more cloud-native approach and new teams? In the first case, it is more likely you will want the configuration of servers to be the same with a shared code base, and having a single pane of glass could be advantageous for the team's managing infrastructure. In this case, deciding whether the infrastructure should be located privately or publicly will come down to cost and whether you intend to take advantage of cloud-native features such as the flexibility of availability sets and load balancers, which could allow compilers to be added on demand.

In the second case, where a fresh start with new teams looking for a new approach is being made, having a separate infrastructure will make the most sense, reviewing what is useful in the current build and what is only relevant to the private data center. As was addressed in the *a platform engineering approach* section, this involves finding out the requirements of the teams working in the cloud and ensuring Puppet is used as part of a platform to meet these needs. The cloud teams should then be able to use the APIs to self-service while gaining the advantage of Puppet providing the audit and security requirements of your organization in the cloud.

It can also be an opportunity to move from heavily customized standards used in the traditional organization and even consider adopting compliance to implement CIS standards, which was mentioned in *Chapter 14*. This will be a cost consideration as to whether it makes sense to have your own team maintaining these standards.

Summary

In this chapter, we discussed how to look beyond just pure technology consideration and make the adoption of Puppet a success. We reviewed how to choose a focus and scope to allow Puppet to be delivered in iterations of continuous improvement, using a regular delivery cadence and methods such as sprints with a small focus for the team to work together on. We talked about breaking down this focus into deliverables that can be collaboratively worked on and demoed on a regular cycle. We also covered allowing the Puppet team to build confidence and learn as decisions are made together and coding practice is established while showing meaningful returns and progress to management and stakeholders. We discussed how the use cases of Puppet should be outlined and ensure that the temptation to maximize the return from Puppet does not result in unsuited tasks trying to be shoehorned in, which can destabilize the reliability and performance of the Puppet infrastructure and give general maintenance headaches.

The approach to adopting a heritage estate was then reviewed showing how even with estates fractured by changes in standards and strategies as well as company mergers and acquisitions, can follow a progressive adoption pattern to slowly reduce configuration drift over time and gather information about the estate.

We looked at rolling out an agent with no configuration first and gathering facts to create an asset view, which could feed into a CMDB and, in the case of Puppet Enterprise, using built-in integrations to manage patching and packaging. We then showed that orchestration could be considered to wrap up current scripts and give better automation. This could be done using fully licensed Puppet Enterprise via PCP or WinRM/SSH connections on an agentless license in Puppet Enterprise, or using a Bolt server with WinRM/SSH connections. This would depend on your license considerations and the need for RBAC and logging. We then covered forming a baseline of stateful Puppet code looking for mandatory settings to enforce and using no-op where appropriate to get a current view of the estate, slowly building a profile where exceptions could be accepted into Hiera or drift could be remediated. Having established this baseline, we discussed repeating this process with application teams to bring heritage applications under control and then use the configuration data to automate audit reporting.

We discussed how to work across multiple teams, establishing a Puppet platform team that championed the platform and provided APIs and self-service to teams, setting standards for teams to adopt and use Puppet well, but not gatekeeping their delivery.

Deployment into regulated environments was shown to be something that worked well with Puppet by communicating how Puppet worked to key process teams, such as change and risk management, and addressing the processes your implementation would use to develop and deploy code into production while taking on board how best to integrate with the current process. Winning the confidence of the process stakeholders and operation teams can lead to changing processes in the future to further automation.

Finally, public cloud adoption was reviewed, discussing two of the biggest issues suffered in the public cloud with either a *cloud-first* policy resulting in a poorly thought out lift and shift of technology and process, or application teams going it alone and forgetting the lessons of automation for security and audit made in private data centers. We explained that you should consider the purpose of your public cloud. Is it an extension of the data center and something you would want to be brought into view of an in-house Puppet server, something you are moving to? Moving Puppet infrastructure into the public cloud can be something that starts to adopt the flexibility of the cloud. Is it something different with a new approach and an opportunity to take new Puppet infrastructure with code optimized for cloud adoption or taken from compliance as old customized in-house standards can be left for industry-standard CIS approaches? The key action was shown to be meeting application teams and ensuring the APIs and platform approach is available to them so they do not worry about core infrastructure build and security configuration but only what is useful to them and that they should be able to manage via self service.

Throughout this book, it has been shown how Puppet's stateful approach can reduce drift and technical debt, automating audit reporting and giving a standard way to deliver change, even in heavily regulated environments, and providing a platform that users can trust to meet their infrastructure requirements and freeing up teams to work on delivering their products to customers. Configuration management is a complex problem with no silver bullet solutions but we have shown with a considered iterative approach how, by working with the processes of your organization and involving everyone, Puppet can bring transformational change to your organization.

Index

Symbols

1Password 338

A

abstract data types 89
patterns 91
prefixes 90
abstract resource type 61
Abstract Syntax Tree (AST) 253
ad hoc commands
running, with Bolt 292-294
Admin API
environment cache 239
JRuby pool 239
adoption approaches
adoption, in regulated environments 374
cloud, moving to 375, 376
heritage servers, managing with no-op mode 370, 371
platform engineering approach 372-374
scope and focus 368-370
aggregate facts 110
alias function 213
anti-patterns 61
abstract resource types 61
defaults 61
schedule 63
Any type 94
Apache module
reference link 170
reviewing 171
APT-based Linux desktop
Puppet lab, deploying on 27, 28
arithmetic operators 77, 78
array index
accessing 82
array operators 84
append 84
concatenate 84, 85
remove 85
splat 86
arrays 82, 86
assigning 82
nested array 83
subset, accessing of 83
arrays of titles 59
attribute splat (*) 60
audit metaparameter 57
Augeas type 55, 56
reference link 56
Azure Key Vault 338

B

backend storage service (BSS) 336
Betadots Hiera Data Manager 228
BODMAS rules 78
Bolt 289, 290
ad hoc commands, running 292-294
debugging 295, 296
logging levels 294, 295
output 294-296
plugins 313
reference link 290
Bolt project 296
configuring 297, 298
creating 316
structure 296
system-level settings 300, 301
transports, configuring 298-300
using 316
Booleans 80
conversion 81
built-in backends, Hiera
using 203-205
built-in functions
change case 116, 117
comparison 115, 116
data handling 121
hash/array 119-121
lambdas 118
sizing 115, 116
string manipulation 117
templating 118, 119

C

ca_extend module
reference link 240
capture variables 162
case statement 160, 161
catalog compilation 330
catalog statements 113
Centre for Internet Security (CIS) benchmarks 362
certificate authority (CA) 234, 348
certificate signing logging
monitoring 250, 251
certificate signing request (CSR) 237
chained functions 115
Chief Information Officers' (CIOs) 375
classes 34, 35
including 35
resource declaration 35
syntax 34
Clojure 282
Clojure application 352
Code Manager service 350
Collection type 94
collectors 63, 64
comment tag 148
common performance and capacity issues
catalog compilation 330, 331
catalog runtimes 331, 332
identifying 330
PuppetDB and PostgreSQL tuning 332
tuning sizing 333
compiler
scaling with 256, 257
viewing 257
compliance enforcement modules (CEM) 363
conditional statements 159
capture variables 162
case statement 160, 161
if statement 159, 160
selectors 161
unless statement 160

Configuration Management Databases (CMDBs) 140, 335
container transports 291
containment 135-140
Content and Tooling Team (CAT) 196, 281
Continuous Delivery for Puppet Enterprise (CD4PE) 285, 362
core facts
reference link 99
core resource types 52
group type 52, 53
user type 52, 53
custom backends, Hiera
data_dig backend type 214
data_hash backend type 214
hiera3_backend type 214
lookup_key type 214
using 214-216
custom facts 105, 106
aggregate facts 110
confining 106-108
rescue blocks 109
structured facts 111
timeouts 109
weighted resolutions 108, 109

D

data security 221-223
secret, storing with eyaml 223, 224
data transformation 157, 158
data types 70, 94
Booleans 80
numbers 76, 77
reference link 71
regexp 81
strings 71
undef 80
deep merge 210
defaults 61
resource body 62
resource default syntax 62, 63
deferred functions 124, 125
defined types 36, 37
syntax 36
Desired State Configuration (DSC) Puppet modules 197
DevOps
relationship 4, 5
Directed Acyclic Graph (DAG) 131
disaster recovery (DR) 357
Docker 291
Domain-Specific Language (DSL) 279
double-quoted strings 72, 73
dynamic data
usage criteria 220, 221

E

Embedded Puppet (EPP) templates 145-150
versus ERB templates 152
Embedded Ruby (ERB) templates 145, 151, 152
reference link 146
ENC scripts 12, 271, 272
Enum data type 91
environment 238
exec type 53-55
executable external facts 104
exporters 63, 64
expression printing tag 148
external data provider pattern 335
1Password integration 338
Azure Key Vault integration 338
components 336, 337
hiera_vault 338

implementations 337
Puppet Data Service (PDS) 338, 339
Satellite 337
ServiceNow integrations 338
Splunk 339

external facts 102
executable external facts 104
static external facts 103, 104

External Node Classifier (ENC) 35, 94, 203, 261

F

Facter 12, 98, 348
examples 98-102
facts 23

Facter 2 102

Facter 3 98, 99

Facter 4 98, 102

facts 98-100, 309
custom facts 105, 106
external facts 102

facts hash 23

Filebeat 324

file type 44-47

flat facts 103

float data type 79

fully qualified domain name (FQDN) 188, 266

functions 113
built-in functions 115
chained functions 115
prefix functions 114
reference link 115
statement functions 113

G

getvar function 162

group type 52, 53

H

hashes 82, 88, 98
assigning 86, 87
mixing, with arrays 89

hash operators 88
merging 88
removal 88

hash values
accessing 87

heredocs 73, 74

Hiera 11, 202
built-in backends, using 203-205
custom backends, using 214-216
data, accessing 206-214
issues 224-228
nodes, classifying with 268-271
pitfalls 224
troubleshooting 228, 229

Hiera layers 216
data, adding to module 219
environment layer 216, 217
global layer 216
module layer 217-219

hiera_vault 338

high availability (HA) 358

HOCON-formatted file 359

I

idempotent 53

if statement 159, 160

include function 35

indirectors 238
Infrastructure-as-a-Service (IaaS) 15
Infrastructure as Code 5
inline template 146
integer data type 79
iteration 153-155
functions 153
iterative loops 156

J

Java Virtual Machine (JVM) 236, 353
jobs 354
running 354, 355
JRuby 243

L

lambda 153
functions 153
LDAP solutions
Active Directory (AD) 349
legacy Puppet patterns 23
Linux Container Hypervisor (LXD) 291
Linux Containers (LXC) 291
literal function 214
load balancer configuration
viewing 257
Logback library
URL 322
logging
adding, to plans 307
logging statements 114
Logrotate 324
logs
agent logs 322
console and console services logs 321
database logs 321
primary server logs 320, 321
logs and current status
finding 320
log locations, exploring 320-322
report processors 324
server logs, forwarding 322, 323
status APIs, accessing 325, 327
log tooling 322
lookup function 213
loops 153-155

M

Mac desktop
Puppet lab, deploying on 25
manifest order 134
metaparameters 57
for creating, dependencies 130
metrics 327
metrics dashboards
configuring 334, 335
exploring 328-330
module 165, 166
contents 166
directory and file structure 167-170
testing, with PDK 181
writing, with PDK 177-179

N

naïve signing 242
named scope 141
namespaces 37
namevar attribute 44
nested array 83
nested data 158
nested hashes 87

node classification 266
best-practice approaches 275
node definition 266-268
node groups 350
nodes
classifying, with Hiera 268-271
non-printing tags 149
noop mode 51
notify type 56
numbers 76, 77
arithmetic operators 77, 78
float data type 79
integer data type 79
numeric to string conversion 79
string to numeric conversion 79
Numeric type 94

O

Open Source Puppet 281
Optional data type 91
orchestrator 289
orchestrator services 352
orchestrator services, components
ace service 353
bolt service 353
inventory service 353
orchestrator service 352
PCP broker 353
PXP agent 353
ordering 130-135
overview 142, 143
organization ID (OID) 248
out-of-memory (OOM) resources 355

P

package type 42, 43
parameters 35
overriding 59
parameter tag 147
parent data types 94
Any type 94
Collection type 94
Data type 94
Numeric type 94
Scalar data type 94
Pattern data type 92
patterns 91
Enum data type 91
Pattern data type 92
Variant data type 92
PCP Execution Protocol (PXP) 351
PE classifier 272-275
plan functions
using 306, 307
plans 289, 301, 305
data sources, managing 309, 310
errors, handling 308, 309
logging, adding to 307
metadata, documenting 310
results 308
testing 311
YAML plans 311-313
plugin hooks 298
plugins 166
plugins, Bolt 313
Puppet library 313, 315
reference 313-315
secret 313, 315
Pod Manager (Podman) 291
PostgreSQL 252
PowerShell cmdlets 291
prefixes 90
Optional data type 91
Sensitive data type 90, 91

prefix functions 114
Promtail 324
providers 6, 37, 44
public key infrastructure (PKI) 239
pull request (PR) 285, 362
Puppet
add-ons 16
best practices 142
change management 16
declarative 5, 6
defining 16
history 3, 4
installation 235
integrations 16
learning 16
pitfalls 142
running locally, with multiple resources 49, 50
templating formats 146
versioning 235
Puppet 5 21
Puppet 6 22
Puppet 7 22
Puppet agent-to-server lifecycle 246-250
certificate signing logging, monitoring 250
Puppet Application Manager (PAM) 363
Puppet code
classifying 285, 286
deploying 277-286
managing 277-283
workflow, creating 284
Puppet Code Validator
reference link 131, 143
Puppet Communication Protocol (PCP) 290
Puppet Comply 362
Puppet Data Service (PDS) 275, 338, 339, 347
installing 339, 340
PuppetDB 251-253
directories 252
performance tuning 253-255
querying 255
reference link 252
Puppet development
IDEs and tools, using 23, 24
Puppet Development Kit (PDK) 24, 131, 175, 362
best practices 176
gem list 176
module, testing with 181
module, writing with 177, 179
used, for testing RSpec 182
workflow 179, 180
Puppet Enterprise
Bolt, using with 351
Code Manager service 350
compliance, ensuring 362, 363
configuration 359, 360
console and services 347
database components 350
deployment, automating 359, 360
deployments, managing 362
extensions and configurations 364
jobs, running 354, 355
long-term support (LTS) version 346
monitoring 361, 362
nodes, types 351
orchestrator services 351-353
overview 346, 347
performance settings, configuring 355
primary server 347
projects and tooling 361
Puppet Server service 348
Puppet web console components 348, 349
standard architectures 356, 357
status_check module 361

supported architectures 356-359
support_tasks module 361
transport 290
troubleshooting 361, 362
Puppet Enterprise Administration Module (peadm) 347
Puppet environments 262
configuration files 264, 265
directories and paths 263
validation and deployment 265
Puppetfile 15
Puppet Forge 194
module, creating 198
module, testing 199
using 195-197
Puppet lab
deploying 24
deploying, on APT-based Linux desktop 27, 28
deploying, on Mac desktop 25
deploying, on RPM-based Linux desktop 26, 27
deploying, on Windows desktop 26
resources and references 30-32
tools, configuring 28-30
Puppet language
key terms 6-12
Puppet library plugins 313-315
Puppet Metrics collector module
reference link 329
Puppet native type refresh options 134
Puppet Operational Dashboards 328
reference link 328
Puppet plans
creating 305
Puppet platform 12-15
Puppet Query Language (PQL) 15, 234, 353
Puppet runs 276, 277
puppet run scheduler module
reference link 331
Puppet Server 236
Admin API 239
certificate authority (CA) 239-242
configuration files 243-246
embedded web server 236, 237
JRuby interpreters 243
logs 243-246
Puppet API service 237-239

R

RBAC service 349
Red Hat Package Manager (RPM) 40
Red Hat Satellite 337
reference plugins 313-315
regexp type 81
relationships 130-135
overview 142, 143
reserved variable names 69
reserved words
reference link 70, 71
resource body 38
resources 37
current system state, examining 41
Puppet, running locally 49, 50
title 39
resources metatype 58
role-based access control (RBAC) 282
roles 349
administrators 349
code deployers 349
custom roles 349
operators 349
project deployers 349
viewers 349
roles and profiles method 142, 171-175

RPM-based Linux desktop
Puppet lab, deploying on 26, 27
RSpec 182
context keyword 184, 185
coverage reports 193
data, from Hiera and facts 189-192
dependencies, managing with fixtures 192
describe keyword 184, 185
examples 185
expectations 186
matchers 186, 187
parameters 187
preconditions 187, 188
relationships 188
research 194
Serverspec 194
testing with 182-184
tools 193
Ruby 166, 290
basics 16
run interval 330

S

SAML solutions
Okta 349
Scalar data type 94
scaling 327
schedule 63
scope 94-96, 140, 141
overview 142, 143
scope function 214
scope, levels
local scope 140
node scope 140
top scope 140
secret plugins 313, 315
Secure Shell (SSH) 290
Secure Sockets Layer (SSL) 236
selectors 161
Sensitive data type 90, 91
Serverspec 194
ServiceNow 338
service types 47-49
single-quoted strings 72
Software-as-a-Service (SaaS) 15
SOScleaner 361
split function
reference link 163
Splunk 339
installing 339
Splunk HTTP event collector (HEC) module
reference link 329
Splunk Plugin 329
reference link 329
statement functions 113
catalog statements 113
logging statements 114
static code
usage criteria 220, 221
static external facts 103, 104
stdlib module functions 122
for arrays, and strings 122, 123
for file information 123
string data type parameter 76
strings 71
double-quoted strings 72, 73
heredocs 73, 74
single-quoted strings 72
to numeric conversion 79
unquoted strings 71
struct 93
structured facts 111
style guides 39
examples 38
reference link 38

substrings
accessing, in variables 74-76
supported architectures, Puppet Enterprise
extra-large installation 356
large installation 356
standard installation 356
symbolic link (symlink) 240
system transports 290

T

tag parameter 57, 58
tarball file 330
targets 291
constructing 305, 306
used, for connecting to clients 291
tasks 289, 301
creating 301-304
technical account managers (TAMs) 347
templates, with loops and conditions
creating 162
testing 162
templating formats 146
EPP templates 146-150
ERP templates 151, 152
termini 337
Thundering Herd 331
tokens 350
top-level variables 23
transports 290
configuring 298-300
reference link 291
used, for connecting to clients 290, 291
troubleshooting
Hiera 228, 229
tuning 327
tuple 92, 93
types 37
Augeas type 55, 56
core resource types 52
exec type 53-55
file type 44-47
notify type 56
package type 42, 43
service types 47-49

U

undef 80
Uniform Resource Identifier (URI) 298
unless statement 160
unquoted strings 71
user acceptance testing (UAT) 272
user type 52, 53

V

variables 68, 69
interpolation 70
names 69
reserved variable names 69
substrings, accessing in 74-76
Variant data type 92
Vault module 150
vertical scaling 331
Vox Pupli community 195
URL 195
vRealize Orchestrator (VRO) 242

W

web console 348

Windows desktop

Puppet lab, deploying on 26

Windows Remote Management (WinRM) 290

Y

YAML plans 311-313

reference link 313

Other Books You May Enjoy

If you enjoyed this book, you may be interested in these other books by Packt:

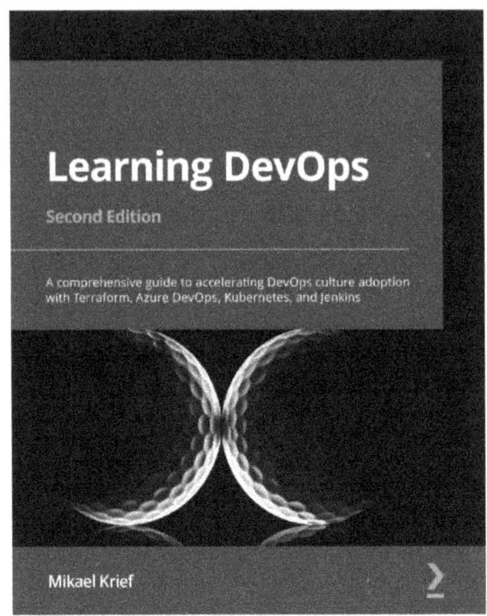

Learning DevOps - Second Edition

Mikael Krief

ISBN: 9781801818964

- Understand the basics of infrastructure as code patterns and practices
- Get an overview of Git command and Git flow
- Install and write Packer, Terraform, and Ansible code for provisioning and configuring cloud infrastructure based on Azure examples
- Use Vagrant to create a local development environment
- Containerize applications with Docker and Kubernetes
- Apply DevSecOps for testing compliance and securing DevOps infrastructure
- Build DevOps CI/CD pipelines with Jenkins, Azure Pipelines, and GitLab CI
- Explore blue-green deployment and DevOps practices for open sources projects

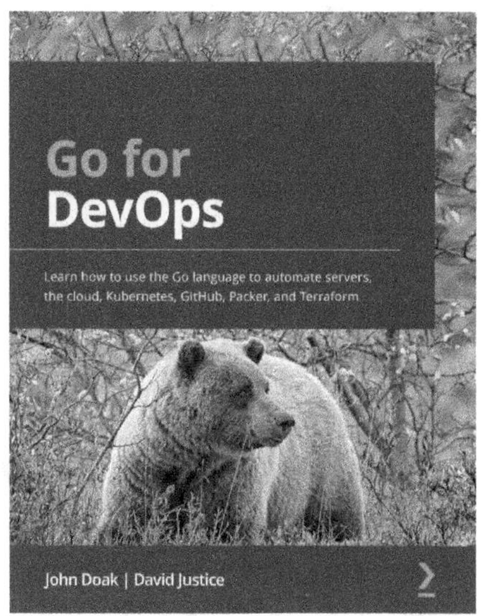

Go for DevOps

John Doak, David Justice

ISBN: 9781801818896

- Understand the basic structure of the Go language to begin your DevOps journey
- Interact with filesystems to read or stream data
- Communicate with remote services via REST and gRPC
- Explore writing tools that can be used in the DevOps environment
- Develop command-line operational software in Go
- Work with popular frameworks to deploy production software
- Create GitHub actions that streamline your CI/CD process
- Write a ChatOps application with Slack to simplify production visibility

Packt is searching for authors like you

If you're interested in becoming an author for Packt, please visit `authors.packtpub.com` and apply today. We have worked with thousands of developers and tech professionals, just like you, to help them share their insight with the global tech community. You can make a general application, apply for a specific hot topic that we are recruiting an author for, or submit your own idea.

Share Your Thoughts

Now you've finished *Puppet 8 for DevOps Engineers*, we'd love to hear your thoughts! Scan the QR code below to go straight to the Amazon review page for this book and share your feedback or leave a review on the site that you purchased it from.

`https://packt.link/r/180323170X`

Your review is important to us and the tech community and will help us make sure we're delivering excellent quality content.

Download a free PDF copy of this book

Thanks for purchasing this book!

Do you like to read on the go but are unable to carry your print books everywhere?

Is your eBook purchase not compatible with the device of your choice?

Don't worry, now with every Packt book you get a DRM-free PDF version of that book at no cost.

Read anywhere, any place, on any device. Search, copy, and paste code from your favorite technical books directly into your application.

The perks don't stop there, you can get exclusive access to discounts, newsletters, and great free content in your inbox daily

Follow these simple steps to get the benefits:

1. Scan the QR code or visit the link below

`https://packt.link/free-ebook/9781803231709`

2. Submit your proof of purchase
3. That's it! We'll send your free PDF and other benefits to your email directly

www.ingramcontent.com/pod-product-compliance
Lightning Source LLC
Chambersburg PA
CBHW081504050326
40690CB00015B/2914
9781803231709